principles of **PLANT BREEDING**

principles of **PLANT**

R. W. ALLARD

Professor of Agronomy
University of California
Davis, California

NEW YORK · LONDON, *John Wiley & Sons, Inc.*

BREEDING

33285

Preface

This book has been written primarily for undergraduate students of agriculture. It has been my experience that students of agriculture, whether or not they intend to pursue careers in plant breeding, prefer instruction based on principles rather than on breeding procedures related to specific crops. At the same time, principles cannot be taught in a vacuum, as their importance and applicability may not be appreciated by those with little prior knowledge of breeding methods or problems. In preparing *Principles of Plant Breeding*, therefore, the goal was set to produce a textbook which, although placing principles foremost, would also provide specific examples in order to avoid abstractness. Since a textbook of plant breeding obviously cannot encompass all of agriculture, it was necessary to assume the general acquaintance of the reader with agricultural practices and problems.

To achieve this goal required that the organization of *Principles of Plant Breeding* should depart from tradition in two respects. First, it was assumed that the student possessed a background in genetics equivalent to that obtained from a course of instruction based on one of the standard textbooks in that field. The elementary facts and principles of genetics are therefore not considered. Instead, certain aspects of inheritance are developed to a level beyond that generally attempted in a first course in genetics. For example, quantitative genetics, population genetics, systems of mating, heterosis, the genetics of pathogenic organisms, and several other topics have been singled out for amplification because of their importance to understanding not only principles but also practices in plant breeding.

The second departure required that lesser emphasis be placed on the breeding of specific crops. This permitted bringing to the forefront the basic unity of methods used in breeding self-pollinated species on the one hand, and cross-pollinated species on the other. Examples were drawn from as many different crops as possible to widen interest. In

v

selecting examples, primary consideration was given to suitability in illustrating points at issue, and only secondary consideration was given to economic importance.

The above considerations led to an organization in which the subject matter of plant breeding is divided into nine major areas (sections), each of which is further divided into a number of chapters. This detailed breakdown of topics is intended as an aid to instructors in emphasizing or omitting subject matter as appropriate to the backgrounds and interests of their students. An attempt was made to enable each section to stand on its own. Sections devoted primarily to principles precede sections on methods. Thus subject matter pertaining to the genetic basis of breeding self-pollinated crops is collected in Section Two, and methods of breeding these crops follow in Section Three. Sections Four and Five are concerned with the genetic basis and methods of breeding cross-pollinated species, respectively, and within Sections Six, Seven, and Eight chapters on principles come before related chapters on practices. Chapters on principles are for the most part not essential prerequisites to corresponding chapters on methods; however, they should provide for better understanding and greater insight if they are studied prior to or in conjunction with the sections on methods.

The most difficult problem encountered arose in connection with the important subjects of biometry and the design of experiments. Since the knowledge of these subjects required of students of plant breeding is too extensive to be acquired as a mere adjunct to a course in plant breeding, the trend is to require students to take them as prerequisites to plant breeding. Moreover, there are several excellent books which cover the biometrical background essential to plant breeding. It was therefore thought redundant to include biometry and experimental design here.

A few words on the way this book might be used may be helpful. Sections One, Three, and Five, together with Chapters 27, 32, 34, 35, and 36, provide elementary coverage of the subject matter included in practically every general course in plant breeding. They might serve as the basis for a one-quarter course concerned primarily with practices. Supplementation from chapters devoted primarily to principles should be possible in more rigorous one-quarter courses or in courses given over a full semester. The entire book could serve as the basis for a two-semester course.

Special attention is directed to Chapters 9, 10, 15, 16, 17, and 19, which may be too difficult for some undergraduate students. These chapters were included as a challenge to undergraduates with special interest in plant breeding, and to graduate students taking their first course in the subject.

I am conscious of my debt to plant breeders and geneticists from many nations who have either discussed with me various ideas that appear in this book or who have read and criticized parts of the manuscript. Although these contributors are too numerous to mention, I would be remiss if I did not acknowledge the aid of Drs. W. E. Nyquist and E. H. Stanford, who read and criticized the entire manuscript.

R. W. ALLARD

Davis, California
July, 1960

Contents

1

Nature and Goals of Plant Breeding

Man is almost absolutely dependent on plants for his food. The things he eats, virtually without exception, are either plant materials or derived rather directly from plants as are, for example, meat, eggs, and dairy products. Plants are also the major source, directly or indirectly, of most clothing, fuel, drugs, and construction materials. Moreover, as ornamentals, they are both useful and aesthetically pleasing. Considering the prime importance of plants, it is not surprising that men have long been concerned with developing types better suited to satisfying their needs. Only recently, however, and largely in conjunction with the growth of genetics, have these attempts been systematized to the point where they can be called a science. This comparatively new science, the science of plant breeding, is the concern of this book.

Augmentation of Food Supplies

Much of the emphasis in plant breeding has been placed on increasing agricultural productivity. This has been in response to the pressure on an adequate food supply caused by constantly increasing population in a world of limited acres. Except during brief periods of abundance, human populations, like populations of other animals, have suffered from hunger. Each period of abundance has brought a jump in population, followed by hunger and its aftereffects, pestilence, high infant mortality, and meager sickly lives. This pattern has been described by Harrison Brown in his readable book *The Challenge of Man's Future,* in which he discusses the interrelations of food supply and other factors in problems of populations. Brown believes that this

1

pattern persists into the twentieth century because many nations have not learned to escape famine consistently.

Man's concern about expanding agricultural production to keep pace with need has not been confined to the present century. The best-known writer on this popular subject, the Reverend Thomas Robert Malthus, postulated in 1798 that human populations, unless checked by wars or disaster, increase until hunger takes control. He foresaw catastrophe because he believed that population was capable of increase at a geometric rate and food supply only at an arithmetic rate. On this basis he predicted that Britain would be in disaster by the midnineteenth century. This dire prediction was not realized despite the fact that population did increase at nearly the Malthusian rate. The failure of the prediction resulted from underestimating man's ingenuity in increasing agricultural productivity. As Malthus was writing, his countrymen about him were ushering in the age of scientific agriculture. From it came tremendous increases in food supplies as a result of better methods of production and improved varieties of plants and breeds of livestock. About the same time, the machine age started. It had a profound effect not only on the efficiency of production but also on the efficiency of utilization of agricultural products through better methods of preservation, storage, and transportation. The advent of machines was also an important factor in the vast expansion into new agricultural areas which has occurred since the time of Malthus. Many are concerned that, in the broadest sense, Malthus may still be right. Modern Malthusians maintain that catastrophe has only been delayed and that the misery associated with over-population is still the fate of modern civilization. If this is not to be true, constantly increasing supplies of foodstuffs will be required. What opportunities remain to further increase food supplies?

The supply of new agricultural land is of course decreasing steadily. This is not to say, however, that opportunities no longer exist for the development of new areas. For example, there are some who regard the agricultural potentialities of the tropics as favorable. Many believe that the sea has great potential for the production of food-stuffs. Many opportunities of more conventional and proven types also remain, but they are increasingly associated with comparatively expensive reclamation projects such as irrigation of deserts and drainage of swamps. Unfortunately, most of the new sources of sustenance that have been suggested, such as algae and plankton, are uninspiring.

Another opportunity to increase production lies in improved agricultural practices, including better fertilization, more effective crop rotations, improved tillage methods, and more efficient weed, disease, and insect control. If improved methods of crop production now at our disposal were used throughout the United States, and more particularly in many nations less agriculturally advanced, the productivity of the earth would doubtless be greatly increased. It also seems certain that the inventiveness of crop-management scientists will continue to provide better and better environments for domestic species so that we can expect further improvements in total production and in quality of product. There is, of course, no way of estimating how close modern production methods have brought us to the maximum yields possible. It does appear certain, however, that great advances remain to be made.

Finally, we can expect plant breeding to contribute substantially to greater agricultural productivity. This will be accomplished not only by the breeding of basically higher-yielding varieties but also by the development of varieties that help to stabilize production through resistance to disease, drought, heat, cold, and wind. The probable future of plant breeding can perhaps best be illustrated by an examination of the contributions it has made in the past.

Goals of Plant Breeding

Increased yield has been the ultimate aim of most plant breeders. Sometimes this has been accomplished by providing varieties basically more productive, not because of specific improvements such as in disease resistance but as a result of generally greater physiological efficiency. This type of improvement can be illustrated with sugar beets. Early attempts to produce sugar from beets were unsuccessful largely because of the low sugar content, generally less than 7 per cent, in the varieties then available. During the subsequent 175 years varieties have been developed which are capable of producing 15 to 18 per cent of sugar consistently. Examples of varieties superior because of their generally more efficient physiological processes can be cited in practically any major crop species.

One of the most important contributions of plant breeding has been the development of better varieties for new agricultural areas. Frequently this has been accomplished by adjusting the growth cycle of the variety to better suit the available growing season. A good example is the modification of grain sorghums since their successful introduction to the United States about 100 years ago. This species, a tropical grass, was originally confined to the warmer parts of the southwest and southern-plains area. Gradually, earlier and earlier varieties have been developed until grain sorghums are now an important crop as far north as South Dakota.

Another contribution of plant breeding has been the improvement of plants in agronomic or horticultural characteristics. Again the grain sorghums make a good example. The types originally introduced grew taller than a man's head and had to be harvested by hand. The breeding of dwarf varieties, which grow to a height of 3 to 4 feet,

FIGURE 1-1. Modification of grain sorghum through plant breeding. In the background are tall, late-maturing varieties similar to the types introduced to the United States from Africa about 100 years ago. In the foreground are dwarf and double-dwarf varieties. What importance did the breeding of the short, early types have in American agriculture?

has made combine harvesting a practical undertaking. Probably no factor was more important in the ultimate success of the grain sorghums in attaining major status as an agricultural crop in the United States than the breeding of these dwarf varieties.

One of the most dramatic and certainly the best-known contribution of plant breeding has been the development of crop varieties resistant to diseases and insects. For some crops these resistant varieties have provided the only feasible control for such pests. With stem rust of wheat, for example, the benefits have been spectacular, representing the difference between a good harvest and crop failure. In other crops only slightly higher tolerance to diseases or insects, discernible only to competent observers, has made substantial differences in productivity. Perhaps the most important feature of resistant varieties is the stabilizing effect they have on production. This is important not only to the individual farmer but to his nation as well. Reasonable

FIGURE 1-2. The benefits of disease resistance illustrated by onions resistant and susceptible to pink root fungus. (After Jones and Perry, *Jour. Hered. 47* : 33. Photograph courtesy United States Department of Agriculture.)

harvests every year are to be preferred to the economic hardships associated with extreme fluctuations in yield. The benefits of varieties tolerant to heat, cold, or drought are similar to the benefits attending disease-resistant varieties.

Thus far the only contributions of plant breeding considered have been ones concerned with increased production. Also of importance to human well-being are those advances leading to improved quality of agricultural products. Certainly the breeding of stronger and longer staple cotton has added to human welfare. Similarly stringless green beans and apples with superior flavor are worth-while improvements. Nor can it be denied that wheat with more protein or tomatoes with higher vitamin content have contributed to health. The desires of food processors as well as consumers have been considered. Canners are concerned whether the liquor in a can of green Lima beans remains clear or becomes murky, with the result that varieties more satisfactory in this aspect of quality have been developed. During the last few years plant breeders have given constantly increasing attention to the needs and tastes of man.

Examples such as the foregoing illustrate some improvements that have occurred in agricultural varieties within the last 150 years. From the historical record we know that steady gains have been made over long periods of time so that the more important domestic plant species represent vast improvements over their wild progenitors. It is now known that most organisms have tremendous plasticity in their hereditary make-up. There is every reason to believe that the exploitation of the variability of most domesticated species that has occurred up to the present has not taken us even near to the maximum productivity that is theoretically possible. If this is true, plant breeding has greater contributions to make in the future than it has made in the past. In one way the contribution to human welfare made by superior varieties is the most satisfying of all methods of increasing production. Normally such varieties add nothing to the cost of production beyond that required to handle the additional increment of yield. The appeal of this situation to human nature is undeniable.

REFERENCES

Brown, Harrison. 1954. *The challenge of man's future.* Viking Press, New York.
Malthus, T. R. 1798. *An essay on the principle of population.* Oxford University Press. Reprinted by Macmillan and Co., London, 1926.
Yearbook of agriculture, U. S. Dept. Agric. 1936, 1937, 1943–1947. (These yearbooks include many essays on plant and animal breeding. For a general survey of plant breeding see pp. 118–164, 1936 *Yearbook.*)

2

*Patterns of Evolution in Cultivated Species**

The earliest physical evidence of the antiquity of the cultivation of plants goes back about 5000 to 6000 years to plant remains left by the lake dwellers of Switzerland and those found in the ruins of ancient Mesopotamian and Egyptian cities and in the barrows of the British Isles. It is improbable that these remains represent the earliest cultivated types of plants. Both barley and wheat, for example, are Asian in origin, and they are represented in Asia by a tremendous range of types. In Egypt and Europe these cereals are represented by a paucity of types. Apparently only a handful of the immense range of types of Asia reached Egypt and Europe, and even the earliest types of these areas were almost certainly recent developments in the history of cereal breeding. According to Harlan (1957), barley types have changed little in Egypt during the past 5000 years, and he believes the ancient Egyptians and Europeans obtained their cultivated plants from earlier plant breeders.

Evidence from other crops also suggests that most of plant breeding was accomplished before the earliest dates set by archeological records. For example, the Lima beans found in the ruins of some of the oldest Indian civilizations of Peru have seeds that are nearly 100 times as large as those of wild Limas of that area. Apparently the city-dwelling pre-Incan Indians of Peru obtained their beans from still earlier plant breeders who left no archeological record, because this increase in seed size is inconsistent with any short period of domestication.

* This chapter is based in part on an unpublished manuscript made available to the author by the late Professor R. E. Clausen. This manuscript, *The Origin of Cultivated Plants,* was the basis of a lecture delivered to the faculty of the University of California by Professor Clausen in 1956.

How long has it taken to develop from wild species the remarkable arrays of modern cultivated forms? Consider the characteristics of the wild relatives of wheat and barley. These species have a fragile rachis and many other features which equip them to survive in nature, and, all told, they are very different from their cultivated relatives. If we assume that the wild progenitors of cultivated barley and wheat had these same adaptive features, and this is not unlikely, it is difficult to reconcile the changes necessary to convert them to the cultivated types with short periods of selection under domestication. Similarly, selection experiments to increase seed size in beans indicate that many dozens or more probably some hundreds of generations of the most intense selection would be necessary to increase seed size from that of wild beans to that of the earliest known cultivated types. Using such measuring sticks, we can see that the time needed to produce the cultivated plants of the world was indeed great.

For the person interested in the origins of cultivated plants, Oakes Ames's stimulating book, *Economic Annuals and Human Cultures,* is required reading. His ideas and interpretations have influenced subsequent students of the subject, and the pages of his book include a wealth of information on the domestication of our annual crop species.

Natural and Artificial Selection

Although we may not know when man became a plant breeder, we can be sure that nature has always been one. Heritable variations certainly occurred both before and after the beginning of cultivation. Strains carrying these variations eventually were grown near each other, natural hybrids occurred, and greater numbers of combinations of variants resulted. Some were advantageous, some not. Some steps were forward, some back, but usually the types poorly adapted to local circumstances were eliminated, and the opportunity for better matings increased. This process was not always local. Trade routes have existed for a long time, and they were certainly important in distributing seed. Transportation from place to place provided opportunity for amalgamation of types so that modern varieties probably have tenuous threads extending into the past in dozens of ways. What modern plant breeders have received are the end products of a long period of "natural" selection under the conditions of cultivation. Man no doubt aided this process and, in some instances, acted contrary to natural selection by preserving variants useful to him that would have expired but for human interference. In any case natural and artificial selection acting together have provided the modern plant breeder with a liberal heritage of plant materials.

Although written records show that artificial selection has been practiced throughout historical times, it is a matter of interpretation as to the precise point at which man became a plant breeder. It could be argued, for example, that breeding started as early as cultivation, because the choice of some particular wild plant to cultivate could be considered to be plant breeding. There is some evidence that primitive peoples differed in their skill as plant breeders. Mangelsdorf, for example, believes that the archaeological remains discovered in the Bat Cave of New Mexico indicate that little artificial selection was practiced by the peoples who inhabited the cave from approximately 2500 B.C. to A.D. 500. Although the remains show that average size of ear and kernel increased during this period, the small primitive corn present at the beginning had not become extinct at the end of the period. This, he argues, must surely have occurred had selection been practiced for large ears. Other American Indians, however, must have practiced such selection, because the corn cultivated at the time of the discovery of America—about 500 to 1000 years after the end of the story told in the Bat Cave—uniformly approached present-day corn in kernel size. Furthermore, that some American Indians were able to maintain two or more types of this cross-pollinated species in a single locality demonstrates considerable skill in plant breeding.

Patterns of Evolution

Even though there are many unanswered questions concerning the origin and domestication of cultivated species, the main patterns in which they evolved are, fortunately, limited in number. They fall into three main categories: (1) Mendelian variation, (2) interspecific hybridization, and (3) polyploidy. This does not exhaust the methods by which plants have been molded into forms useful to man. But they are the chief ways in which cultivated species have evolved, considering both the pre- and postdomestication periods. Also, these patterns are not mutually exclusive; in fact, much of the difficulty in dealing with the subject lies in the way in which different methods have been interwoven in a given group of plants.

Interest in these patterns of evolution extends beyond the historical because they are also the methods currently used by plant breeders. The contribution of modern man to plant breeding has been not so much to develop new methods as to accelerate progress, a process that has gained momentum proportionately with greater understanding of the nature of the genetic systems that govern heritable variability. For these reasons, discussion of the evolutionary patterns in which our cultivated plants have evolved is important for understanding modern plant-breeding methods. At least one example will be given of each of these modes of origin of cultivated plants. In the main, the evidence is circumstantial, based on records written in the genes and chromosomes of the plants themselves of alterations impressed there long before the existence of written records.

Mendelian Variation

Mendelian diversity arises basically from gene mutations, which are the building blocks of evolution. Mendelian recombinations resulting from hybridization among types carrying different mutations provide further diversity among individuals upon which both natural and artificial selection may act.

Sometimes a single mutation with drastic or multiple effects may considerably enhance the usefulness of a species to man. The single macromutation that differentiates pod corn from the important commercial types falls in this category. Heading cabbage, cauliflower, broccoli, Brussels sprouts, and kohlrabi have all been derived fairly directly from the wild cabbage that still grows in the coastal regions of Europe and North Africa. The great morphological differences separating these cultivated types depend on few gene differences. They represent botanical monstrosities that would have had little chance to survive in nature.

The Lima bean, an American species brought to a high level of perfection in pre-Columbian times, affords a good example of another and more complex type of development. Gray-brown seeded Lima beans, probably very similar to the ancestors of the cultivated Lima, have been found growing as wild trailing plants in the subtropical brushlands of Central America and also in similar environments in the Andes from Peru to central Argentina. The seeds of these wild beans, little larger than wheat kernels, are one tenth to one hundredth as large as the seeds of cultivated types. Seeds found in Peruvian tombs prove that modern types were cultivated as long ago as 2600 years, and there is ample evidence that the species was widely cultivated throughout tropical and temperate America in pre-Columbian times.

Study of native Lima beans over the regions of its commercial production in the Americas suggests that this species spread out in prehistoric times from two centers, one in Central America (Guatemala) and one in Peru. Modern types in those two regions exhibit great diversity in size, shape, color, and pattern of seed, habit of growth, and various other features. On moving away from the two centers of diversity, variation decreases and specialization develops along characteristic lines:

1. *The Hopi Line* extended northward from Guatemala along the west coast of Mexico into Arizona.

2. *The Carib Line* extended eastward from Guatemala to Yucatan, across the West Indies, thence northward along the Atlantic coast of North America to the Potomac and spread southward into coastal South America from northern Colombia to the region of the Orinoco. The North American varieties of this line were the progenitors of the baby Limas of modern trade.

3. *The Inca Line* extends from Peru northward into coastal Ecuador and southward into subtropical parts of northern Argentina. Many varieties of this line are characterized by large seed size, and they provide the source of the modern large, high-quality Lima beans of commerce.

It is interesting to note that the distribution of the wild Lima bean contacts two ancient centers of civilization in the Americas, the pre-Mayan and pre-Incan centers. There is, in fact, considerable evidence suggesting that the species was domesticated independently in these widely separated localities, and that development remained largely independent subsequent to domestication.

The development of the Lima bean exemplifies a pattern of transformation characteristic of cultivated species. Two distinct features emerge: (1) there is an accumulation of a store of diversity, primarily by spontaneous mutations, with the preservation by selection of types desirable to man; and (2) varieties are produced from this diversity which are suitable for different purposes and adapted to different conditions. This process, in total, can be called Mendelian variation. The accumulation of diversity, other things being equal, is proportional to the length of time the plant has been under cultivation and the scale upon which it has been grown.

Other important crops which appear to have developed by variation within a single species are barley, rice, corn, flax, common beans, and tomatoes. Each of these species has developed in its unique way, but the general pattern for each crop is that of Mendelian variation. In some of the crops mentioned above there is evidence of introgression, a subject which will be discussed later. Divergence in cultivated plants has continued neither long enough, not even during long periods of domestication, nor far enough, despite conspicuous morphological differences in some cases, to produce the kind of isolation barriers

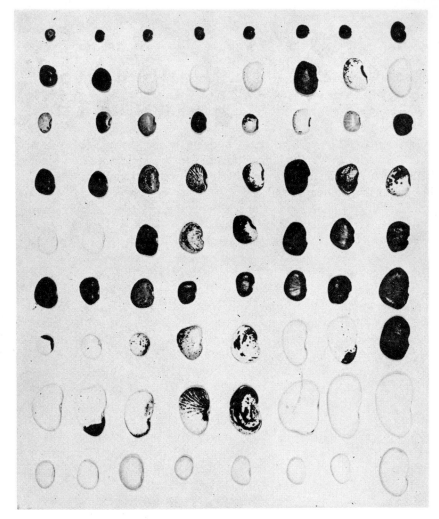

FIGURE 2-1. Mendelian variation in the Lima bean. Top row, wild Limas Second row, Hopi Indian types. Third row, Mexican varieties. Rows 4 to 6, Carib types (Central America, Caribbean Islands, Northern South America). Rows 7 and 8, Incan types from Southern Ecuador and Peru. Bottom row, cultivated varieties of the United States. The smallest seed shown weighs about one eightieth as much as the largest seed. Seed coat colors are white, green, and various shades of brown, red, and purple to black. Seed coats of many types are multicolored in various patterns.

found in natural evolution. As a consequence, the original taxonomic species are still recognizable. What is more significant, however different various local races may be in regard to gene frequency or morphology, they usually hybridize freely and produce fertile offspring,

FIGURE 2-2. Transformation of the Lima bean under domestication. *A*, perennial wild Lima growing as a trailing plant on brush in El Salvador, Central America. *B*, primitive, perennial cultivated Lima being grown on poles as a garden plant in northern South America. *C*, modern, annual, high-yielding, high-quality, determinate-habit variety developed to meet the requirements of large-scale field production and mechanical harvest for quick-freezing.

thus establishing the failure of the transformation process to affect their taxonomic status.

Interspecific Hybridization

A second method of evolution of cultivated plants depends on the crossing of distinct taxonomic species with the preservation of improved types from the segregation products. The large number of gene differences likely to occur in interspecific hybrids, combined with differences in chromosome organization, usually lead to bewildering complexity in the segregating generations. Moreover, many of the recombinations are disharmonious ones, neither likely to pass the test of natural selection nor to be selected and preserved by man. For this reason it is difficult to cite instances in which two or more natural species have contributed to the development of a series of cultivated varieties, apart from plants propagated by the horticultural arts such as budding or grafting. Vegetative propagation preserves the exceptional vigor which characterizes many F_1 interspecific hybrids and avoids the necessity of attempting to fix the good types likely to occur in extremely low frequencies in subsequent generations. Certain varieties of pears, plums, cherries, and grapes have arisen from interspecific hybridization, as have many ornamentals such as certain irises, roses, lilies, and rhododendrons.

The garden strawberry is one of the best authenticated cases in which interspecific hybridization has contributed to the development of a series of cultivated varieties. Prior to the discovery of America, European gardeners had domesticated their native species, but progress toward developing superior types was slow, apparently because the necessary variability was not present in the germplasm of the European species. Real progress in strawberry breeding was delayed until early in the nineteenth century, when hybrids were made by English gardeners between *Fragaria virginiana,* an eastern North American species, and *F. chiloensis,* a Pacific Coast species. Crossing and recrossing between these two species and their segregation products over the last century and a half have combined in some measure the large size of the Pacific Coast species with the fruitfulness, quality of flesh, and excellent flavor of the eastern species.

There is another form of interspecific hybridization, namely *introgressive hybridization,* which may have been operative in some instances. In this form, hybridization is followed by recrossing with the parental species in such a way that certain features of one species become transferred to the other species without impairment of taxonomic integrity. In essence this amounts to one species becoming enriched to a small degree with genes derived from another species, thus broadening its base of variability and increasing the variety of recombination products which may be secured from it. Introgression is difficult to detect, especially if the contamination is slight, and the whole matter is thus rather speculative. It does deserve attention, however, because of the opportunities it offers for improvement in some crop species.

There is some reason to believe that the distinctive features of North American corn are the result of contaminating South American corn by introgressive hybridization with genes of *Tripsacum,* the related wild gamagrass of Central America. This theory, in brief, visualizes that corn was domesticated in the Andean region of Bolivia and Peru. Spreading northward it eventually was carried to Central America. There it encountered maize from an independent domestication in Mexico which was characterized by features foreign to maize as it existed in South America but included those peculiar to North America. Gamagrass possesses the essential morphological characters which distinguish North American varieties of maize. The maize-*Tripsacum* cross is highly sterile, so if *Tripsacum* is a progenitor of maize, the fertile progeny of maize-like plants likely arose from repeated backcrosses to maize.

Polyploidy

Another form of evolution that has been important in the development of cultivated plants has been polyploidy, variation arising through reduplication of chromosome sets. The normal plant has the diploid number of chromosomes, that is, two complete sets of chromosomes, one derived from each of its parents. Misadventures in the reproductive processes may give rise to haploids with one set of chromosomes, triploids with three, tetraploids with four sets, or even higher levels of polyploidy. If identical genomes, those of a single species, are reduplicated, it is designated as a simple polyploid or *autopolyploid.* If the genomes are dissimilar, having arisen from two or more species, it is designated as a hybrid polyploid or *allopolyploid.*

Autopolyploids have not been very important in natural evolution probably because they are usually characterized by aberrant chromosome behavior and a high degree of sterility. Because of their limited seed production, autopolyploid species are usually propagated by asexual means. Frequently autopolyploids have larger flowers, fruits, or other plant parts than their diploid counterparts and for this reason are interesting and useful, despite their lower fertility and generally slower rate of development.

The simplest form of autopolyploid is the triploid which, because of its unbalanced chromosome sets, is highly sterile and usually must be reproduced asexually. Although of little importance in nature, triploidy has played a role in the evolution of cultivated species. Approximately one fourth of American varieties of apples are triploids, and the same is true of pears in Europe. Commercial bananas have **33** instead of the **22** chromosomes of ordinary diploid varieties. Perhaps the most important characteristic of these banana triploids is the seedlessness of the fruit, although the triploidy imparts increased vigor and size of plant and fruit as well.

An outstanding example of the next level of simple polyploidy, namely tetraploidy, is found in the potato. It has **48** chromosomes whereas ordinary diploid species have only **24**. Although the origin of the tetraploid potato is predominantly by simple polyploidy, it appears to have been complicated to some extent with hybrid polyploidy. Collections of potatoes from Bolivia, the probable center of origin, include $2N$, $3N$, and $4N$ types, all regarded as members of the cultivated species, *Solanum tuberosum*. The cultivated varieties grown in the United States and Europe are exclusively tetraploid.

It should be noted that simple polyploidy does not always confer superiority. Tetraploids of spontaneous origin have been discovered in at least fifteen varieties of diploid grapes. Not one of the tetraploids has become important in commercial production, because along with the favorable features of tetraploidy, there have been disabilities which have outweighed the advantageous effects.

It must be concluded that simple polyploidy has been of limited importance as a source of cultivated plants. Likewise its use to the plant breeder in the future is likely to be confined largely to ornamental and horticultural types which can be propagated vegetatively.

Allopolyploidy depends on hybridization between species followed by doubling of the chromosome number. The rationale of this method is simple in principle. Smooth functioning of the reproductive mechanism requires the simultaneous presence of two chromosomes of each kind. As a result of certain evolutionary processes, related species frequently become so distinct that on hybridization their chromosomes fail to participate normally in the reproductive processes in the hybrid, and, as a consequence, the hybrid is usually more or less sterile. An outstanding discovery in the field of interspecific hybridization was that such sterile hybrids may become reasonably fertile when their chromosome number is doubled. Not only does fertility increase, but such amphiploids may become reasonably constant for the hybrid condition, producing progeny that are more or less uniformly like the original hybrid. In this form of evolution we must, as in simple polyploidy, deal with Mendelian variation superimposed on the polyploid condition. Polyploidy then has the effect of broadening the potential

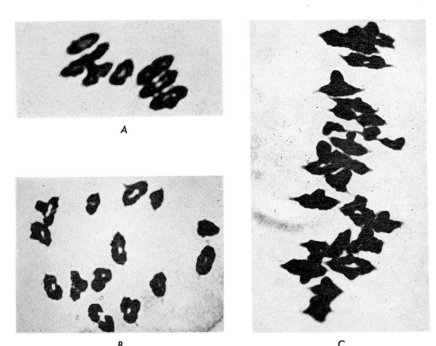

FIGURE 2-3.   Allopolyploidy in wheat.   *A, Triticum monococcum,* seven pairs of chromosomes at meiotic metaphase.   *B, T. dicoccum,* fourteen pairs in polar view. *C, T. vulgare,* twenty-one pairs.   The origin of each of the three sets of seven chromosomes combined in bread wheat is discussed in detail in Chapter 31.

base of variation by increasing the number of genes which can mutate. With hybrid polyploidy, the potential for variation of two species is combined into one. A large portion of cultivated plants, possibly as many as half, are hybrid polyploids. A recent survey among grasses sets the proportion of polyploidy, mostly hybrid polyploidy, in the neighborhood of **70** per cent.

The role of hybrid polyploidy in the development of wheat, tobacco, and cotton has been the object of a great deal of research with the result that it has been determined in considerable detail. The origin of cultivated tobacco will serve to illustrate the main features of this type of evolution.

Commercial tobacco, *Nicotiana tabacum,* does not occur as a natural species, nor is there any dependable evidence that it ever did. It is a tetraploid species with 48 chromosomes, whereas related tobaccos usually have 24 chromosomes. From two of these twenty-four chromosomes species, *N. sylvestris* and *N. tomentosa,* a sterile hybrid may be obtained which on doubling of the chromosome complement, becomes reasonably fertile and constant, and equivalent in many respects to commercial tobacco. Evidently *N. tabacum* arose as a hybrid polyploid between these two species. This probably was followed by a long period of evolution on the pattern of the Lima bean, bringing into existence the numerous varieties of commercial tobacco. All of this happened in South America long before Columbus, because, by the time of the discovery of America, *N. tabacum* was a highly developed cultivated species, widely distributed and far removed from its probable place of origin.

REFERENCES

Ames, Oakes. 1939. *Economic annuals and human cultures.* Botanical Museum of Harvard University, Cambridge, Mass.
Anderson, E. 1949. *Introgressive hybridization.* John Wiley and Sons, New York.
Harlan, H. V. 1957. *One man's life with barley.* Exposition Press, New York.
Hutchison, J. B., R. A. Silow, and S. G. Stephens. 1947. *The evolution of Gossypium and the differentiation of cultivated cottons.* Oxford University Press, London.
Mackie, W. W. 1943. Origin, dispersal and variability of the Lima bean, *Phaseolus lunatus. Hilgardia* 15: 1–29.
Mangelsdorf, P. C. 1952. Evolution under domestication. *Amer. Nat.* 86: 65–77.
Muntzing, A. 1936. The evolutionary significance of autopolyploidy. *Hereditas* 21: 263–278.
Roberts, F. 1926. *Plant hybridization before Mendel.* Princeton University Press.

3
Plant Introduction and Domestication

Wherever man has gone, his plants have gone with him, and, as we shall see, this carrying of plants from place to place has been one of the most important features in the development of agriculture throughout the world. The acquisition of superior varieties by importing them from other areas accomplishes the same purpose as developing superior varieties in deliberate breeding programs of the type we shall consider later. For this reason, plant introduction deserves consideration as a method of plant improvement.

Originally, plant introduction must have attended the wanderings of nomadic tribes. Later it accompanied the expansion of more sedentary agricultural peoples into new areas as human populations increased. When trade routes were developed, there was probably a great increase, not only in the scope of plant introduction, but also in the distances plants were transported. An inscription of about 2500 B.C., found in Mesopotamia, tells of Sargon crossing the Tsurus Mountains in Asia Minor and mentions incidentally that he brought back figs, vines, and roses with him. The earliest recorded account of an expedition organized expressly for the collection of plants is that of Queen Hatshepsut of Egypt who sent ships to East Africa in 1500 B.C. to procure the incense tree. In Japan there is a monument to Taji Mamori, who went to China in A.D. 61 on imperial orders to bring citrus fruits back to Japan. Numerous other examples of early plant introductions can be cited. If we consider that the center of production of nearly all major crops is now far removed from the native areas of the parent species, the success and importance of plant introduction can hardly be overestimated.

Undoubtedly the most important event in the movement of agricultural plants was the discovery and settlement of the Americas. The

19

early explorers returned to Europe with New World species that greatly influenced European agriculture. For example, the European diet was changed drastically by the acquisition of the potato, which almost immediately on its introduction achieved rank among the most important energy foods. European colonists to America, on the other hand, brought with them the crops of their native lands. Many of the introduced crops were immediately successful and others ultimately found areas where they were adapted. As settled community life gradually replaced the hazardous early days in the colonies, increasing attention was given to bringing the best of food and fiber plants, fruit and shade trees, and ornamentals from the Old World. The development of present-day crops is due in considerable measure to the importation in the last half of the eighteenth century of varieties of various crops that were suited to special needs or growing conditions. The importance of these Old World species to the agriculture of the New World is shown in Table 3–1. The ten most important crops grown in the United States in 1951 were about equally divided between Old and New World species, taking into account both number of crops and their value.

TABLE 3–1.   Farm Value of the Ten Most Important Crops Grown in the United States in 1951

(Data from 1952 United States Census)

Crop	Farm Value $1000	Place of Origin
Corn	4,981,156	Tropical America
Cotton	3,349,315	Tropical North America
Wheat	2,091,535	Asia Minor
Tobacco	1,190,920	Tropical America
Oats	1,099,202	Asia Minor or North Africa
Soybeans	735,613	China
Potatoes	535,304	South America
Sorghums	343,030	Africa
Barley	318,157	Asia Minor
Tomatoes	259,839	Tropical America

Native American crops account for 69 per cent of the total farm value.

Nor did important introductions end with the Colonial period. For example, Mennonite immigrants, who settled in Kansas in the decade 1870–1880, were largely responsible for establishing the hard red winter-wheat industry of the United States. The wheat these settlers brought with them from southern Russia came to be known as Turkey, and it has provided, along with similar introductions, most of the germplasm for this wheat area, one of the great grain producing areas of the world. It is probable that the varieties introduced by the colonists and later immigrants were highly heterogeneous, giving them great potential to fit into new environments. Even the varieties of self-pollinated crops were undoubtedly a complex mixture of types, providing the flexibility necessary to meet the particular conditions of new circumstances.

Public Exploration and Experimentation

According to Nelson Klose, in his book *America's Crop Heritage*, the earliest record of organized public efforts to encourage crop cultivation in the Americas is found in the annals of an experimental farm established in 1699 in South Carolina. This farm was set up by the Lord Proprietors of the colony to test the adaptation of various introduced crops. The Royal Botanical Garden at Kew in West London is an example of an early public institution that made important contributions to plant introduction. Established in 1760, the garden assisted in the spread of valuable plants among the British colonies. It sponsored the first professional plant hunter, Francis Masson, dispatching him to Africa in 1772 and later to the West Indies and South America. These explorations marked the first of many sponsored by the British and the French governments. Many of America's early statesmen were aware of the importance of introducing useful plants, but the new government, occupied with other pressing problems, did not immediately engage in plant explorations. The attitude of the time was expressed by Thomas Jefferson, who wrote: "In an infant country, as ours is, we are probably far from possessing, as yet, all of the articles of culture for which nature has fitted our country. . . . to find them will require an abundance of unsuccessful experiments. But if, in a multitude of these, we make one successful acquisition, it repays our trouble." It was not the responsibility of the Federal Government, but "perhaps it is the peculiar duty of associated bodies, to undertake these experiments." Jefferson was referring to the agricultural societies which had been formed in many states by that time.

As a result of the activities of these agricultural societies, a large number of plant introductions were made in the early days of the United States. The first of these societies to be established was the South Carolina Society for the Promotion of Agriculture, founded in 1785. It provided a sum of $200 annually for the purpose of corresponding with consuls of the United States, officers of the United States

Navy, and persons in foreign lands. The society also provided seeds gratis to its members. Excerpts taken from the minutes of other societies, for example, the Albemarle Agricultural Society of Virginia, the Columbian Agricultural Society of Washington, and the Berkshire Society of Massachusetts, show that shipments of seed were frequently received from abroad and distributed among the members for trial.

Although it was late in entering the field of plant introduction, the United States Government was destined to become a major contributor in plant exploration and introduction. It was recognized that the Navy had greater opportunities in this work than other departments of government, and early this organization was encouraged to make the most of its opportunities. One result was the introduction of large Lima beans from Peru by Captain John Harris, U.S.N., in 1824. This bean soon became popular and remains so to the present. The attitude of the Navy is exemplified in the orders issued to Commander William Crane in 1827: "It will probably be in your power, while protecting the commercial, to contribute something to the agricultural interests of the nation, by procuring information respecting valuable animals, seeds, and plants, and importing such as you can, conveniently, without inattention to your more appropriate duties, or expense to the Government."

Subsequently a number of major expeditions were sent out by the Navy for the purpose of opening trade and exchanging agricultural implements and seeds. The best known of these expeditions was the Perry expedition to Japan, which was instructed to "collect all indigenous vegetable products within your sphere of operations, with a view to their introduction into the United States, preserving seeds and dried specimens of as many plants as possible." In 1839 an Agricultural Division of the Patent Office was formed to aid in publishing agricultural statistics and collecting and distributing seeds. Although this organization aided in the introduction of many plants, it was the establishment of the Colleges of Agriculture and Agricultural Experiment Stations, following the Morrill Act in 1862, that gave new and important emphasis to plant introduction.

The importance of the experiment stations to plant introduction was primarily in providing the means for testing new plants. Their locations represented a great variety of environmental conditions, with the result that the local adaptation of varieties could be determined much more quickly and accurately than was possible previously. Soon materials were being exchanged, both nationally and internationally, on a scale never before attainable. For the most part, however, these exchanges depended on personal arrangements among individuals, and organization and purpose in introducing new types were still largely lacking. This was remedied in the last decade of the nineteenth century when the U. S. Department of Agriculture actively entered the field of foreign plant introduction. The mechanism for testing had been available for some time in the state experiment stations, but now direction and stability were introduced into the program by the active participation of the U. S. Department of Agriculture in introducing new types and overseeing their distribution.

Because of his extensive explorations, writings, and leadership, David Fairchild stands in the forefront of plant explorers. His remarkable career with the Department of Agriculture, which began about 1890, carried him to all parts of the world. Among the thousands of collections he sent back were Hairy Peruvian Alfalfa, Berseem Clover, and a great range of fruits and vegetable varieties. Niels Hansen made his first exploration in 1897, going to Siberia to search for alfalfa and other forage plants able to thrive in the cold and arid prairie regions. He turned out to be a highly successful collector, introducing crested wheat grass and bromegrass and in later explorations many varieties of cereals. Mark Carleton, named wheat specialist of the "Section of Seed and Plant Introduction" when it was organized in 1898, collected wheat in Russia. Among the varieties he introduced were Kharkov, a hard red winter variety which at one time occupied 21 million acres annually, and varieties which formed the basis of the durum wheat industry of the north plains area.

Commercial Varieties Originating from Introductions

As a result of the work of such plant explorers as Fairchild, Hansen, Carleton, and their successors, large numbers of introductions became available for testing by plant breeders. In general, commercial varieties that originated from these introductions were developed in three different ways: (1) directly by increase en masse from the introduced stock, (2) from selections made from introductions, and (3) from hybrid offspring of introductions crossed with adapted varieties.

The most important commercial types that arose from direct increase en masse of introduced materials were developed so long ago that the details are now obscure. Sometimes the introductions became domesticated and escaped into uncultivated areas to the extent that they are now commonly regarded as native species. Some have even acquired local names such as Kentucky and Canada bluegrass. Other thoroughly domesticated forage species include, among the grasses, Bermuda grass, Dallis grass, meadow fescue, orchard grass, and redtop, and among the legumes, alfalfa, white clover, red clover, and alsike clover.

Ladino clover, bunyip wheat, and rose clover are more recent and better documented examples of commercial types which arose from the direct increase *en masse* of an introduced plant stock. Ladino clover is a large form of white clover believed to have originated by natural selection in the area near Lodi, Italy. It was introduced to the United States in 1891 and is now the basic pasture legume in most of the humid or irrigated regions of the United States.

Bunyip wheat was selected from a cross made in Australia in 1897 and was introduced to the United States at the International Exposition in San Francisco in 1915. The Sperry Flour Company increased and distributed seed of this variety.

Rose clover was introduced to the United States from Turkey in 1944. Among three collections one proved to be outstanding for range reseeding. It was increased and released as a commercial variety in 1948.

Numerous examples can be cited of commercial varieties which were selected from introduced stocks. Acala cotton is an outstanding example. The original stock of Acala came from the village of this name in southern Mexico. Its performance aroused interest in the stock when it was tested in Texas in 1907. After several generations of selection, a reasonably uniform type of Acala was obtained and released for commercial production in 1911. Strains of Acala now occupy almost all the acreage of irrigated cotton in the southwestern states and account for about one fourth of the total cotton production of the United States.

Numerous examples can also be cited of use of an introduced variety as a parent in a cross. One of the most spectacular examples is that of Victoria oats, a variety introduced from South America in 1927. Victoria itself was not suitable for commercial production due to various agronomic deficiencies but it was found to be resistant to many races of crown rust and smut. Crossed with Richland, a stem-rust resistant variety adapted to the central corn belt, it produced a number of varieties which were outstanding agronomically and resistant to crown rust, stem rust, and smut. The Victoria-Richland derivatives at one time occupied millions of acres in various parts of the United States.

It has gradually become more rare for varieties to be directly useful upon introduction to new areas. In the development of varieties to meet precisely the needs of their own areas, plant breeders may be turning out types less useful to other areas. These new types are probably, but not necessarily, less widely adapted than the older types. It may be that they are increasingly unable to compete outside their range of particular adaptation because simultaneous progress toward more specialized types has occurred in nearly all producing areas. This is particularly true of major crops that have been the object of intensive plant breeding wherever they have been grown. The role of introduction in the future would appear to be less in providing varieties directly useful for commercial production and more in providing a reservoir of germplasm at the disposal of plant breeders.

World Collections

Even in the past, only a very small proportion of introductions became successful varieties. The fate of most introductions has not been production on a vast scale, but humble inclusion in the great repositories of plant types known as the world collections. As a result of the activities of various collectors in introducing new types, particularly in the last half century, these collections have become very large, at least for certain crop species. There are, for example, more than 12,000 entries in the wheat collection and more than 5000 in the world barley collection maintained by the U. S. Department of Agriculture. The collections of many other crops maintained by this organization are also very large, and the number of entries is being increased constantly. In discussing modern plant procurement and exploration, W. H. Hodge and C. O. Erlanson of the "Section of Plant Introduction" of the U. S. Department of Agriculture state that

. . . the bulk of plant material is obtained through correspondence, exchange, purchase, or gift. A great part of it is obtained by deliberate intent, to be used in specific breeding programs or tested for use in the diversification of the agriculture of some region, but some also comes in unsolicited from travelers, foreign service workers, and friendly governments as international exchange. Actual explorations are conducted when the plant materials needed cannot be obtained by other means. . . . Through explorations come the most valuable introductions, the wild relatives of our cultivated crops, the locally grown strains and varieties which may have characters useful to the breeder, and occasionally plants which may be the basis of completely new crops for the United States . . . most explorations do not develop on short notice but arise from cooperative planning between workers at state and federal experiment stations, who bring together suggestions as to geographical areas to be explored and special crop groups to be collected, as well as to the order or priority in which explorations are to be made. The Section then undertakes the exploration work based on these recommendations.

Clearly the task of plant introduction has made great gains in precision since its haphazard beginnings. This is the result of the experience gained in foreign exploration coupled with increased appreciation of the agricultural requirements of successful varieties. Also the great advances that have been made in genetic and cytological aspects of taxonomy have provided a much improved basis for judging the potential value of exotic types.

Centers of Origin of Cultivated Plants

Nikoli Ivanovich Vavilov stands in the forefront of contributors to our knowledge of the global dispersal of crop plants and their wild relatives. The tremendous mass of material assembled by the institute which he directed for about 20 years, beginning in 1916, led Vavilov to the belief that most of the varietal wealth in our crop plants was concentrated in eight great centers of diversity: China, Hindustan, Central Asia, Asia Minor, the Mediterranean Region, Abyssinia, Central America, and west-central South America (Peru-Ecuador-Bolivia). Vavilov proposed that the centers of origin of species coincide with the areas where the greatest diversity exists in the species. Whether or not such areas are actually centers of origin or only topographically or otherwise suited for the preservation of variation has little practical concern for plant explorers. The centers of diversity located by Vavilov have turned out to be fertile collecting areas and remain promising areas for future explorations. Vavilov also recognized secondary centers of origin and was careful to point out that valuable forms are found far removed from the primary area of origin. He cited the Washington Navel Orange, discovered in Brazil, whereas the basic center of diversity for citrus is in southeastern Asia.

Another of Vavilov's contributions concerned the "law of homologous series in variation." It had been noted by several early botanists that characters found in one crop species may also be expected in similar species. Vavilov expressed this concept in genetic terms and used this principle as a clue to characters remaining to be discovered.

It has frequently been found that certain characters occur much more frequently in collections made in some areas than in others. In the world barley collection, for example, about 75 per cent of the types resistant to net-blotch disease come from Manchuria, a country that contributed only about 12 per cent of the total collection. Similarly, Hessian-fly-resistant varieties abound in North Africa, greenbug-resistant barleys are likely to originate in China or Korea, and the Crimean wheats are the most common source of resistance to bunt or stinking-smut disease. On the other hand resistance to powdery mildew in barley comes from types originating in many different countries. As experience has been gained it has become increasingly possible to predict where materials valuable for specific areas or purposes are likely to be found, and this information has aided in planning expansion of the collections.

J. R. Harlan, during a plant exploration trip to Turkey in 1948, was impressed with the tremendous plant diversity found in small areas. These areas Harlan referred to as *microcenters* since plant evolution appeared to proceed in them at a more rapid rate than in other areas, particularly larger geographic regions. These microcenters seemed to offer an excellent opportunity, not only to collect valuable types, but also to study evolution of cultivated types experimentally.

In a later discussion (1956) of natural variability in plants, Harlan points out that the geographic centers of diversity upon which we have depended so much in the past for our sources of germplasm are in danger of extinction.

Modern agriculture and modern technology are spreading rapidly around the world. New, uniform varieties from the experiment stations are replacing the old mixed populations that have been grown, in some cases, since the neolithic. In southern Turkey in 1948, I found great acreages of flax planted to a single variety. It was a selection of Argentine origin resistant to rust. From one end of Celicia to the other I could find not a single indigenous variety although this very area had at one time been a center of diversity for flax. . . . We now find ourselves in a predicament in which technologically backward countries cannot afford to keep their great varietal resources and we cannot afford to let them be discarded. The only answer is an extensive exploration and collection program devoted to assembling as much of the germplasm of the world as possible and the diligent maintenance of the material once it is obtained.

Maintenance and Use of World Collections

Once a plant introduction has been received, it must be inspected for pests and diseases and identified before it is ready for the next steps—propagation, testing, and distribution. The original introduction, whether in form of seed or clonal material, is usually limited in quantity and must be increased before it can be tested or distributed to plant breeders. In the United States, these various steps are carried out by an Inspection House through which every introduction must pass, a quarantine station to which suspected materials are referred, four plant-introduction gardens, and four regional plant-introduction stations.

The task of maintaining large collections and making them available to plant breeders is an enormous one. It is even a formidable task for a plant breeder to survey a collection in search of some specific character. Thus to examine the entire world wheat collection of the United States Department of Agriculture requires about twenty-five working days, if only one minute is spent on each entry. Obviously, to package, sow, maintain, and examine this collection is a task not to be undertaken lightly.

Schaller and Wiebe (1952) illustrated a typical use of the world collections in their work with barley. Net blotch was a serious disease of barley in California and one for which no source of resistance was known. The obvious place to turn was to the world collection of barleys from which 4526 types were grown under epiphytotic conditions in 1948. Only 218 types were sufficiently promising to warrant further testing. Rigid screening in 1949 and 1950 further reduced the number to 75 types. These 75 types provided an ample base from which to initiate a breeding program designed to produce varieties combining acceptable quality, agronomic worth, and net-blotch resistance. The value of the world barley collection needs no emphasis when one contemplates an attempt to breed for net-blotch resistance without the collection. The world collections of the various crops represent treasure houses of variability whose exploitation for the benefit of agriculture is still almost entirely in the future.

Domestication

Domestication can be loosely defined as the bringing of a wild species under the management of man. It is a method of plant breeding in the sense that, when successful, it provides domestic types that are superior to ones previously available.

Among the higher plants the greatest opportunity and progress in domestication at present are perhaps with timber and forage species. The species which occur in natural stands of timber can, with some logic, be regarded as wild species. But when these trees are harvested and the forest replanted using selected strains, a case can be made for regarding the same species as now domesticated. The situation is perhaps even less clear cut among the forage species, especially the ones which provide feed on range or wild lands. Forage production on the ranges is almost entirely from resident species which are present naturally. When, however, management is practiced to encourage certain of these species, it might be argued that their domestication has started. Certainly a start toward the domestication of such species has been made when plant breeders exploit the variability existing in them by selecting desirable types for use in range improvement.

Although early man did a remarkable job of domesticating those species most useful for food, fiber, and medicinal purposes, advances in technology have created demands for new plant products sometimes unavailable in the present domesticated species. For example, the discovery of the usefulness of cortisone for medicinal purposes created demand for this drug, and a search was begun for plants to satisfy this demand. The tropical wild yams, *Dioscoria*, are potentially rich sources. Domestication is not new to this family, as some of its species have been cultivated for food for hundreds of years.

The most dramatic example of domestication in historical time has been that of microorganisms for the production of antibiotics. Improved strains of *Penicillium* in use today far exceed the wild types in productivity. Microorganisms very largely escaped the attention of our early ancestors, and understandably so, although they inadvertently utilized yeasts and some other lower organisms in much the same way they are used today.

One important aspect of domestication at present is the use of particular genes from wild relatives in the improvement of cultivated species. When a plant breeder transfers one or a few desirable genes from a wild relative to a cultivated type he is, in a sense, domesticating the wild species in part.

REFERENCES

Ames, Oakes. 1939. *Economic annuals and human cultures.* Botanical Museum of Harvard University, Cambridge, Mass. (Includes a discussion, species by species, of the areas of origin of cultivated annuals.)

Anderson, Edgar. 1952. *Plants, man and life.* Little, Brown and Co., Boston. (This stimulating book covers a range of topics related to the effect man has had on plants.)

Fairchild, David. 1938. *The world was my garden.* Charles Scribner's Sons, New York and London. (The entertaining and informative diary of a noted plant explorer.)

Harlan, J. R. 1951. New world crops in Asia Minor. *Scientific Monthly* 72: 87–89.

Harlan, J. R. 1951. Anatomy of gene centers. *Amer. Nat.* 85: 97–103.

Harlan, J. R. 1956. Distribution and utilization of natural variability in cultivated plants. *Brookhaven Symposia in Biology,* No. 9: 191–208.

Hodge, W. H., and C. O. Erlanson. 1955. Plant introduction as a federal service to agriculture. *Adv. in Agron.* 7: 189–211.

Klose, Nelson. 1950. *America's crop heritage*. Iowa State College Press. (Detailed description of the introduction of plants to the United States.)

Moseman, A. H. 1956. New crops from strange plants. *Crops and Soils* 8: 22, 42.

Ryerson, K. A. 1933. History and significance of the foreign plant introduction work of the U. S. Department of Agriculture. *Agric. History* 7: 110–128.

Schaller, C. W., and G. A. Wiebe. 1952. Sources of resistance to net blotch in barley. *Agron. Jour.* 44: 334–336.

Vavilov, N. I. 1951. The origin, variation, immunity and breeding of cultivated plants. Translated from the Russian by K. S. Chester. *Chronica Botanica* Vr. 1/6. (The most important of Vavilov's contributions to biological thought have been brought together in this important volume.)

4

Reproductive Systems in Cultivated Plants

From the standpoint of the plant breeder, agricultural species can be divided into two groups, depending upon whether they are predominantly self-pollinating or largely cross-pollinating. This distinction is one of primary importance because the methods of breeding applicable to the self-pollinated group are for the most part distinct from those that apply to the cross-pollinated species. The important difference between the two groups is related to the influence of inbreeding versus outbreeding on the genetic structure of populations.

All plants in populations of outcrossing species are highly heterozygous and, almost without exception, enforced inbreeding results in deterioration in general vigor and in other adverse effects. Heterozygosity appears to be an essential feature of commercial varieties of these species and, as a result, it must either be maintained during the breeding program or restored as a final step of the program. On the other hand, populations of self-pollinated plants usually consist of mixtures of many closely related homozygous lines, which, although they exist side by side, remain more or less independent of one another in reproduction. Individual plants in such populations are likely to be fully vigorous homozygotes. With these species, the goal of most breeding programs is a pure line. The general pattern of the breeding program appropriate to any particular species is therefore as much influenced by its mode of reproduction as by any other single feature of the species. When vegetative parts of the plant can be used to produce new individuals (asexual reproduction), additional patterns become possible in plant-breeding programs.

Besides its influence in determining the general features of the breeding program, the reproductive biology of a species has an important role in determining the more specific procedures which are likely to be successful. In modern plant breeding, hybridization between deliberately chosen parents has become increasingly important. The ease with which controlled hybrids can be made, or selected individuals selfed, therefore frequently has an important role in determining not only the goal of the program but also the details of its execution. Thus the type of program most successful with maize, in which selfing is easy and hybridization on a large scale is economical, turns out to be different in both goal and execution from the type of program natural and suitable for alfalfa, also a cross-pollinated species, but one in which both selfing and controlled crossing are difficult. Similarly the ease with which either selfed seeds or reasonably large numbers of hybrid seeds can be obtained in a self-pollinated species such as tobacco permits more flexibility in choice of breeding program than for oats, also a self-pollinated species but one in which artificial hybrids can be produced only laboriously and with great expense. Many species are so highly self-fertilized that measures to protect against outcrossing are unnecessary even in breeding nurseries where many different types are grown in close proximity, whereas in other species outcrossing is so frequent that some form of pollination control is essential.

Knowledge of reproductive systems is so clearly fundamental to plant breeding that discussion of them must precede any consideration of breeding methods themselves. In the remainder of this chapter we shall discuss some of the more common systems by which cultivated plants reproduce themselves. The impact of these reproductive systems on plant breeding procedures will be considered, in outline form, in Chapter 5 and in more detail in later chapters.

Devices of Pollination Control

Devices for the control of the breeding system, whether to encourage inbreeding or outbreeding, occur in a variety of forms in the higher plant species. In general, efficient inbreeding schemes are less difficult to secure than equally efficient outbreeding schemes because the former do not require genetic diversity among individuals. Restrictions on outbreeding are almost always imposed either by cleistogamy, or by devices with similar effects. True cleistogamy is exemplified in the basal inflorescences of California oatgrass, *Danthonia californica.* These inflorescences are borne between the culm and sheath, and, since they never emerge from the sheath, seeds produced in them are clearly self-pollinated. In certain of the annual fescues, for example, *Festuca megalura,* the inflorescences emerge from the boot but the florets do not open, also insuring selfing. Lettuce is also cleistogamous in that the stigma is pollinated before the flower opens. It should be recognized that the precision with which inbreeding devices operate is subject to modification by both genetic and environmental forces. Degree of outcrossing thus varies from one genetic background to another and varies in different seasons and locations.

Wheat flowers are structurally similar to those of rye (a strong outbreeder). But in wheat the anthers burst before extrusion so that the empty anthers hanging out are only a mockery. This amounts to effective if not morphological cleistogamy. The chasmogamous behavior of wheat flowers does, however, permit a low incidence of outcrossing. In the tomato, pollination follows the opening of the flower, but the stamens form a cone enclosing the stigma in such a way that self-pollination is almost insured. Among cultivated types of tomatoes there is some variation in the length of the style, thus affecting the position of the stigma within the cone. Other variations in floral morphology can also occur so that considerable outcrossing takes place in some types. The point is this: Variations in characteristics such as floral morphology and time of pollen shed, which may be under simple genetic control, can drastically alter the mating system of a species; this in turn is likely to produce dramatic and far-reaching effects on the genetic structure of populations and hence on the breeding procedures that are natural and appropriate to the species.

In both wheat and tomatoes, inbreeding appears to have supplanted outbreeding. The same is true for peas, where the showy entomophalous flowers are wasted. There is some evidence that this situation is common and that inbreeders arise constantly from outbreeders, but rarely the reverse. It may be both interesting and profitable to speculate about this situation.

Mather believes it may result from the greater flexibility provided by the genetic structure associated with outbreeding populations, permitting outbreeders to adapt better to long-range changes in environment than genetically less flexible inbreeders. According to his hypothesis, inbreeders arise constantly from outbreeders because inbreeders, by virtue of the uniformity accompanying homozygosity, are capable of high immediate fitness. This allows them to meet particular short-term requirements of environment better than their more variable outbreeding ancestors. This line of reasoning appears to fit the situation of domestic species, many of which are inbreeders, perhaps because prime importance is attached to high immediate fitness under cultivation. In nature inbreeding may be suicidal in the long run, according to Mather, because it destroys the flexibility theoretically necessary to meet the challenges of long-term changes in environment. There is some recent evidence that inbreeding species may not be as inflexible as traditionally regarded. Although this argues against the ideas presented above, it cannot be denied that far more devices exist to discourage inbreeding than to encourage it, and that there are more outcrossers proportionately among natural species than domesticated species.

It has been noted previously that inbreeding devices can be simply controlled genetically because genetic diversity among individuals is unnecessary. There are also a number of outbreeding devices that do not require genetic diversity. Among the most common of such devices are ones causing the gametes to be separated in space. Maize, walnuts, strawberries, and pecans are examples of monoecious crop plants. In protandry and protogyny, the anthers and stigma of a flower mature at different times. Carrots and raspberries are examples of protandrous species (pollen shed first); avocados and walnuts are examples of protogynous species (stigma receptive first).

Another example of an outbreeding device is provided by alfalfa. In this species the stigma does not become receptive until a protective membrane enclosing it is ruptured. This is accomplished by tripping. In alfalfa the stamens and stigma grow inside the keel and are held by the keel under considerable tension. When this tension is released by mechanical pressure, supplied in nature by bees, the stigma snaps against the standard, the membrane is ruptured, and pollen is released. Some of this pollen adheres to the hairy body of the bee and serves to pollinate the next flower visited.

Comparatively little is known about the effectiveness of such devices because measurements have been made in only a few species. Certainly they can be very effective, as in maize, which is highly outcrossed (90 per cent or more) by virtue of a combination of monoecism and protandry. Beets, which feature protandry, are also highly outbred. The same is true for red clover. The high degree of outcrossing in these species, however, may not be due to dichogamy alone. Red clover, for example, is highly self-incompatible in addition to being protandrous. Alfalfa, besides its elaborate entomophalous pollinating mechanism, has some sort of incompatibility system. Even further, Brink and Cooper have found this species to feature what they call somatoplastic sterility, in which embryos derived from selfing survive less frequently than embryos produced by outcrossing. Somatoplastic sterility obviously favors heterozygosity. If little is known about the effectiveness of these outbreeding devices, even less is known about their genetic control. Obtaining this information on the genetic control of mating systems is one of the most important areas of research in biology.

In monoecism, protandry, and protogyny, all individuals of the population are genetically alike with regard to the outbreeding mechanism. Such schemes can only hinder, not preclude, self-fertilization, however, and for systems that are absolute or nearly so in their effects, some sort of genetic diversity is necessary.

The most striking of the systems employing genetic diversity is, of course, dioecy or sexual differentiation. It is clearly an outbreeding device, obviously prohibiting selfing, but it does not preclude brother-sister mating or less severe forms of inbreeding. Dioecy, the great mating system of animals, is sporadic in higher plants, probably because it is wasteful of gametes in non-mobile organisms. Among cultivated plants the important dioecious species are date palms, hemp, hops, spinach, papayas, and asparagus. In fact, some of these species commonly produce hermaphroditic individuals and cannot be regarded as inviolately dioecious. The genetic control of dioecy in higher plants is reasonably well understood (Westergaard, 1958).

Self-incompatibility is far more important in higher plants than dioecy. There are two general schemes of self-incompatibility in the higher plants: (1) gametophytic or haplo-diplo schemes, in which the incompatibility depends upon the genotype of the gametophyte; and (2) sporophytic or diplo-diplo schemes, where the incompatibility is impressed upon the gametophyte by its sporophytic parent. In the gametophytic system the effects of the incompatibility are easily classified. Upon selfing no seed is produced. Upon crossing either all of the pollen grows (fully compatible), one half grows and one half fails (half-compatible), or no pollen grows. Classification is even more simple for the sporophytic system: upon selfing no seed is produced, and upon crossing either all or none of the pollen grows. The genetical interpretation of these results is also straightforward but will be deferred to Chapter 20, where it will be discussed more fully. It should be pointed out before leaving this important subject temporarily, that self-incompatibility is not necessarily absolute, and the plant breeder who wishes to self-pollinate highly self-incompatible species can usually do so by one stratagem or another. Also, the incompatibility mechanism is poorly developed in many species, serving to discourage rather than prohibit self-fertilization.

Asexual Reproduction

Reproduction by asexual means is common among higher plants. The best-known means of asexual reproduction are by corms, bulbs, rhizomes, stolons, tubers, or other vegetative organs. Plants normally propagated in agricultural practice by these means or by budding or grafting include nearly all fruit and nut trees, strawberries, blackberries and raspberries, grapes, pineapples, almost all ornamental shrubs and trees, and a few field crops such as sugar cane, potatoes, and sweet potatoes. The experience of plant breeders indicates that agricultural varieties of asexually propagated species are highly heterozygous and segregate widely upon sexual reproduction. This is not unexpected, because the types selected as commercial varieties are likely to be vigorous ones, and a positive correlation exists in most species between vigor and heterozygosity.

It should be noted that some species not normally propagated vegetatively can be increased by vegetative means when considerable amounts of seed of certain genotypes are required for experimental purposes.

Asexual reproduction leads to perpetuation of the same genotype with great precision. It can be a great advantage in plant breeding since an indefinitely large number of genetically identical individuals can be obtained irrespective of the degree of heterozygosity of the genotype. The breeder can therefore take advantage of outstanding individuals occurring at any stage in a breeding program. For this reason breeding these crops is in some respects less troublesome than breeding other species.

Besides seeking desirable recombinants following sexual reproduction, breeding in asexual plants also makes use of the search for desirable mutants (sports), both natural and artificially induced.

In addition to the types of vegetative reproduction commonly used in the horticultural arts, there are a number of other types grouped under the term apomixis. In apomictic reproduction seeds are produced, but through processes other than normal meiosis and fertilization. The main effect of apomixis is an increase in the proportion of maternal individuals through prohibition or modification of genetic segregation and recombination. Apomixis is often not stable, however, with the result that apomictic species are notoriously variable. Detailed descriptions and classifications of the many types of apomixis have been given by several authors (e.g., Stebbins, 1941), and types of apomixis will not be considered here.

Determining Mode of Reproduction and Prevalence of Natural Crossing

Mode of reproduction and prevalence of natural crossing are reasonably well known for most important agricultural species. This information is incomplete in varying degrees, however, particularly regarding the prevalence of natural crossing under varying environmental conditions. Successful conduct of a breeding program may therefore require determination of rates of crossing for the particular area in which the breeding program is to be conducted.

Experiments to determine whether a plant is self- or cross-pollinated are ordinarily simple and straightforward. Examination of floral structure is an obvious first step and for some plants, dioecious species, for example, will settle the question forthwith. Similarly, dichogamy, monoecism, or other outbreeding devices furnish presumptive evidence that the species is cross-pollinated, just as cleistogamy provides evidence of self-pollination.

Usually the next step is to isolate single plants and observe whether or not seeds are produced. Isolation in space is the preferred procedure, since isolation by bags or cages introduces the possibility of imposing environmental conditions adverse to seed production. Failure of seed set in isolation is an almost certain indication that the species is cross-pollinated. The reverse is not necessarily the case, however, because many cross-pollinated species such as maize are highly self-fruitful. Thus, although outcrossing species are fairly easily identified, self-pollination is more difficult to establish. One good indicator of self-pollination is the effect of inbreeding. If inbreeding can be carried out without adverse effects, the species is probably normally self-pollinated.

Once it has been established that a species is largely self- or cross-pollinated, the question arises as to the amount of natural crossing that occurs when different genotypes are grown in proximity to one another. The extent of natural crossing is ordinarily determined by interplanting strains carrying a recessive marker gene with strains carrying the dominant alternative allele. Seeds are harvested from the recessive type and the amount of natural crossing calculated from the proportion of recessive and dominant progeny. From the standpoint of efficiency, seed or seedling characters are preferred in studies of this sort. Such genes as the ones governing starchy versus glutinous endosperm in grasses or white versus green cotyledon in legumes make desirable markers, since large numbers of individuals can be handled with little effort or expense.

A number of factors must be taken into account in conducting studies of this sort. Time of flowering must be considered, because varieties flowering at different times will not hybridize in the same amount as varieties flowering concurrently. These experiments can frequently provide information about the insect vectors of pollen, the effect of wind direction, temperature, and the interactions of such factors on the extent of natural crossing. Studies conducted over a number of seasons and locations are particularly valuable, because amounts of natural crossing in different genotypes can be greatly influenced by environmental conditions.

Experiments of the type described above can also be used to establish the necessary distance between planting areas, both with and without intervening crops, to prevent unacceptable amounts of contamination.

Self- and Cross-Pollinated Crop Species

At the end of this section appear lists of the more important self- and cross-pollinated crop species. Information on some of the species included is fragmentary, or even inconsistent, and it is likely that some reclassification will be necessary as more accurate information on mating systems is developed. The lists should, nevertheless, prove useful as a general guide to the breeding behavior of the major agricultural crops.

In the preparation of these lists, the only criterion of classification was degree of self-pollination versus cross-pollination. Self-pollination is of course only one of many mating systems and only one way by which populations can become inbred. The cucurbits are a case in point. Although monoecious, they do not suffer inbreeding depression, and in many ways their population structure is more similar to that of inbreeders than outbreeders. There are several possible explanations for this, both biological and economic. Cucurbit plants are large and expensive of space. Also a few plants are likely to satisfy the wants of any one grower. Cucurbits may consequently have existed in small colonies both in nature and under cultivation, and this restriction of population size may have produced inbreeding despite the floral mechanism favoring outcrossing. Similar situations may exist in other cross-pollinated species less fully investigated than the cucurbits. For this reason it has seemed necessary to classify all species known to be cross-pollinated in the outbreeding group, although this classification may not accurately predict response to inbreeding or appropriate breeding methods. In this respect, the self-pollinated species are much more homogeneous. None of these species suffers conspicuous inbreeding depression, and the same breeding methods are for the most part applicable to all members of the group.

Within the self-pollinated group the amount of outcrossing is important largely because it affects contamination of breeding stocks. There are great differences in amounts of outcrossing among different species of the group. In fact different varieties within the same species may show widely different amounts of natural hybridization. Also the amount of crossing for a given variety can be considerably influenced by environmental changes. Generalizations about the amount of natural crossing to be expected in self-pollinated species are therefore valid only within limits defined in terms of specific varieties and delimited environmental conditions.

Nevertheless, in certain species of this group—especially cotton, sorghums, and annual sweet clover—some type of control of outcrossing will almost always be required, since at least 5 per cent of natural crossing occurs under most environmental conditions and as much as 50 per cent under some conditions. In other species, such as barley, oats, rice, lettuce, and tomatoes, the amount of natural crossing is rarely more than a fraction of 1 per cent, regardless of variety. This is too little to be a factor in most breeding programs. Other species are more variable in their behavior. In Lima beans, for example, 25 per cent or more of natural crossing is usual and 100 per cent is sometimes approached under the humid conditions of the eastern seaboard of the United States. Less than 1 per cent is characteristic of most varieties in California, but as much as 20 per cent may occur occasionally between adjacent plants of a few varieties. It is clearly desirable for studies of amounts of outcrossing to be conducted, so far as practical, for every set of conditions and with as many genotypes as feasible.

In the list of cross-pollinated species an attempt is made to indicate the type of mechanism by which outcrossing is encouraged or enforced. Some species are difficult to classify because the outcrossing mechanism varies from strain to strain or even within strains. In many instances this is at least partly the result of selection under domestication, as exemplified in the changes which have occurred in the breeding behavior of the garden strawberry. Both of the wild ancestors (*Fragaria virginiana* and *F. chiloensis*) are dioecious. Dioecious varieties are less desirable than types with perfect flowers because the latter are fruitful when planted alone. Consequently strawberry breeders have favored hermaphroditic types and have selected toward them until nearly all modern strawberries are fully self-fruitful. This process started long ago, since perfect-flowered *F. chiloensis* was being cultivated in western South America at the time of Columbus. There has been a similar tendency in many other crops to select away from dioecy, monoecy, and self-incompatibility, toward fully self-fruit-ful types.   This process is particularly noticeable in grapes and in fruit and nut crops, where the proportion of self-fruitful types has increased markedly in recent years.

Self-Pollinated Crop Plants

Cereal Grasses

Barley
Foxtail millet
Oats
Rice
Sorghum*
Wheat

Legumes

Broadbean*
Chick pea
Common bean
Cowpea
Lima bean
Mung bean
Peanut
Pea
Soybean
Sweet pea
Urd bean

Forage Grasses

Annual fescue
Foxtail barley
Mountain bromegrass
Slender wheatgrass
Soft chess

Forage and Green Manure Legumes

Annual sweet clover
Bur clover
Crotalaria juncea
Hop clover
Strawberry clover (common)
Subterranean clover
Velvet bean
Vetch (common, hairy, and pannonica)

Fruit Trees

Apricot
Nectarine
Peach
Citrus

Other Species

Cotton*
Eggplant
Flax
Lettuce
Okra
Pepper (Capsicum annuum, C. frutescens)
Tobacco
Tomato
Parsnip
Endive

*Frequently more than 10 per cent outcrossed.

Cross-Pollinated Crop Plants

Cereal Grasses

Maize[b]
Rye[c]

Forage Grasses

Annual ryegrass[d]
Buffalo grass[a]
Orchardgrass[d]
Meadow fescue[d]
Perennial ryegrass[d]
Smooth bromegrass[d]
Tall fescue[d]
Timothy[d]

Forage Legumes

Alfalfa[d]
Alsike clover[c]
Birdsfoot trefoil[d]
Crimson clover
Red clover[c]
Strawberry clover (Palestine)[d]
Sweet clover[d]
White clover[c]

Legumes

Scarlett runner bean[d]

Fruits

Apple[d]
Avocado[d]
Banana[b,f]
Cherry[c]
Date[a]
Fig[e,f]
Grapes[b]
American grapes[a,b]
Mango[d]
Olive[d]
Papaya[a]
Pear[d]
Plum[d]

Nuts

Almond[c]
Chestnut[b]
Filbert[b]
Pecan[b]
Pistachio[a]
Walnut[b]

Other Species

Artichoke	Celery	Pumpkin[b]
Asparagus[a]	Chard	Radish[c]
Beet[d]	Collard[c]	Raspberry
Blackberry[b]	Chicory[c]	Rhubarb
Blueberry	Chinese cabbage[c]	Rutabaga[c]
Broccoli[c]	Cucumber[b]	Safflower
Brussels sprouts[c]	Hemp[a]	Spinach[a]
Cabbage[c]	Kale[c]	Squash[b]
Cauliflower[c]	Kohlrabi[c]	Strawberry[b]
Carrot	Mangel[d]	Sunflower
Castorbean[b]	Muskmelon[b]	Sweet potato[d]
	Onion	Turnip[c]
	Parsley	Watermelon[b]

[a]Dioecious.   [b]Monoecious or monoecious strains occur.   [c]Strongly self-incompatible.   [d]Self-incompatible in some degree or self-incompatible strains occur.   [e]Effectively dioecious.   [f]Parthenocarpic.

REFERENCES

Brink, R. A., and D. C. Cooper. 1939. Somatoplastic sterility in *Medicago sativa*. *Science* 90: 545–546.

Darlington, C. D., and K. Mather. 1949. *Elements of genetics*. The Macmillan Co., New York. (Includes chapters on breeding systems in plants.)

Lewis, D. 1942. The evolution of sex in flowering plants. *Biol. Rev.* 17: 46–67.

Mather, K. 1943. Polygenic inheritance and natural selection. *Biol. Rev.* 18: 32–64. (A general discussion of the consequences of inbreeding and outbreeding on population structure.)

Stebbins, G. L., Jr. 1941. Apomixis in the angiosperms. *Bot. Rev.* 7: 507–542.

Westergaard, M. 1958. The mechanism of sex determination in dioecious flowering plants. *Adv. in Gen.* 9: 217–281.

Yearbook of Agriculture, U. S. Dept. Agric. 1936, 1937. (Gives detailed descriptions of floral structure and mode of reproduction in crop species.)

5

Reproductive Systems and Plant Breeding Methods

The general pattern of the breeding program appropriate to a particular species, as we saw in Chapter 4, is determined in a large part by the reproductive system of the species. We shall now develop this idea further. In the present chapter the development will be brief and in outline form to provide general orientation to plant breeding methods. Later chapters will give detailed descriptions of each method. Special methods, such as mutation breeding, will not be considered in the present brief outline of breeding procedures.

Self-Pollinated Species

The breeding methods that have proved successful with self-pollinated species fall into the following categories:

1. Pure-line selection
2. Mass selection
3. Hybridization, with the segregating generations handled by the
   (*a*) Pedigree method
   (*b*) Bulk method
   (*c*) Backcross method

All of these methods are based on the fact that selfing, or backcrossing to a homozygous parent, leads to homozygosity.

Pure-line selection has been widely used to select new varieties from the old "land" varieties that have passed down from generation to generation of farmers. Although they may be reasonably similar in gross morphology, lines within a farmer's variety may be different in agricultural value. Most plants selected from such varieties can be expected to be homozygous and hence the starting point of a new

43

true-breeding variety. The usual procedure in pure-line breeding has been to select a large number of single plants, compare their progeny in field trials, and save the single most valuable progeny as a new variety. Many valuable varieties, however, are not the result of an organized program, but trace back to a single chance variant noticed and selected by a farmer.

Mass selection differs from pure-line selection in that a number of plants, rather than just one, are selected to make up the new variety. Varieties developed by this method thus include fewer genotypes than the parent population but more than the "single" genotype of varieties developed by pure-line selection. The number and variability of types included depend on the variability of the original population and the kind and intensity of selection practiced. As a breeding method, mass selection now appears to have its greatest applicability in underdeveloped countries where land varieties still exist. There the method may be useful in eliminating types of low agricultural value without the dangers associated with the selection of single genotypes. In highly developed agricultural areas, mass selection now finds its greatest use in the purification of existing varieties in connection with pure-seed programs. Usually several dozens or even several hundreds of typical individuals are selected from fields of the variety to be purified, each line progeny tested, and "on-type" progenies bulked to form the pure-seed lot.

As plant breeding progressed and more and more of the natural variability existing in populations of self-pollinated species was exploited, it became increasingly important for the plant breeder to set up his own variability by making artificial hybrids. Selfing hybrids leads to the separation of large numbers of homozygous types, combining in various proportions the genes contributed by the parents. During the period while homozygosity is being attained by selfing, most plant breeders have handled the hybrid swarm in one of two ways, the pedigree method or the bulk method.

The pedigree method is widely used by modern-day plant breeders. It derives its name from the records that are kept of the ancestry or pedigree of each of the progenies. Selection for superiority is based on the vigor and other agricultural features of individuals or progenies (families). In the F_2 selection is, of course, limited to individuals. In F_3 and subsequent generations, until homozygosity prevails, selection is practiced both within and between families. Thence selection is practiced among families until the number of progenies has been reduced to the point where comprehensive evaluation trials can be undertaken.

The bulk method differs from the pedigree method in that the hybrids are grown in bulk with no attempt being made to keep track of ancestry of individuals. If artificial selection is practiced during the period of bulk propagation, this selection is usually based on individual plant performance (and not on tests of the progenies of the selected individuals). The period of bulk propagation is usually terminated in the F_6 to F_8 generation by selecting desirable individual plants from the population. These selections are then carried as families which are evaluated in the same way as in the pedigree method.

The backcross method is particularly suited for transferring specific genes to a good variety which is deficient in one or a few characteristics. In this method recurrent backcrosses are made to the more desirable parent while selection is practiced for the characters being transferred from the donor parent. The final product need not be evaluated agriculturally because its performance will be the same as that of the recurrent parent, except for its improved characteristics.

Outcrossing Species

The outcrossing group of crop species, as has been noted previously, is much less homogeneous than the self-pollinated group. Every plant in such species is certain to be heterozygous, and this heterozygosity must either be maintained during a breeding program or restored at the end of the program if productivity is to be satisfactory. There are wide variations in the agencies and mechanisms controlling cross-pollination and also in the ease with which cross-pollination can be controlled by plant breeders. Some species are self-compatible and can be inbred without undue difficulty; others are highly self-incompatible. There are also great differences in the effects of inbreeding, ranging from little or no deterioration in such crops as hemp and the cucurbits to such drastic effects in others, for example, alfalfa, that few lines can be carried beyond the second or third selfed generation. As a result of this diversity of breeding behavior, the methods of breeding cross-pollinating species cannot be so neatly categorized as can the methods successful in self-pollinated species. The most important methods applicable to outbreeding crop species are mass selection, backcross breeding, hybridization of inbred lines or other suitable materials to form hybrid varieties, recurrent selection, and the development of synthetic varieties from selected genotypes.

Historically, mass selection has been the most important method by which varieties of cross-pollinated crops have been developed. In mass selection a large number of plants are chosen to propagate the next generation, the basis for selection being generally good vigor combined with specific characteristics that appeal to the selector. Seed of the selected plants is bulked together and the next generation produced from the mass. In this procedure selection can be only for maternal characteristics because the pollen source is unknown. Many important varieties of corn, melons, vegetables, and forage species were developed by mass selection.

The backcross method can be used with outcrossing species equally well as with self-pollinated species. The main distinction is the requirement that sufficient plants of the heterozygous recurrent parent be used so that its characteristic gene frequencies will be recovered in the backcross-improved product.

Hybrid varieties depend for their superiority upon the heterosis that characterizes the F_1 hybrids between certain genotypes. The genotypes crossed to produce hybrid varieties may be inbred lines, clones, strains, varieties, or in fact, any stocks, which on crossing produce an F_1 of sufficient superiority to be attractive. It is important that the parental genotypes be maintained without change so that the hybrid will be essentially the same, genetically, year after year. Obviously inbreds or clones meet this requirement best. Another important requirement is that hybrid seed be easily made in large quantities for commercial production. Male sterility is a device receiving increasing attention as an aid to the production of these large amounts of seed needed for field-scale plantings of hybrid varieties.

In recurrent selection, desirable genotypes are selected, and these genotypes, or their selfed progeny, are intercrossed in all combinations to produce populations for reselection. These procedures help maintain the gene frequency of the selected materials. Recurrent selection thus differs from mass selection in that gene frequency is controlled in the male lineage as well as in the female lineage. Additional or recurrent cycles of selection can be made as long as satisfactory improvement continues. The genotypes used as the parents for the next generation in recurrent selection are chosen with the aid of progeny tests when the characters under selection (yield, for example) are difficult to evaluate on the basis of the phenotypic appearance of single plants. Progeny tests are not necessary when selection is practiced for characters which are easily identified by visual observation or simple tests on single plants.

The populations obtained by recurrent selection procedures can be utilized in various ways. Their most common uses are likely to be as source populations for inbred lines to be used in the production of hybrid varieties or as source populations for genotypes to be recombined into *synthetic* varieties. A synthetic variety is one that has been synthesized from all possible intercrosses among a number of selected genotypes; thereby a population is obtained that is propagated subsequently from open-pollinated seed. The essential difference between a variety developed by mass selection and a synthetic variety hinges on the way in which the genotypes to be compounded into the new variety are selected. In mass selection the next generation is propagated from a composite of the seed from phenotypically desirable plants selected from the source population. A synthetic variety is made up of genotypes which have previously been tested for their ability to produce superior progeny when crossed in all combinations. Also, in mass selection the male gametes represent a more or less random sample from the entire previous generation whereas in a synthetic variety pollination is controlled so that the gene frequency of the selected materials is maintained in the male as well as in the female lineage.

Control of Pollination

From the foregoing discussion it can be seen that plant breeders are interested in controlling pollination for two general reasons. The first is to prevent cross-pollination and resultant unwanted hybrids both in breeding nurseries and in the production of pure seed for commercial purposes. Prevention of natural crossing is usually accomplished in one of two ways, either by isolating in space or by the use of bags, cages, or some other artificial barrier to the spread of pollen.

Second, control of pollination is essential to making the particular types of matings required in the several breeding methods. Thus in hybridization methods with self-pollinators, crosses between selected strains are necessary to provide the segregation and recombination which furnish the raw materials for selection. Ordinarily a small number of F_1 seeds, frequently fewer than a dozen, will be adequate for this purpose. Backcross programs usually require somewhat larger numbers of hybrid seeds, although modifications are possible to keep the seed requirements within reasonable limits when hybridization is difficult. With cross-pollinated species, inbred lines are frequently required or useful; their development depends, of course, on ability to control pollination. Again, a few seeds per line are adequate. The amounts of seed required for the production of hybrid varieties are of quite a different order, however, and very large amounts are needed in most instances. For this reason hybrid varieties are feasible in only a few favored species.

Techniques in Hybridization

The essential problem in pollination control, whether in producing hybrids or in inbreeding, is to place functional pollen from the desired male strain on receptive stigmas at the correct time. Protection must usually be provided against selfing and also against chance outcrossing. The former is often accomplished by removing the anthers from the female strain before they mature and the latter by use of bags or other devices to exclude unwanted pollen. Elaborate equipment is usually not required. A pair of forceps and scissors are normal, and a small camel's hair brush is sometimes required. Low-magnification glasses may be necessary when dealing with small flowers.

Species vary in the amount of mutilation they will tolerate in the process of hybridization. For example, small-flowered legumes are among the most difficult species to hybridize because damaging mutilation is almost impossible to avoid as a consequence of the size and structure of the flowers. In the small-flowered legumes, and in fact in most species, emasculation is usually more difficult and time consuming than pollination. However, the latter also can be difficult enough to justify measures to circumvent hand operations.

Special techniques that have been used for emasculation include exposure of the flowers to heat, cold, or chemicals such as alcohol. These techniques are based upon the fact that pollen is usually more sensitive to unfavorable environmental conditions than the stigma. Hence a time of exposure can frequently be discovered that will destroy the viability of the pollen without excessive injury to essential floral organs. When temperature is the emasculating agent, the usual procedure is to immerse flowers in water maintained at the appropriate unfavorable temperature in a wide-mouthed thermos bottle. Another technique to avoid hand emasculation employs suction. In this method a small vacuum nozzle is used to suck adhering pollen from the stigma. Emasculation is, of course, unnecessary with fully monoecious, dioecious, or self-incompatible species. This may also be true of certain self-fruitful species. In Lima beans, for example, pollinations made slightly before or even at anthesis produce nearly all hybrid seeds. Apparently foreign pollen is more effective than pollen from the same strain, even though the species is fully self-fruitful. Recently the effectiveness of genetic emasculation by means of male-sterility genes has become recognized, and this technique seems likely to receive widespread use in the future.

In hybridizing most species, pollinations must be made by hand, but there are some species in which insects have been used as pollinators.

Cages of one sort or another are necessary to keep the desired insects in contact with the plants to be crossed and to exclude other pollinators. This technique has been used successfully with onions, where flies were the pollinators, and with red clover, using bees as the vector, as well as in some other species.

Where artificial hybrids are made only with the greatest difficulty, the final resort may be to natural crossing, even though it is rare. The parents to be crossed are grown together in isolation from possible contamination, and progeny of one, or sometimes both parents, is examined for hybrids. This method is clearly at its best when seed or seedling markers are available, obviating the growing of large numbers of plants to maturity to find rare hybrids. This method also can be used where inbreeding depression is so marked that selfs and hybrids can be identified by differences in vigor.

REFERENCES

Lawrence, W. J. C. 1951. *Practical plant breeding.* Allen and Unwin, London. (Chapters 1, 2, and 3 describe in detail pollination control in flower species.)

Hayes, H. K., F. R. Immer, and D. C. Smith. 1955. *Methods of plant breeding.* McGraw-Hill Book Co., New York. (Includes a chapter on the details of selfing and crossing agricultural species.)

Keller, W. 1952. Emasculation and pollination technics. *Proc. Sixth Intern. Grasslands Congr.* (Reviews hybridization techniques with grasses and legumes.)

Yearbook of Agriculture, U. S. Dept. Agric. 1936, 1937. (Gives details of hybridization for many agricultural species.)

6

Selection under Self-Fertilization

In both evolution and in plant breeding, populations are constantly being sifted for superior types. In this continual sifting the primary force is *selection*, in which individuals with certain characteristics are favored in reproduction. Before we can deal properly with this all-important subject with its many implications in plant breeding procedures, it is necessary to discuss the genetic basis of selection. In this and the next four chapters we will consider the basic attributes of selection. The development of this topic will be in terms of self-pollinated species, but much of what will be said applies equally well to cross-pollinated species.

Among the attributes of selection, two are especially important to the understanding of breeding principles: (1) selection can act effectively only on heritable differences; (2) selection cannot create variability but acts only on that already in existence. It is in connection with this second attribute that inbreeding assumes importance in plant improvement. Inbreeding causes an increase in homozygosity. This effect applies to all loci, so that quantitative characters as well as characters determined by major genes are subject to its influence. The result of sufficiently long and intense inbreeding is fixation of genetic characters—separation of the population into genetically distinct groups, each uniform within itself. Another way of putting the argument, with slight but important shift in emphasis, is to say that inbreeding uncovers genetic variability concealed in heterozygotes and brings it into the open where it can be acted upon by selection. This effect of inbreeding has important implications in plant-breeding practice.

To the present we have referred to inbreeding and outbreeding a number of times without defining these terms precisely. Inbreeding

50

occurs when matings are made between closely related parents; outbreeding, on the other hand, describes matings between individuals not closely related. Technically any mating system in which individuals have fewer ancestors than a population mating at random involves some degree of inbreeding. The system of mating to which we shall confine our attention for the present—self-fertilization—is clearly an inbreeding system, since any line of descent has only a single ancestor in each generation. In later chapters we shall consider some other systems of mating involving inbreeding and treat the subject quantitatively.

Early Observations on Selection

There is ample evidence that the practice of selecting desirable individuals as parents for the next generation is an ancient one. Virgil, for example, wrote that selected seed, improved through years of labor, was seen to run back, unless man selected the largest and fullest ears. Although much observational evidence concerning selection accumulated over the centuries as a result of agricultural practice, the first recorded accounts of the extent to which selection is effective appeared in the late eighteenth century as the result of methodical attempts at plant breeding. Van Mons in Belgium, Knight in England, and Cooper in America demonstrated that selection can lead to worthwhile improvement in varieties of plants, and each of these men, as a result of his plant-breeding activities, made contributions not only to agriculture but also toward an understanding of selection.

More informative work followed, however, in the nineteenth century. John Le Couteur, a farmer of the Isle of Jersey, had noted the diversity of plant types in his wheat field. He discovered that the progeny of single plants was remarkably uniform and that differences existed in the agricultural value of various selections. Le Couteur, who published a summary of his work in 1843, was apparently the first plant breeder to set down clearly the importance of selecting individual plants in the improvement of cereals, although a Scotsman, Patrick Sheriff, somewhat earlier had used the same methods as Le Couteur in breeding wheat and oats and was successful in developing some extensively grown varieties. In midcentury an Englishman, F. Hallett, also used single-plant selection with small grains. However, he believed that acquired characters were inherited and consequently grew his progenies under the best of conditions and continued to select the best plants in each generation. Despite his false premise, he succeeded in isolating some valuable varieties, the most noteworthy being Chevalier barley. In failing to observe that continuous selection in single-plant progenies of autogamous species is ineffective, he contributed less to an understanding of selection than his predecessors, Le Couteur and Sheriff.

The most notable contribution of the nineteenth century to an understanding of the effects of selection was made by Louis de Vilmorin, who took charge of the Vilmorin seed firm of France in 1843. He conceived a method of line selection with progeny testing which came to be known as the "Vilmorin method" or "Vilmorin Isolation Principle." When he applied this method to four varieties of wheat, selecting the best plants in each generation for fifty consecutive years, he was able to detect no changes in appearance. However, selection for high sugar content during the period 1850 to 1862 so markedly improved sugar beets that the beet-sugar industry became economically important. With this work the differences between the effects of selection on self- versus cross-pollinated species had been discovered.

The Vilmorin method was advocated and used by Willit M. Hays in the United States as early as 1888. The best organized and most widespread use of the method, however, was by the Swedish Seed Association, which had been established at Svalof in 1886. The spectacular results obtained there with this method and later with other methods of plant breeding have made that institution internationally famous. Thus, although pure-line selection had become well organized as a method of breeding by the late nineteenth century, it remained for the geneticists of the twentieth century to explain the origin of the variation upon which selection depends and for Johannsen to elucidate the genetic basis of pure lines.

The Pure-Line Theory

W. L. Johannsen (1903, 1926), a Danish biologist, provided a sound scientific basis for selection in self-pollinated crops when he defined "pure lines" and described the genetic mechanism by which they are established. In a notable series of experiments on beans, a highly self-pollinated species, Johannsen studied the effects of selection for seed weight. In his original seed lot of the Princess variety, obtained commercially, there were beans of many different sizes. He observed that progenies derived from the heavier seeds generally were characterized by greater mean seed weight than progenies from the lighter seeds. This result demonstrated that selection for seed weight had been effective.

Johannsen next established nineteen lines by growing progenies derived from nineteen different seeds of the original seed lot. Detailed studies of these lines revealed two main features of behavior as regards seed weight. First, it was discovered that each lot showed a characteristic mean weight of seed. The heaviest line, No. 1, produced seed with a mean weight of 64 centigrams. All intergrades were observed from this value down to 35 centigrams, the mean weight of the smallest line, No. 19. Johannsen concluded that the commercial seed lot had consisted of a mixture of "pure lines." A pure line was defined as the progeny of a single self-fertilized homozygous individual.

Second, it was observed that seeds of various sizes occurred within each progeny, but the variability within progenies was much less than in the original seed lot. Johannsen believed this variability did not have a genetic basis, but depended on slight differences in various environmental factors which affected individual plants to different degrees. This interpretation was easily susceptible to experimental verification.

As a first check the beans in each pure line were separated into classes of 10-centigram size and grown separately. It was found that the classes of different sizes within a pure line produced progeny with the same mean weight. For example, the categories 20, 30, 40, and 50 centigrams occurred in line No. 13. The mean weights of their progenies were 47.5, 45.0, 45.1, and 45.8 centigrams respectively. Thus, within a pure line, seeds of all sizes produce progeny with the weight that is characteristic of the line.

A further proof was achieved by selecting large and small beans in each line for six successive generations. The results are shown in Table 6–1. After six generations of selection in line No. 1 the mean weight was 69 centigrams for the light line and 68 centigrams for the heavy line. It was concluded that the mean weight of each line remained remarkably constant generation after generation, whether or not it was reproduced by its heaviest or its lightest seeds.

Further insight into the forces at work was obtained by determining parent-offspring correlations within pure lines and in the mixed population consisting of all 19 pure lines. Within pure line No. 13, for example, r was $-.018 \pm .038$, indicating no relation between seed weight of parent and offspring. However, the highly significant correlation coefficient of $.336 \pm .008$ for the mixed population showed that here the seed weights of the offspring were related to the seed weights of the parents.

TABLE 6–1.   The Effects of Selection for Six Generations in Line One of the Princess Bean

(After Johannsen, 1926)

Harvest Year	Mean Weight of Parent Seeds			Mean Weight of Progeny Seeds		
	Light Line	Heavy Line	Difference	Light Line	Heavy Line	Difference
1902	60	70	10	63.15	64.85	+1.70
1903	55	80	25	75.19	70.88	−4.31
1904	50	87	37	54.59	56.68	+2.09
1905	43	73	40	63.55	63.64	+0.09
1906	46	84	38	74.38	73.00	−1.38
1907	56	81	25	69.07	67.66	−1.41

In each generation the light line was selected for light seeds and the heavy line for heavy seeds.

The Genetic Basis of Pure Lines

These results were explained by Johannsen on the basis of the joint effects of heredity and environment. The environmental component of the variation depended on slight differences in external and internal conditions of various sorts acting on individual seeds. The important hereditary consideration was the homozygosity, a result of long-continued self-fertilization, which characterized the lines. This effect of inbreeding had been demonstrated mathematically by Mendel himself, who showed that, starting with the heterozygote Aa, continued self-pollination caused a decrease in heterozygosity by one half per generation. Only a few generations of selfing are required to produce a population of equal proportions of AA and aa individuals, with a negligible remaining proportion of Aa individuals.

Each heterozygous gene pair, irrespective of the total number involved in any plant, will be reduced in heterozygosity as Aa above. With n heterozygous gene pairs, the proportion of completely homozygous plants after m generations of self-fertilization is given by the formula $[(2^m - 1)/2^m]^n$. With 5 independent gene pairs, 85 per cent of the population will be homozygous at all 5 loci after only 5 generations of selfing. Even with 100 independent gene pairs, obviously more than are possible in most plant species, 10 generations of selfing will render about 90 per cent of the individuals homozygous at all loci. The results of applying this formula to 1, 5, 10, 20, 40, and 100 gene pairs for 1 to 12 generations are shown in Figure 6–1.

The preceding formula assumes equal survival of all genotypes and the independence of all gene pairs. Failure to meet the first requirement results in a decreased rate of return to homozygosity if the heterozygotes are favored and the reverse if the homozygotes have the higher survival ability. Linkage increases the proportion of homozygous individuals in any generation, though it does not influence the percentage of homozygosity.

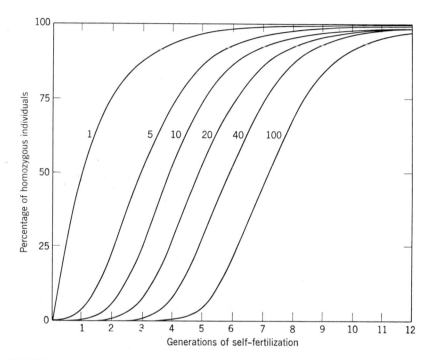

FIGURE 6-1. Percentage of homozygous individuals after various generations of self-fertilization, when the number of independently inherited gene pairs is 1, 5, 10, 20, 40, or 100. The percentage of homozygosity in any selfed generation is given by the curve for one gene pair.

It is evident that self-pollination, excluding special genetic circumstances, rapidly reduces any population to the homozygous condition regardless of the number of heterozygous gene pairs present in the beginning. The final result of selfing is a homozygous but not a homogeneous population, because such populations include different kinds of homozygous families. It will be sufficient to indicate at this time that the number of different kinds of homozygous types possible is 2^n where n represents the number of heterozygous gene pairs. With one gene pair only 2 pure lines are possible. With 10 gene pairs the number increases to 1024 and with 20 gene pairs to more than 1 million. Obviously comparatively few Mendelian differences provide the basis for very complex populations.

The Significance of Johannsen's Experiments

Johannsen's experiments provided the basic facts necessary to clarify the fundamental attributes of selection and its consequences. Since beans are a self-fertilizing species and, furthermore, since continued inbreeding leads to homozygosity, it might be expected that his original seed lot consisted of a mixture of homozygous lines. Progeny from an individual seed normally would not be expected to show genetic segregation. Hence variation within the lines had only one component, environmental in nature. Selection within pure lines was not effective, because all individuals in any one line had exactly the same genotype and hence the same capacity to respond to environmental stresses as did their parents and every other member of that line.

On the other hand selection was effective in the original population because the variation there had an additional component, hereditary in nature. It was upon this component of variation that selection was able to act, resulting in the isolation of lines with different seed weights. Johannsen's brilliant interpretation of these results made clear the essential difference between phenotype and genotype, and selection had been placed on a firm scientific basis.

Sources of Genetic Variation

In the previous section pure lines were, for the sake of simplicity, discussed as fixed and inalterable genetic units. There is a great deal of evidence that, in short-term experiments such as those of Johannsen, this viewpoint is a close approximation to reality. On the other hand, there is ample evidence that genetic variation arises *de novo* in pure lines at rates that can assume importance in plant breeding. In fact, the great diversity of types occurring in the world collections of self-pollinated species is in itself evidence of the long-term magnitude and importance of spontaneous heritable variations. These variations have their origin in mutation; their dispersal through populations and their combination with other mutants are accomplished through natural hybridization and its Mendelian consequences. An appreciation of these processes is therefore important to plant breeders, since the naturally occurring variation in populations of self-pollinated species is the primary basis for improvement of these species.

The ultimate nature of the mutation process or processes is a question as old as the science of genetics, and it is still warmly disputed. For our purpose it is adequate to regard any sudden heritable change as a mutation, recognizing that it may be a specific chemical change in a gene, that is, a so-called point mutation, or that it may result from a change in the number or structure of chromosomes. These chromosomal aberrations include duplication or losses of parts of chromosomes (duplication or deficiency), rearrangements of parts of chromosomes (translocations or inversions), and multiplication of entire chromosomes or sets of chromosomes (polyploidy). In evolution, and in the development of new and different kinds of plants, point mutations appear to have provided the great majority of the raw material for change. Apart from polyploidy, which has a special place in the origin of genetic variability, chromosomal aberrations appear to have been the source of only limited variation useful to plant breeders.

Both the effects and the rate of natural mutations have importance in plant breeding. Mutations presumably can be specific, affecting only a single character. More usually, however, they appear to have manifold effects upon the organism in which they occur, and some mutations, particularly those whose actions are expressed early in the life of the individual, can produce a syndrome of effects. Moreover, a number of different mutation possibilities can exist for any particular gene, leading to allelic series at some loci. Whether or not mutations produce single or compound effects and whether there is more than a single mutation possibility at any locus, both major effects and any secondary effects of the great majority of the mutants which have been recorded have not been favorable to the organism nor have they improved its utility to man.

Mutation is a recurrent process—that is, any mutation observed today has probably occurred many times before in the history of the organism. Hence it is reasonable to suppose that the majority of all possible mutations have occurred in cultivated species during the thousands of years they have been under cultivation, and, as a result of selection by nature and man, most of the alleles now in existence are superior ones. Therefore most current mutations must be in the direction of inferiority. Comparatively little is known about mutations governing metrical characters such as height or maturity because of technical difficulties in studying them. Such genes are rarely individually recognizable, and they are usually not unique in action but rather are capable of supplementing or replacing one another. Application of the terms superior and inferior are therefore largely meaningless in connection with the mutations of such genes.

The rate at which mutations occur is as important to breeders as the effects of mutations, and there is considerable evidence on mutation rates, both observational and experimental. Some important evidence comes from the experiences of plant breeders who have selected and observed pure lines. Atlas barley, for example, was derived from a single-plant selection and hence was presumably homozygous when it was isolated from a well established land variety nearly half a century ago. This variety has subsequently been re-selected a number of times and the progeny of individual plants carefully observed. On painstaking study they are found to differ from each other in quantitative characters as well as in minor mor-phological ways. It is significant that no line has been found that is noticeably superior to the original Atlas.

Although there is no precise information on the point, Atlas appears to be a relatively stable variety. Certainly it is more stable than many varieties that have been investigated, for example, the quite unstable ones described in wheat by Love. Although the great mass of observational data from pure-line varieties is highly suggestive in indicating that mutations of a nondefective type are relatively frequent, it is not critical evidence because of the near impossibility of evalu-ating the contributions of mechanical mixing and natural hybridiza-tion to the total variability. It is therefore desirable to examine the evidence on rates of mutation from certain less comprehensive but more definitive experiments.

East was one of the early workers to obtain reasonably precise infor-mation on mutation rates. His best evidence came from certain matriclinous individuals obtained in connection with attempts to pro-duce certain species hybrids in *Nicotiana*. These individuals arose parthenogenetically from unfertilized eggs and were presumably com-pletely homozygous diploids. According to East the progenies of these plants were remarkably alike, much more so than inbred lines in the same species. After four generations of self-fertilization, however, these lines were as variable as ordinary inbreds. This was presumably due to mutations, since hybridization and mixture had been excluded beyond reasonable doubt. Other investigators have also reported evidence suggesting relatively high mutation rates for nondefective genes.

It remained for Stadler, however, to obtain definitive evidence of the frequencies of particular gene changes. Stadler selected favorable material for study, namely eight genes governing endosperm characters in maize, where examination of the seeds on a single ear suffices to establish the genotype of several hundred individuals. Some of Stad-

TABLE 6–2.   Frequency of Spontaneous Gene Mutations in Maize

(After Stadler, 1942)

Gene	Number of Gametes Tested	Mutations Observed	Mutation Rate per 1,000,000 Gametes
R (color factor)	554,786	273	492
I (color inhibitor)	265,391	28	106
P₂ (purple)	647,102	7	11
Su (sugary)	1,678,736	4	2.4
Y (yellow)	1,745,280	4	2.2
Sh (shrunken)	2,469,285	3	1.2
Wx (waxy)	1,503,744	0	0

ler's results appear in Table 6–2. It is unfortunate that mutation studies in higher plants are almost entirely confined to maize. Nevertheless, the similar results obtained from precise experiments with a few other prolific organisms, together with the observational data in several cultivated species, suggest that usual mutation rates are of the order of one mutation per 100,000 to one mutation per million gametes, although higher and lower frequencies are known for some loci.

These controlled experiments on mutation thus complement and support the observations made by plant breeders. Furthermore, these measured mutation rates appear adequate to explain the variability observed in pure lines. They also indicate that mutation frequencies are generally so low that the chances of a favorable mutant appearing in an adapted variety are small in any short period of time and consequently that selection within pure lines is not likely to be profitable compared with other procedures for plant improvement. On the other hand the rates are high enough, considering the long period since most crops were domesticated, to account for the great variability existing in cultivated species today. The possibility of discovering a favorable mutant in a pure line should therefore not be ignored.

REFERENCES

Brookhaven Symposia in Biology, No. 8, 1956. (A collection of papers on the status of mutation research in 1956.)

East, E. M. 1936. Genetic aspects of certain problems of evolution. *Amer. Nat.* 70: 143–158.

Johannsen, W. L. 1903. *Ueber Erblichkeit in Populationen und in reinen Leinen.* Gustav Fischer, Jena. (One of the great papers in genetics.)

Johannsen, W. L. 1926. *Elemente der exacten Erblichkeitslehre.* Gustav Fischer, Jena.

Love, R. M. 1951. Varietal differences in meiotic chromosome behavior in Brazilian wheats. *Agron. Jour.* 43: 72–76. (Discusses unstable varieties of wheat.)

Stadler, L. J. 1942. *Some observations on gene variability and spontaneous mutation.* Spragg Memorial Lectures in Plant Breeding, Michigan State College.

7
Genetic Consequences
of Hybridization

In the previous chapter it was seen that natural hybridization and its Mendelian consequences, compounded many times over many generations, play an important role in producing variability in self-pollinated species. However, the range and variety of the recombinations that can be expected from natural hybridization are severely limited by the similarity of the types that are most likely to hybridize. It is consequently not surprising that variability existing in land varieties at the beginning of the twentieth century was rapidly exhausted in many self-pollinated crops and that returns from pure-line selection progressively became smaller and smaller. As this happened the emphasis in plant breeding gradually shifted to planned hybridization between carefully selected parents until, at present, hybridization methods completely dominate the breeding of self-pollinated species.

In making hybrids that possess, between them, the characters required to obtain specifically desired recombinations, the breeder creates populations in which selection is most likely to be profitable. However, it should be recognized that the series of events initiated by hybridization can have unfavorable as well as favorable consequences and that all the phenomena of Mendelian genetics—segregation, recombination, linkage, nonallelic interaction, penetrance, expressivity, thresholds, and so on—have a bearing on both the successes and failures of hybridization as a method of breeding. With the advance of genetics, the information about these phenomena has gradually increased until the modern plant breeder is in a much more favorable position than his pre-Mendelian predecessors in recognizing and dealing with their consequences. It is to this background of genetic understanding that we will now direct our attention.

The Gene-Character Relationship

Genes are the building blocks from which new varieties are fashioned, and it is essential in plant breeding to keep in mind the relation between genes and characters—that is, the relation between genotype and phenotype. In the higher plants it is now commonly accepted that virtually all phenotypic effects are not related to the gene in any simple way. Rather they result from a chain of physico-chemical reactions and interactions initiated by genes but leading through complex chains of events, controlled or modified by other genes and the external environment, to the final phenotype. To a greater or lesser degree the functioning of any gene depends on the genotype of which it is a part and on the external environment to which it is exposed.

There are numerous cases where well-defined morphological or physiological characters are controlled largely by single genes and are little affected by either the genetic or physical environment. Important commercial characteristics of many crop species, such as the main difference between field and sweet corn, flower color in some ornamentals, determinate versus indeterminate habit of growth in beans, and resistance to particular races of diseases, are governed by genes with large, easily recognizable, and relatively stable effects. Such genes are particularly valuable to plant breeders, because they are readily analyzed and managed by standard Mendelian techniques.

These familiar "qualitative" genes of classical genetics can, however, have more than a single phenotypic manifestation. The well-known white-eye-color mutant of *Drosophila melanogaster,* for example, affects not only the pigmentation of the eye, but also the testicular membrane, the shape of the spermatheca, length of life, and the general viability of the individual. In a study to determine the frequency of manifold effects, Dobzhansky examined a random sample of twelve mutants of *D. melanogaster* affecting such characters as wing size, body color, and similar differences. Ten of the twelve had the secondary effect of changing the shape of the spermatheca. Evidently secondary effects are common, and they may be quite unrelated to the primary effect of the gene.

A case of particular interest to plant breeders has been described by Suneson, et al. (1948). They found that yield and bushel weight in wheat are influenced by the gene pair governing awned versus awnless lemmas. More specifically they found that the addition by recurrent backcrossing of awns to Onas wheat, an awnless variety, increased the yield by approximately 7 per cent and the bushel weight by about 1 pound.   Comparable decreases attended the removal of awns from Baart wheat by a similar process.

The discovery of beneficial side effects such as those associated with awns in wheat provides a positive method of improving varieties in yield or in other characters that are difficult to identify among the segregation products of hybrids.   Unfortunately only a few instances are known at present, but some investigators have undertaken systematic searches for such effects.   Atkins and Mangelsdorf (1942) suggested the use of isogenic lines for this purpose.   After crossing two varieties, individuals heterozygous at a particular easily identified locus are selected in each successive generation, thus enforcing heterozygosity at this locus while inbreeding occurs at all other loci.   After many generations the stock can be expected to be homozygous at all loci (isogenic) except the one maintained in heterozygous condition. At this point the two homozygotes at the locus in question are selected and any differences between them in quantitative characters can be reasonably ascribed to the gene in consideration or to tightly linked genes.

Isogenic lines can also be obtained by recurrent backcrossing.   This approach has also been used to identify side effects of major genes. In the breeding of a semismooth, awned version of Atlas barley, it was discovered that the semismooth derivatives obtained in the tenth backcross outyielded both the rough-awned recurrent parent, Atlas, and also the rough-awned derivatives.   This difference in yield could be ascribed either to a secondary effect of the smooth-awned allele or to the action of a closely linked gene or genes.   A careful study (Everson and Schaller, 1955) was required to settle this question in favor of the latter explanation.   The linkage proved to be close as might have been expected from the fact that the relation between smooth awns and yield survived ten backcrosses.   In plant-breeding practice it therefore makes little difference whether a favorable side effect results from pleiotropy or linkage, although the genetic basis of the difference is quite different in the two cases.   When side effects are unfavorable, however, pleiotropy excludes the use of major genes with important unfavorable effects, but the situation is not hopeless if the correlated effects result from linkage.

Epistasis

Epistasis is a term originally used by Bateson in 1909 to describe genes whose effects mask or cover the effects of other genes.   The term has since acquired a more general meaning that is synonymous with interactions between genes at different loci; that is, it is used to describe the phenomenon whereby the effect of one gene may change according to the presence or absence of another gene or genes. Epistasis, or interallelic interaction, should be carefully distinguished from dominance, which refers to nonadditivity of alleles at the same locus, that is, intraallelic interaction. In self-pollinated species epistasis is perhaps more important to breeders than dominance, because the latter is necessarily ephemeral in such species. Epistasis, which does not necessarily depend on heterozygosity, makes possible recombinations that amount to something more than simply new ways of putting old characters together. Through gene interactions it is possible for different and unexpected types to appear, and some of them may represent real advances over their parents.

The simplest epistatic situations are the ones leading to such familiar dihybrid F_2 ratios as $9:7$, $13:3$, $15:1$, $12:3:4$, and $12:3:1$. When three genes interact, the possibilities are greater and ratios such as $37:27$, $55:9$, and $27:9:9:19$ can occur. The material basis for epistasis has been analyzed biochemically in a number of cases to the point where the gene interactions involved are understood at least superficially. The most common pattern appears to be one in which one gene produces a substrate that provides the necessary raw material for the action of a second gene. The case of HCN content in Ladino clover, worked out by Atwood and Sullivan (1943), follows this general pattern. High HCN is governed by two dominant alleles, one of which produces a cyanogenic glucoside and the other the enzyme that catalyzes the release of HCN from the glucoside. Individuals with at least one dominant allele at each locus are high in HCN because both the substrate and the enzyme are present. When a plant is homozygous recessive at either or both loci, it will be low in cyanide for lack of either the substrate, the enzyme, or both.

It would indeed be surprising if many of the morphological and physiological characters in which the plant breeder is interested do not have their material basis in reaction systems of this sort. Undoubtedly many are quite complex, and although it may be interesting to speculate about their biochemical basis, such speculation is not likely to be conclusive at the present time.

Modifying Factors

Another way in which the genetic environment can alter the gene-character relationship involves genes with small effects that exert their chief influence by intensifying or diminishing the expression of major genes. These genes are appropriately called modifying factors. In some instances it has not been possible to demonstrate that the modifiers have any effect other than that they produce in modifying the expression of some specific major gene. One of the best documented examples of modifiers is that of the control of spotting in mice. First, whether a mouse is spotted at all depends on a major gene. However, the degree to which the animal is spotted depends on a galaxy of modifiers. Variation may be practically continuous from self-color, through coats with small white spots, to ones that are all white. Although most modifiers have not been studied in so great detail as those of the spotted mice, there is every reason to believe that most major genes have a complement of modifiers.

Most plant breeders are greatly interested in maximum yield. Although it is easy to demonstrate that such observable characters as maturity date, height, and resistance to drought, winter killing, diseases, and insects have a marked influence on yield, it is more difficult to ascertain the number and kinds of genes that contribute directly to yield. It must be assumed, however, that both major and modifying genes are concerned. In a worth-while discussion of breeding for yield, Frankel (1947) expressed the belief that wheat breeding in New Zealand had progressed to the point where gains can no longer be made by adjusting observable characters. The future must then lie in manipulating genes for strictly quantitative characters. Frankel raised the question whether selection based on replicated yield trials will be effective for these nonobservable characteristics. He suggests that progress in breeding will depend on the close cooperation of geneticists, breeders, and statisticians.

In general, plant breeders are interested in the maximum expression of the characters with which they work. Consequently modifiers are likely to become increasingly important as more and more major genes are concentrated in single lines. Plant breeders may thus ultimately be concerned only with modifier genes.

The effects of modifying genes are primarily quantitative and can be described accurately only in terms of measurements. Methods of analyzing and dealing with genes governing metrical characters are subjects of such great importance to plant breeders that they have been set aside for special consideration in Chapters 8, 9, and 10.

Penetrance and Expressivity

Penetrance can be defined as the ability of a gene to be expressed in individuals which carry it. Thus a variety of Lima beans called

Ventura carries a dominant gene that causes the tips and margins of the unifoliate leaves of seedling plants to be partially deficient in chlorophyll. However, rarely more than 10 per cent of plants show this characteristic even though the variety is homozygous for the allele governing the character. This gene may be said to have a penetrance of approximately 10 per cent on the average, although under certain environmental conditions all plants show the character, that is, penetrance is complete. In other environmental situations none of the plants show the character, and penetrance is zero. Additionally, there is considerable variation in the manner in which the character is expressed in different plants. In some individuals the entire unifoliate leaves are devoid of chlorophyll but soon become yellow and shortly thereafter normal green in coloration. In other individuals the leaves fail to develop chlorophyll and abscise. In still other individuals only the tip of the blade may be affected and in some only the margin of the leaf. In other words, this single gene can express itself in a variety of ways that may resemble a number of characters. This gene then evidences incomplete penetrance and variable expressivity. Whether a gene is expressed at all is denoted by the term penetrance whereas the term expressivity is concerned with the manner of expression. Penetrance and expressivity are therefore phenomena that exert confusing effects on the gene-character relationship, and they tend to obscure the correspondence between genotype and phenotype. In general, therefore, incomplete penetrance and variable expressivity complicate the task of the plant breeder.

Thresholds

An additional phenomenon that can obscure the gene-character relationship is the threshold effect. Collins (1927) discovered an albino barley in which the character was expressed only at temperatures below 45° F. When plants of the albino type were grown above 65° F., they developed normal chlorophyll. Most albino plants had small sectors of green in their leaves at low temperatures but made a complete recovery when the temperature was increased. Another example of a threshold is provided by Atlas, a pure-line variety of barley which also exhibits a temperature sensitive type of albinism. Atlas frequently shows a low percentage of albinism when grown in the field. Albino plants tend to occur in patches. This suggests that some kind of genetically controlled threshold is involved in the reaction that produces albinism. Under the variable microclimatic conditions in a field of barley, the threshold is apparently reached for some individuals but not for others. The idea of thresholds is useful as a way of looking at many of the homozygous genotypes that behave irregularly.

Environmental Effects on Gene Expression

The examples discussed under previous subheadings in this chapter were chosen deliberately to demonstrate that the effects of genes can be conspicuously altered by factors of both the genetic and external environment. If these genes represented the norm with respect to stability under environmental stresses, the selection and evaluation of homozygous genotypes derived from hybrids would be almost hopelessly difficult. However, most genes express themselves quite uniformly and predictably under the range of environmental conditions they are likely to encounter. In other words, most genotypes are reasonably buffered to variations in their environment, and their performance under one set of conditions is at least a reasonable guide to their probable performance under all fairly similar sets of environmental conditions. Nevertheless, the complex nature of the gene-character relationship leads us to expect that the pattern of inheritance can be obscured by environmental agencies, even though these are subtle and apparently trivial in some instances; and the plant breeder must consequently remain alert to unpredicted deviations from previous performance in the genotypes he selects from his hybrids.

Segregation and Recombination of Genes

The two laws of heredity announced by Mendel that form the foundation of the modern theory of particulate inheritance were developed from studies of single differences. Mendel worked with well-defined characters such as yellow versus green and smooth versus wrinkled seeds, and only when he understood their individual behavior was their joint behavior investigated. His success depended upon treating the simplest ones first. Once the $3:1$ ratio was understood, the interpretation of ratios such as the $9:3:3:1$ and $27:9:9:9:3:3:3:1$ followed much as a matter of course. Carried to its conclusion this approach theoretically permits differences of any order of complexity to be analyzed as long as each gene produces a large enough effect for it to be treated as a uniquely recognizable unit. In practice gene-by-gene analyses of differences between parents have not been carried out, and plant breeding has not become an exercise in assembling favorable genes as was optimistically predicted by some early geneticists. The first reason this has not been done is that many genes are not unique in action, and their effects cannot be identified individually. The second reason is related to the rate at which complexity increases as the number of segregating allelic pairs become more numerous. This is illustrated in Table 7–1.

TABLE 7–1. Numerical Characteristics of Hybrids between Parents Differing in n Allelic Pairs

Number of Allelic Pairs	Kinds of* Gametes Possible in F_1	Kinds of† Genotypes Possible in F_2	Smallest Perfect Population in F_2	Kinds of Phenotypes in F_2 Assuming	
				Full Dominance	No Epistasis and No Dominance
1	2	3	4	2	3
2	4	9	16	4	9
3	8	27	64	8	27
4	16	81	256	16	81
10	1,024	59,049	1,084,576		
21	2,097,152	10,460,353,203	4,398,046,511,104		
n	2^n	3^n	4^n	2^n	3^n

*Also gives the number of genotypes occurring in backcrosses and the number of homozygous genotypes.

†Assuming coupling and repulsion heterozygotes to be equivalent.

The values in the first four lines in Table 7–1 are easily obtained by reasoning from Mendelian principles. The generalized prediction, given in the last line, can be deduced from the simple situations. It is seen from the general prediction that the complexity of genetic situations increases exponentially with increase in the number of segregating gene pairs. Since both components of the formulas of the general prediction, that is, the base number and the exponent, are positive integers, an inevitable result is enormous complexity when the exponent, representing the number of gene pairs, becomes at all large.

The consequences of hybridization can be illustrated with a hypothetical cross between two wheat varieties assumed to differ in 21 gene pairs or one gene pair per chromosome. This cross is probably unrealistically simple because it is unlikely that such closely similar varieties exist. Table 7–1 shows that such an F_1 hybrid has the potentiality of producing more than 2 million different kinds of gametes, and that the ways in which these gametes can combine permits the formation of more than 10 billion different genotypes in the F_2 generation. The great majority of these genotypes—that is, $3^n - 2^n$—will be heterozygous at one or more loci. With continued selfing, these heterozygotes disappear; consequently they are interesting to the plant breeder only because they (1) tend to confuse early-generation progeny tests by producing variable offspring and (2) have the potential of producing homozygotes that can be the basis of new varieties. Because the kinds of gametes produced in the F_1 also represent the kinds of homozygous genotypes the hybrid is capable of producing, they have particular significance to the plant breeder. As we have seen, our hypothetical hybrid has the potential of producing a very large number of kinds of gametes and also of pure-line varieties, even though it is probably more simple than any real hybrid.

If an F_2 population large enough to permit every genotype to occur with a frequency based on Mendelian expectancy were to be grown from this hypothetical hybrid, it would number 4,398,046,511,104 individuals, among which only two are the same as the parental types. At spacings normally used by farmers, this population would occupy about 50 million acres or approximately 60 per cent of the total area devoted to wheat in the United States in 1949. Upon backcrossing the hybrid to either parent, the situation is much less complex. The total number of genotypes possible is 2,097,152, each occurring with equal frequency. The smallest perfect backcross population therefore includes only 2,097,152 plants, rather than 4 trillion, made up of more than 10 billion different genotypes, as required to produce the smallest perfect F_2 population. The only homozygous genotype in the first backcross generation will be one with the same genotype as the recurrent parent. These relations can be seen by examining the accompanying simple illustration.

F_2	Backcross
AaBb × AaBb	AaBb × aabb
1AABB	
2AABb	
1AAbb	
2AaBB	
4AaBb	1AaBb
2Aabb	1Aabb
1aaBB	
2aaBb	1aaBb
1aabb	1aabb

These considerations suggest that backcrossing is a useful procedure for dealing with complex hybrids when most of the desired characters are concentrated in one of the parents.

The Composition of Populations Derived from Hybrids

The composition of populations derived from hybrids depends not only on the number of genes by which the parents differ but also depends on the number of generations the population has been self-pollinated. There are various ways of calculating the changes that occur with the progress of selfing; for example, the proportion of homozygous and heterozygous genotypes can be determined from columns 4 and 3 in Table 7–1. A more precise picture of the composition of any inbred generation can be obtained by expanding the binomial $[1 + (2^m - 1)]^n$, where n is the number of gene pairs involved and m represents the number of generations of selfing. In the expanded binomial, the first exponent in each term gives the number of homozygous loci, and the second the number of heterozygous loci. As an example, consider a population originally heterozygous for three gene pairs and selfed for four generations—that is, $n = 3$ and $m = 4$. Substituting in the binomial, we obtain

$$[1 + (2^4 - 1)]^3 = [1 + 15]^3 =$$
$$1^3 + 3(1)^2(15) + 3(1)(15)^2 + [15]^3.$$

Translated into genetic terms this indicates that our hypothetical F_5 will include

$$\begin{array}{r} 1 \text{ plant with 0 homozygous and 3 heterozygous loci} \\ 45 \text{ plants with 1 homozygous and 2 heterozygous loci} \\ 675 \text{ plants with 2 homozygous and 1 heterozygous loci} \\ \underline{3375} \text{ plants with 3 homozygous and 0 heterozygous loci} \\ \hline 4096 \end{array}$$

Considering only the second term of the expanded binomial, we see that the coefficient 3 indicates that there are three ways in which an individual can be homozygous at one and heterozygous at two loci, namely, $AABbCc$, $AaBBCc$, and $AaBbCC$. The value of this term is 45, indicating that these three genotypes make up 45/4096, or 15/4096 each, of the total population.

Let us consider the somewhat more realistic case of a population derived from a hybrid heterozygous for 10 gene pairs. Since the ultimate aim is to obtain the single best homozygous line, we are particularly interested in the 1024 homozygotes that the hybrid is capable of producing. In order for each of these homozygous lines to be represented only once in the F_2, a minimum population of 1,084,576 individuals would be required. Even if it were possible to grow and process a population of this size, there would be no sure way of identifying the homozygous types. It is therefore not practical to attempt to obtain the desired type in the F_2 generation. On inbreeding for 5 generations, however, the population would be expected to have the composition shown below, provided it were large enough to include at least one representative of each genotype.

Number of Individuals	Heterozygous for Indicated Number of Loci	Homozygous for Indicated Number of Loci
1	10	0
310	9	1
43,295	8	2
3,547,920	7	3
193,939,410	6	4
7,329,062,560	5	5
186,375,773,010	4	6
3,305,513,693,320	3	7
38,382,096,684,845	2	8
264,396,221,666,710	1	9
819,628,286,980,801	0	10
1,125,906,021,359,278		

The effect of the inbreeding is a drastic reduction in the frequency of the more heterozygous sorts and a corresponding increase in the proportion of the more homozygous types in which the breeder is most interested. The foregoing calculations show that the F_5 generation is expected not only to be dominated by homozygous genotypes but also to include many plants heterozygous at one or two loci. These heterozygous individuals are interesting to the breeder because they are capable of producing in the next generation homozygous sorts that might not have been included previously. Plant selections made in the F_5 are therefore likely to be homozygous, or nearly so, and the performance of their progeny consequently more predictable than the performance of the progeny of the heterozygotes that earlier dominated the population. For this reason many plant breeders prefer not to emphasize selection on a single plant basis until the later generations when the confusing effects of dominance and heterozygous types of interallelic interaction have largely been dissipated.

In actual practice, the calculated expectancies will not hold unless each genotype is equally productive. There are, in fact, good reasons for believing that selective differences frequently upset the expected progress of inbreeding; much of the evidence suggests that heterozygotes are favored, so that homozygosity is attained more slowly than simple theory predicts. Calculations of the theoretical composition of populations, therefore, at best only approximate the actual situation. Nevertheless such calculations have at least indirect usefulness in contributing to the background of understanding that guides the judgment of breeders in managing their operations. Despite the lack of precise information about the number of genes segregating and the effects of selection, the conclusion is inescapable that the number of possible combinations from most hybrids is so large that only a small proportion can be dealt with at any one time. Populations should therefore be as large as practicable in the early generations, and the management should be directed toward obtaining in the later generations a manageable number of pure lines for final evaluation.

Linkage

In the previous examples the consequences of hybridization between hypothetical varieties differing at a single locus per chromosome were examined. Because it is more likely that any two varieties will differ by several or even many loci per chromosome, consideration must be given to the effects of linkage on recombination. First, it should be understood that linkage does not influence the percentage of homozygosis. The effect of linkage is rather to influence the frequency of the various possible combinations of genes. With independent inheritance, assortment is free, and all combinations are equally frequent. The general effect of linkage is to upset this equality and to cause an overabundance of parental combinations and a corresponding deficiency in recombinations. The magnitude of this effect is suggested in Table 7–2, where the recovery of parental and recombinant types for the digenic case is given for several different linkage intensities. The numerical relationships become quite complicated when three genes are linked and incredibly so if the segregation of several linked genes is considered simultaneously.

In general, linkage can be regarded as a conservative influence, tending to hold together existing combinations of genes. In recently domesticated species or in species that have not been subjected to long and continued improvement, it is unlikely that any single type has concentrated in it the alleles best suited for its domesticated state. Therefore linkage can be regarded as a hindrance to the wide segre-

TABLE 7–2. **The Effect of Linkage upon the Proportion of** *AB/AB*
Genotypes Expected in F_2 **from the Double Heterozygote**

Recombination Value	Per Cent of *AB/AB* Individuals in F_2 if the F_1 Is	
	AB/ab	*Ab/aB*
.50 (independence)	6.25	6.25
.25	14.06	1.56
.10	20.25	0.25
.02	24.01	0.01
.01	24.50	0.0025
p	$\frac{1}{4}(1-p)^2$	$\frac{1}{4}p^2$

gation desired. On the other hand, highly developed varieties are
likely to have the best alleles at many loci. Here linkage can be
regarded as an aid, because it tends to hold together the existing
favorable combinations.

Examination of Table 7–2 shows that close linkages between desirable and undesirable genes can drastically delay the progress of
breeding programs, and many examples of this have been described.
There is, for example, a tight linkage between the genes governing
stem-rust resistance and late maturity in certain common wheat strains
derived from *Triticum timopheevi*. Only after very large populations
had been grown and much effort expended were rust-resistant strains
with early maturity recovered. Conversely, linkages between favorable
genes can aid breeding programs. Consider the case of the Rio and
Turkey genes for bunt resistance in wheat, located about fifteen crossover units apart. With these genes entering a cross in coupling phase,
91 per cent of the resistant plants in F_2 are expected to carry both
genes. Each gene imparts resistance to races of the organism not
covered by the other so that lines with both alleles for resistance in
coupling phase are particularly valuable parents in breeding for bunt
resistance.

Although the number of cases of this type that has yet been
described is still distressingly small, close linkages between favorable
genes provide a valuable approach to plant-breeding problems in much
the same way as the favorable pleiotropic effects discussed earlier.
The importance of favorable linked genes in plant breeding cannot be
measured in terms of recognizable cases only. In the broad sense
each step forward in plant breeding builds up the number of favorable

linkages. Often this must escape detection because the effects are slight or not sharply defined. In fact, then, the building of favorable linked combinations of genes is automatic and continuous regardless of the system of breeding used, although the approach is more precise and direct in some breeding systems than in others.

REFERENCES

Atkins, I. M., and P. C. Mangelsdorf. 1942. The isolation of isogenic lines as a means of measuring the effects of awns and other characters in small grains. *Jour. Amer. Soc. Agron.* 34: 667–668.

Atkins, I. M., and M. J. Norris. 1955. The influence of awns on yield and certain morphological characters of wheat. *Agron. Jour.* 47: 218–220.

Atwood, S. S., and J. T. Sullivan. 1943. Inheritance of a cyanogenetic glucoside and its hydrolyzing enzyme in *Trifolium repens. Jour. Hered.* 34: 311–320.

Collins, J. L. 1927. A low-temperature type of albinism in barley. *Jour. Hered.* 18: 331–334.

Everson, E. H., and C. W. Schaller. 1955. The genetics of yield differences associated with awn barbing in the barley hybrid (Lion × Atlas[10]) × Atlas. *Agron. Jour.* 47: 276–280.

Frankel, O. H. 1947. The theory of plant breeding for yield. *Heredity* 1: 109–120.

Hayman, B. I., and K. Mather. 1953. The progress of inbreeding when homozygotes are at a disadvantage. *Heredity* 7: 165–183.

Stern, C. 1949. Gene and character. In *Genetics, palentology and evolution.* Princeton University Press. (A short discussion on the relation between gene and character.)

Suneson, C. A., B. B. Bayles, and C. C. Fifield. 1948. Effects of awns on yield and market quality of wheat. *U. S. Dept. Agric. Circ.* 783.

Wagner, R. P., and A. K. Mitchell. 1955. *Genetics and metabolism.* John Wiley and Sons, New York. (An excellent discussion of the ways genes act and the effects of their actions.)

8
Quantitative Inheritance

Anyone who has observed the segregation in hybrids between tall and short races of peas, one of the trait differences studied by Mendel, has probably been struck by two aspects of the variation in height among the progeny. In the first place, there is such a gap between the tall and short segregates that the difference can be treated as a qualitative one, and accurate classification into tall and short categories can be made by a quick visual estimation of height. As a result, the analysis can proceed along familiar genetic lines, and the difference is found to be governed by a single gene pair, with tall being dominant to short as Mendel first demonstrated. Many characters of economic importance are of this type, and they are appropriately referred to as *qualitative* characters.

The second and perhaps less obvious aspect of the variability is the variation within the short and tall groups. If a number of short individuals are examined carefully, all intermediate degrees between the shortest "shorts" and the tallest "shorts" will be found, the middle expression being the most common with the frequency falling off toward either extreme. A similar graduated series for tallness is found among the tall individuals. Many of the characters of practical importance with which the plant breeder must work vary in this same way, in that easily distinguished alternatives do not exist. For example, if a survey is made of varieties of wheat with respect to such commercial characteristics as height, time of maturity, size and shape of kernel, protein content, cold or drought tolerance, and ability to tiller, all degrees of intermediacy between one extreme and the other will be found, and the segregants from hybrids involving such characters will usually be distributed in a continuous series. Because these characters can be specified accurately only in terms of metrics such as length, time, weight, or proportion, they are referred to as *quantitative* or *metrical* characters.

Mendel himself, and the rediscoverers of Mendelism, deliberately neglected the continuous variation in their material, presumably because they recognized that it would only confuse their analyses. This sort of variation had, however, been the object of active and vigorous study in the last part of the nineteenth century, principally in England by Galton and his students. Although these workers failed to discover the mode of transmission of continous variation from parent to offspring, they were able by the application of statistical methods to their data to demonstrate that it was at least partly heritable. When Mendel's work was brought to light in 1900, this group was understandably reluctant to accept the simple Mendelian ratios and the discontinuous or particulate mode of inheritance they implied as anything more than trival exceptions to their continuous or "blending" type of inheritance. The early Mendelians, on the other hand, regarded the blending inheritance of the biometrical group as incompatible with discontinuous genetic variation and some, notably Hugo de Vries, regarded the existence of continuous phenotypic variation as a criterion of nonheritability. One of the sharpest controversies in the history of biology grew out of this difference of opinion.

The first step in reconciling the differences between the two groups was taken by Yule in 1906 when he suggested that there need be no conflict between particulate inheritance and continuous inheritance if many genes having small and similar effects were postulated. About the same time Johannsen showed that seed weight in beans varied continuously as a result of the joint and more or less equal effects of heritable and nonheritable agencies. Nilsson-Ehle, a Swedish plant breeder, soon found a naturally occurring model for the type of inheritance suggested by Yule. In wheat and oats there are three genes for red versus white kernels. Any one of these genes gave a ratio of 3 red to 1 white when segregating alone. Two segregating together gave a 15:1 ratio and all three a ratio of 63:1. When more than one gene was segregating, differences in intensity of color were observed, and these Nilsson-Ehle was able to associate, by progeny tests, with differences in the number of genes for redness. It was thus established that different genes could have similar and cumulative effects and therefore that Mendelian genes have the properties necessary to account for continuous variation. Conclusive evidence that plural segregating genes with similar effects explain the inheritance of quantitative characters was provided by E. M. East in 1916, and since his experiments so nicely reveal the essential features of this type of inheritance, we will now consider one of them in some detail.

The Multiple-Factor Hypothesis

Flower size in tobacco is a particularly favorable character for studies of quantitative inheritance because the size is relatively constant under a wide range of environmental conditions. For this reason East selected two true-breeding varieties of *Nicotiana longiflora* for one of his investigations of the nature of quantitative inheritance, obtaining the data summarized in Table 8–1. It was immediately apparent that the continuous nature of the variability precluded an analysis of the genetic situation in the standard Mendelian manner. East postulated, however, that the data would meet several requirements if a considerable proportion of the variation was under the control of Mendelian genes. East's reasoning in correlating his observations with Mendelian expectancies was as follows:

1. The parents differed strikingly in corolla length when they were grown under similar conditions. The biometrical constants in Table 8–1 indicate that the difference was a real one, having its basis in genetic differences between the parents.

2. Although the parents had been long inbred, and must have approached complete homozygosity, there were differences within the parental groups. This variability presumably resulted from subtle environmental differences within the experimental garden.

3. The variability within the F_1 population was comparable in magnitude to that within the two parental populations. This conforms to Mendelian expectations, since all plants in a hybrid between two pure lines should be identical genetically.

4. Judging from the frequency distributions, standard errors, and coefficients of variability, the F_2 generation was more variable than the parental and F_1 generations. Although the F_2 generation cannot be expected to be immune from environmental effects, neither can it be expected to be drastically more sensitive than the other generations, and the excess of variability in the F_2 cannot reasonably be accounted for on the basis of environmental effects alone. The segregation and recombination of Mendelian genes is expected in the F_2, however, and East postulated that the excess of variability in that generation resulted from these causes.

The results observed in the parental, F_1, and F_2 generations thus conform to Mendelian expectations, and this raises the question of the number and nature of the genes governing corolla length. If the genetic portion of the variability were controlled by a single allelic pair displaying no dominance, three genotypes would be expected in the F_2 generation: AA, Aa, and aa. Phenotypes similar to those of the

TABLE 8–1. Frequency Distributions for Corolla Length in the Parents, F₁, and F₂ Generations in a Cross between Varieties of *Nicotiana longiflora*

(After East, 1916)

Generation and Year Grown	Class Centers in Millimeters																						n	$\bar{x}$	S.E.	C.V.
	37	40	43	46	49	52	55	58	61	64	67	70	73	76	79	82	85	88	91	94	97	100				
P₁ 1911		13	80	32																			125	40.5	1.75	4.33
P₁ 1912	1	4	28	16																			49	40.6	2.00	4.92
P₁ 1913		4	32	1																			37	39.8	1.09	2.74
F₁ 1911							4	10	41	75	40	3											173	63.5	2.92	4.60
P₂ 1911																	6	22	49	11			88	93.2	2.29	2.46
P₂ 1911																	2	16	32	6	1		47	93.4	2.23	2.39
P₂ 1911																	5	7	10	2			24	92.1	2.70	2.93
F₂(1)1912*						1	5	16	23	18	62	37	25	16	4	2	2						211	67.5	5.91	8.75
F₂(2)1912*					2	4	2	24	37	31	38	35	27	21	5	6	1						233	69.8	6.79	9.73

*F₂ progenies derived from two F₁ plants were grown in 1912.

parents and the F_1 should be associated with these genotypes. Consequently about one fourth of the F_2 should fall in the range of 34 to 43 mm., one half in the range of 58 to 70 mm., and the remaining one fourth should vary from 88 to 100 mm. In other words, all of the F_2 plants should fall into one of three discrete groups. Reference to Table 8–1 shows that no such simple hypothesis can account for the distribution curve that was observed in the F_2 generation.

But suppose that the genetic difference between the parents is represented by two allelic pairs instead of one. Furthermore, let us suppose for the sake of simplicity that the parental values are 40 and 90 mm., that there is no dominance, and that each allele has an equal and cumulative effect. Under this hypothesis the F_1 should have corollas 65 mm. long on the average. Instead of three classes in the F_2, there should be five classes with means of 40, 53, 65, 78, and 90 mm., and they should appear in the ratio of $1:4:6:4:1$. The means of these phenotypic classes are separated by 13 mm., and since the more variable parent varied over a range of only 15 mm., these five classes should be discrete or nearly so. Thus the five genotypes could still be identified with only small possibility of misclassification. Although the F_2 distribution curve constructed on this basis does not correspond to the curve actually observed, the assumption of an extra allelic pair seems to be a step in the right direction. This can be confirmed by constructing curves based on three and then four genes, and it will be found that the frequencies expected approximate more and more closely the normal distribution curve observed in the F_2 generation.

An examination of the F_2 frequency distribution shows that only one individual reached the lower size limit of the larger parent and that none came within two classes of the upper size limit of the smaller parent. Thus, in a population of 444 F_2 plants, the parental genotypes were not recovered. Since with four genes there is about an even chance of recovering the parental types, it is reasonable to conclude that more than four genes were involved in the present case. If only five genes were involved and our previous assumptions apply, the means of the expected classes should differ by about 4.5 mm. Since environmental effects are likely to cause any genotype to vary over a range of at least 11 mm., it is readily seen that the genotypic classes would overlap phenotypically and that the distributional curve should approach smoothness. Although no hypothesis is possible regarding the precise number of genes involved or their exact mode of action, the observed results do appear to be of the type expected if a number of genes, possibly as few as five, govern corolla length in this hybrid.

This explanation fits the observed facts neatly, but it does not prove beyond doubt that plural segregating genes are the correct explanation for the variation observed. Confirmatory evidence is always reassuring, and this East obtained by predicting the consequences of such an explanation in the F_3 and subsequent generations and by testing these predictions against observation. On the basis of the Mendelian scheme, East made the following additional predictions:

5. Individuals from various parts of the distribution curve of an F_2 population should produce F_3 progenies differing markedly in mean size.

6. Different F_2 individuals should produce F_3 progenies with different variabilities. The possible range of variabilities should extend from progenies as variable as the F_2 to ones no more variable than the original parents.

7. In the generations succeeding the F_2, the variability of a family can be the same or less than that of the family from which it came, but not greater.

Data from the F_3, F_4, and F_5 generations, shown in Table 8–2, make it possible to test these predictions. The data are clear in indicating that different F_3 families were characterized by real differences in mean size and furthermore that these differences were associated with the size of the F_2 parents. Although the greatest difference in mean size amoung the F_3 families was only 30 mm. or 23 mm. less than the difference separating the original parents, East was able by selecting the largest and smallest individuals in each generation to obtain two F_5 families with corollas 42 and 88 mm., or only 2 mm. longer than the shorter parent and only 5 mm. shorter than the longer parent. These facts provide strong evidence favoring the validity of prediction 5 above.

The two remaining predictions relate to comparative variabilities. It can be seen, both from the frequency distributions of the various F_3 families and from their coefficients of variability, that they differed widely in variability. F_3 families comparable in variability to either parental variety did not appear, but the number of families grown was rather small to expect this to have occurred. However, two F_4 families out of four did show as little variability as the smaller parent.

Because homozygosity within families should increase with continued self-fertilization, a concomitant decrease in genetic variability is expected in each succeeding generation. It can be seen that the

TABLE 8–2. Frequency Distributions of the F_3, F_4, and F_5 Generations in a Cross between Varieties of *Nicotiana longiflora*

(After East, 1916)

Class Centers in Millimeters

Family	Generation	Parent Size	34	37	40	43	46	49	52	55	58	61	64	67	70	73	76	79	82	85	88	91	94	97	n	x	S.E.	C.V.
1-2	F_3	46				1	4	26	44	38	22	7	1												143	53.5	3.74	6.99
1-3	F_3	50				6	20	53	49	15	4														147	50.2	3.17	6.31
2-5	F_3	50					7	25	55	55	18														160	53.0	3.04	5.74
1-4	F_3	60				2	3	9	25	37	70	19	10												175	56.3	4.07	7.22
1-1	F_3	72										4	20	25	59	41	19	2							170	70.1	3.82	5.22
2-1	F_3	77						1	0	1		2	2	16	33	43	34	20	6	1					159	73.0	5.00	6.85
2-4	F_3	80										2	8	14	21	39	39	32	10	1					166	74.0	4.85	6.55
2-3	F_3	81										1	1	8	16	20	32	41	17	3	3	1			143	76.3	5.06	6.63
2-6	F_3	82												3	5	12	20	40	41	30	9	2			162	80.2	4.76	5.93
1-2-1	F_4	44			8	42	95	38	1																184	45.7	2.37	5.18
1-3-1	F_4	43			2	23	122	41	1																189	46.3	1.87	4.04
2-6-1	F_4	85														4	9	38	75	59	6	3	1		195	82.2	3.30	4.01
2-6-2	F_4	87												4	5	6	11	21	23	41	29	8	5	1	164	82.9	5.83	7.04
1-3-1-1	F_5	41	3	6	48	90	14																		161	42.0	2.30	5.49
2-6-2-1	F_5	90														2	3	8	14	20	25	25	20	8	125	87.9	5.52	6.28

TABLE 8–3.   Pedigrees of Families and Their Coefficients of Variation

(After East, 1916)

F_2	F_3	F_4	F_5
8.75	5.22		
	6.99	5.18	
	6.31	4.04	5.49
	7.22		
9.73	6.85		
	6.63		
	6.55		
	5.74		
	5.93	$\left\{ \begin{array}{l} 4.01 \\ 7.04 \end{array} \right.$	6.28

coefficient of variability for the F_2 generation (Table 8–3) was higher than for any of the F_3 families.   In general, the coefficients of variability of F_4 families were lower than those of F_3 families, and a similar trend continued into the F_5 generation.   This trend is shown schematically in Table 8–3.   Among the fifteen families studied, only two were exceptional.   These two exceptions are not to be taken seriously because the families were small, and the biometrical constants describing them cannot be regarded as accurately established.

It was from such reasoning and observations that the Mendelian scheme was established as capable of explaining the inheritance of quantitative characters.   This explanation for quantitative inheritance was called the *multiple-factor hypothesis.*   Although the multiple-factor hypothesis viewed the genotypes for a given quantitative character as simply the sum of the effects of some definite number of gene differences, and did not visualize coordinated genetic systems of the type now regarded as basic to quantitative inheritance, it was one of the most important of all contributions to genetic thought.

Qualitative and Quantitative Characters

That both continuous and discontinuous variation is observed in characters such as stature indicates that the distinction between qualitative and quantitative characters is not absolute.   Stature, it is true, is usually a quantitative character; but dwarf or giant strains dependent upon single-gene differences have been found in all or nearly all plant species in which a search has been made. In some species—for example, corn—several different genes affecting stature are known, each producing a different degree of shortness or tallness. In certain instances, the effects produced by such single Mendelian alternatives vary with environment so that clear-cut categories appear under some conditions, but considerable difficulty is encountered in the classification of the same gene under other environmental circumstances. If identical environmental conditions could be provided for all the progeny of such monogenic hybrids, as, for example, might be possible in controlled environment chambers, segregation into discrete classes should occur even though the gene had only a very small effect. Conversely, when the effects of environment are large, even genes with major effects may be intractable to classical methods of genetic analysis.

In practice, then, the difference between qualitative and quantitative characters depends not so much on the magnitude of effect of individual genes as on the relative importance of heredity and environment in producing the final phenotype. It is therefore apparent that the key to progress in the analysis of quantitative characters lies in evaluating the relative contribution of these two casual agents to the variability, and it is to this subject that we must now turn.

The Concept of Heritability

In a strict sense, the question of whether a characteristic is hereditary or environmental has no meaning. The genes cannot cause a character to develop unless they have the proper environment, and, conversely, no amount of manipulation of the environment will cause a characteristic to develop unless the necessary genes are present. Nevertheless, we must recognize that the variability observed in some characters is caused primarily by differences in the genes carried by different individuals and that the variability in other characters is due primarily to differences in the environments to which individuals have been exposed. It would therefore be useful to have a quantitative statement of the relative importance of heredity and environment in determining the expression of characters.

In order to develop such a quantitative statement, we must first have a clear idea of the effect of environment on variability and also a clear idea of the effect of genes on variability. Let us first postulate a *constant* genetic situation and study the effect of changing the proportion of the total variability caused by environment. This situation is illustrated in Figure 8-1. In the figure a 12-unit difference ($P_1 = 50$

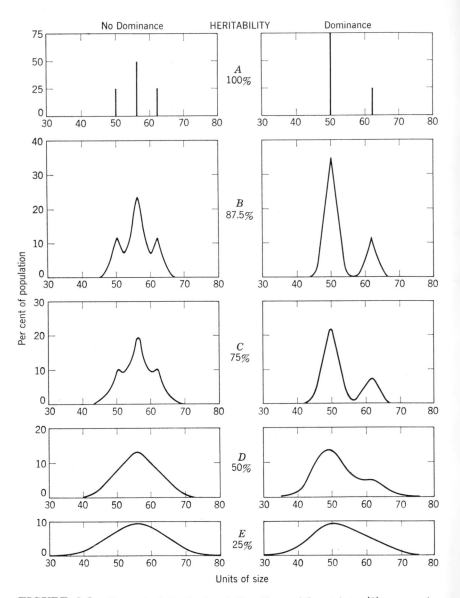

FIGURE 8-1. Theoretical distributions in F_2. The model postulates (1) monogenic inheritance, (2) a twelve-unit difference between the parents, and (3) that the effect of environment varies from nil (100 per cent heritability) to the point where environmental effects account for three fourths of the total variability (25 per cent heritability). The left column depicts no dominance; the right column, full dominance. Note that the scale for A differs from the scale for B through E.

units, $P_2 = 62$ units) in mean value between the parents of a hybrid is assumed to be governed by a single Mendelian gene pair. If there is no dominance, the genetic variance can be calculated to be

$$\frac{1(6)^2 + 2(0)^2 + 1(6)^2}{4} = 18.$$

In similar fashion, the genetic variance for the dominance model is found to be **27**.

In Figure 8–1a all the variability is assumed to be genetic. Since the genotype completely determines the phenotype, heritability is 100 per cent, and each phenotype can be represented accurately by histograms such as the ones shown in the figure. If, however, environment also contributes to the variability, the genotypes of the F_2 can no longer be represented by histograms but are more accurately portrayed by curves. For example, the two curves of Figure 8–1b were constructed on the assumption that one eighth of the total variance was environmental and seven-eighths genetic in origin; in other words, it was assumed that the heritability was 87.5 per cent. Heritability therefore specifies the proportion of the total variability that is due to genetic causes, or the ratio of the genetic variance to the total variance. This can be expressed quantitatively:

$$H = \frac{\sigma_G{}^2}{\sigma_G{}^2 + \sigma_E{}^2}$$

where $\sigma_G{}^2$ is the genetic variance and $\sigma_E{}^2$ is the environmental variance. Later we shall see that the term heritability is also used in more specific ways, but for the present this general definition of heritability will be adequate.

In Figure 8–1, it will be noted that the curves representing the three possible genotypes are discrete only at the highest heritabilities. With lower heritabilities, the degree of overlap increases progressively, and more and more difficulty is encountered in distinguishing among genotypes on phenotypic grounds even though genetic control is monogenic. Had larger differences been postulated between the parents, discreteness would of course have been maintained longer. It should also be noted that the generation curves obtained by summing the curves for each of the genotypes expected in a generation approximate more and more closely to normality as the heritability decreases. For example, even though the skewness due to dominance is striking with high heritabilities, skewness is hardly detectable for this particular model with 25 per cent heritability. These curves illustrate the tremendous difficulties that might be expected in selection when heritability is low. As a standard for comparison, it can be pointed out that the heritability for yielding ability, as measured in actual experiments, is often not significantly different from zero and that of the corolla-length character in tobacco, selected by East because of its remarkable insensitivity to environmental influences, usually has a measured heritability of the order of 75 per cent.

Figure 8–2 illustrates the effect on the F_2 distribution of varying the genetic situation while holding the effect of environment constant. With no dominance the distributions are symmetrical. It can be deduced from Figure 8–2 that, even when there are very few genes, these distributions would be difficult to distinguish from normal distributions whenever some blurring effect is produced by environmental agencies. The right column of Figure 8–2 shows how the skewness produced by isodirectional dominance (dominance all in one direction) becomes increasingly difficult to detect as gene numbers increase. This effect is even more conspicuous when environmental influences are incorporated into the model.

The models upon which Figure 8–2 are based assume that all gene pairs are equally effective, that dominance is isodirectional, that gene frequencies are 0.5 at all relevant loci, and that inheritance is independent. If all the different gene pairs do not have the same effect, the distribution tends to resemble ones produced by smaller numbers of genes. The same effect is produced by coupling-phase linkages. The Mendelian mechanism predicts that the zygotes in the population will be distributed according to the square of the ratio of the gametes, that is, according to the distribution corresponding to $[p(A) + q(a)]^2$. If $p \neq q$, the variance will be smaller—that is, the population will be less variable than if $p = q = 0.5$ for all loci. A similar dampening effect on the variability is caused by repulsion-phase linkages. The distributions can also be modified by epistasis, thresholds, and so on. Such effects are, however, much too complicated to permit generalizations about the way they modify distributions in various generations other than the generalization that whenever many genes interact with one another, the result is likely to be a normally distributed population. Despite these complications Figures 8–1 and 8–2 make clear the point that either moderately low heritability or moderately large numbers of genes will result in continuous and more or less normal distributions. This is even more true when many genes, each more or less influenced by environment, govern the variability.

From the foregoing considerations we can see that the hypothesis of plural segregating genes is capable of explaining the inheritance of continuously varying characters and, furthermore, that this hypothesis

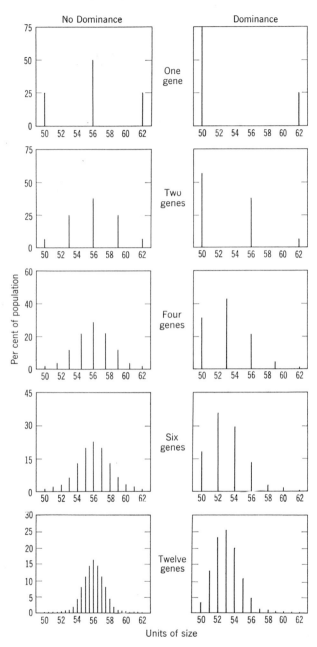

FIGURE 8-2. Theoretical distributions in F_2. The model postulates (1) 100 per cent heritability, (2) that a twelve-unit difference between the parents is governed by various numbers of genes of equal effects, (3) no linkage, and (4) that dominance is isodirectional. The scale is not constant.

tells us in a general way what we can expect when we cross two varieties and select for yield or other quantitative characters. This in itself is useful as a guide to plant breeders, but if selection is going to proceed on an entirely logical basis, more precise knowledge is required of the genetic and environmental forces at work. In the next two chapters we shall be concerned with some of the methods that have been proposed to obtain the required information, subjects we are prepared to consider now that the general features of the inheritance of quantitative characters have been established.

REFERENCES

East, E. M. 1916. Studies on size inheritance in Nicotiana. *Genetics* 1: 164–176. (One of the classic papers of genetics.)

Mather, K. 1943. Polygenic inheritance and natural selection. *Biol. Rev.* 18: 32–64. (Treats many of the basic concepts of quantitative inheritance.)

Mather, K. 1948. The genetical theory of continuous variation. *Proc. Eighth Intern. Congr. Gen.* 376–401.

Mather, K. 1949. *Biometrical Genetics,* Dover Publications, New York. (This little monograph includes a good review of the development of concepts in quantitative genetics and discusses the status of the field in 1949.)

Smith, H. H. 1944. Recent studies on inheritance of quantitative characters in plants. *Bot. Rev.* 10: 349–382. (A concise review with an extensive list of references.)

Yule, G. V. 1906. On the theory of inheritance of quantitative compound characters on the basis of Mendel's laws—a preliminary note. *Rept. Third Intern. Conf. Gen.* 140–142.

9

Roles of Genotype
and Environment
in Continuous Variation

Decisions about the most effective method of solving a practical problem involve ability to predict the consequences of various possible procedures. In plant breeding, such decisions have their basis in knowledge of the structure of the germplasm and the laws governing its behavior. Said in another way, an accurate choice among the many ways in which plant-breeding programs can be varied is possible only in the light of genetic understanding. The science of genetics has been most successful in providing plant breeders with the information they need in connection with genes that have large enough effects to be treated as uniquely recognizable units. Many important commercial characteristics of our crop species, such as fruit color in tomatoes, determinate habit of growth in beans, the difference between field and sweet corn, and resistance to many diseases, are governed by such genes. If a commercial type is required with some particular combination of these major genes, the plant breeder can determine exactly the procedures necessary to satisfy the need, whether few or many genes are involved. In so far as success in breeding operations depends on the manipulation of genes with large and easily recognizable effects, plant breeding can now be regarded much as a matter of routine.

However, in the selective improvement of plants, quantitative characters such as yield, adaptation, and many aspects of quality are also very important. Unfortunately, means of analyzing and managing this type of variability are, owing to technical difficulties, not as well developed as they are for qualitative characters. In dealing with quantitative characters the study of discrete classes is precluded by the very nature of the data, and it is apparent that an entirely different approach is needed to obtain the type of genetic information that will allow prediction of genetic advance when selection is practiced under various systems of breeding.

This need was recognized by a number of early geneticists, particularly Fisher and Wright. Since their pioneer studies, many purely theoretical investigations of quantitative genetics have been conducted, and the principles derived from them have been gradually tested by observation and experiment. The means of study have been experiments in which the system of mating has been controlled, just as in classical genetics. But the observations have had to take another form—means, variances, and covariances—statistics that can be calculated conveniently from the type of data obtained from continuously distributed observations. These statistics provide estimates of certain numerical attributes (parameters) of the genotype-environmental complex that influence breeding results, thus providing a basis for making choices among breeding procedures. In this approach the familiar symbols of classical genetics are abandoned, and genetic significance is attached to a different set of symbols. This does not mean that the new approach is non-Mendelian. It merely means that we recognize the impracticability of attempting to analyze the genetic system in terms of individual genes, symbolized by Aa, Bb, Cc, . . . , and that it is necessary to specify the effects of these genes in ways that are more convenient for the purpose. In the remainder of this chapter and in the next chapter we shall be concerned with the *genetic content* of the parameters that influence genetic advance under selection and the forms in which information about these parameters is most useful in plant improvement.

Factors Affecting Phenotypic Expression

With Johannsen's demonstration of the distinction between genotype and phenotype, and the proof by Nilsson-Ehle and East that quantitative characters were inherited according to Mendel's laws, it became clear that variation arises from the joint action of the genotype and environment. In describing the phenotype it is convenient to express this joint action in a linear fashion so that the phenotypic expression of any particular character, say yield, denoted by A, can be represented as

$$A = \mu + a + e + (ae). \tag{9–1}$$

In this expression the numerical value of the phenotype is visualized as the sum of a general population mean (μ), a genotypic effect (a), an environmental effect (e), and an interaction effect (ae). If it were possible to catalogue all genotypes and all environments, the value obtained would represent the general population mean (μ). Any particular genotype (or particular environment) adds or subtracts from the mean depending on whether its effect on yield is above or below the general population mean. The interaction term (ae) will be zero only when all genotypes behave consistently in all environments, that is, there are no genotypic-environmental interactions. Since all measurements are necessarily on phenotypes, any practical definition of μ, a, e, and ae must also be in phenotypic terms. In this system the effect of a genotype therefore is defined only relative to other genotypes, and this relationship, if it is to have any general utility, must be averaged over a number of environments. If the relationship is not constant from environment to environment, this variability will be reflected by the interaction term.

From the point of view of the plant breeder, genotypic values must be measured with reference to some particular group of environments, usually the ones occurring over a period of years at a number of locations within some comparatively homogeneous geographical area. Experimental measurements of genotypic values, no matter how extensive (within practical limits), can only sample the population of environments occurring within this geographical area. Hence the plant breeder must usually be satisfied with an attempt to identify the effects of the environmental components that are likely to be important in determining the agricultural value of genotypes, for example, soil fertility, rainfall, and temperature. In practice, these items are usually identified with years and locations. If the analysis is conducted in terms of different locations and different seasons, equation 9–1 can be expanded to:

$$A = \mu + a + r + l + y + (al) + (ay) + (lg) + (alg) + e. \quad (9\text{--}2)$$

In this expression, μ and a remain the population mean and the genotypic effect, r, l, and y are the direct effects of replicates, locations, and years, respectively, and e is a composite-error term of the remaining effects (including the effect of the plot compared to others in the same replication, error due to sampling among plants in the same family, and errors of measurement). Combinations of symbols indicate interactions between or among these main factors; (al), for example, represents the effect of interaction between a specific genotype and location. Equation 9–2 therefore identifies the principal environmental factors that bear on phenotypic expression, but does so in terms of locations and years, without attempting to specify in any precise way exact environmental influences, such as temperature, prevalence of diseases, and so forth, that can cause genotypes to vary relative to one another in different locations and seasons. If the breeder wishes to exercise control over some particular component of environment such as soil fertility, he can design experiments to identify and measure such specific components of environment and their interactions with genotype.

Genetic Advance under Selection

According to Comstock and Robinson (1952), the essential aspects of most, if not all, breeding programs are: (1) selection within a base population of genetically variable individuals or families, and (2) utilization of the selected material for the creation of new populations to be employed either as potential new commercial varieties or as the base for a new cycle of selection. It is apparent from equations 9–1 and 9–2 that genetic advance under selection (as represented by improvement in a, the genotypic value) in the new population as contrasted to the base population will depend on (1) the amount of genetic variability, that is, the magnitude of the differences in a among different individuals (or families) in the base population, and (2) the magnitude of the masking effect of the environmental and interaction components of variability on the genetic variability, that is, e and (ae) of 9–1 or l, y, (al), etc., of 9–2. A third factor, the intensity of the selection that is practiced, will also influence the rate of genetic advance under selection.

Against this background, suppose that selection is to be practiced among a large number of homozygous lines of a self-pollinated species. Furthermore, suppose that this selection is based on the information gained from the equivalent testing of each of these n lines for yielding ability in replicated-yield trials conducted at several locations in each of several years. From the data obtained it will be possible to calculate a mean yield (denoted by A) for each line. Assuming that the values of A are normally distributed with standard deviation σ_A, the frequency distribution of the mean values can be represented by Figure 9–1. If it is decided to select a portion q of the total of n lines, say all lines having yield A greater than some fixed value A', the selected families will be the ones falling in the shaded area of Figure 9–1. It can be shown by the methods of the calculus that the expectation of genetic advance for the selected families is

$$G_s = (k)(\sigma_A)(H). \tag{9–3}$$

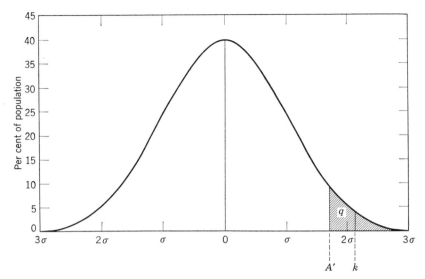

FIGURE 9-1. Theoretical distribution of mean yields of lines compared in repli-cated-yield trials. When the top 5 per cent of the lines are selected, that is, those with yields larger than A'(= 1.65σ), the mean of the q selected families is expected to be k (in standard measure) and kσ when expressed in terms of the actual units in which the measurements were taken.

In equation 9–3, the symbols have the following significance:

G_s represents the expectation of genetic advance under selection. It measures the difference between the mean genotypic value of the q selected lines, that is, $\bar{a}_s$, and the mean genotypic value of the n original lines, $\bar{a}$. Thus $G_s = \bar{a}_s - \bar{a}$.

σ_A is the phenotypic standard deviation of the mean yields of the n original lines.

H is the heritability coefficient, estimated as the ratio formed by dividing the genotypic by the phenotypic variance. Thus

$$H = \frac{\sigma_a{}^2}{\sigma_A{}^2}. \qquad (9\text{–}4)$$

k is a selection differential which takes into account the mean pheno-typic value of the q selected families ($\bar{A}_s$), the mean phenotypic value for all n families tested, the phenotypic standard deviation ($\sigma_A{}^2$), and the stringency of the selection q/n. Since k is expressed in standard-deviation units—that is, in terms of a unit normal curve—k does not vary except as the selection intensity, q/n, is varied. Thus if the highest-yielding 1 per cent of the lines are saved, k can be shown to have the value of 2.64. Other likely selection intensities for plant materials are 2, 5, 10, 20, and 30 per cent of the lines saved, for which selection intensities k takes the values of 2.42, 2.06, 1.76, 1.40, and 1.16, respectively.

If the heritability is 100 per cent (i.e., if $\sigma_A = \sigma_a$), the phenotype provides a perfect measure of the genotypic value. But even in this event the heritability value is an inadequate measure of the genetic progress that would result from selecting the highest-yielding lines, because the heritability could be 100 per cent when σ_a has a small value or a large value, whereas genetic progress would be larger with the larger genotypic variance. Genetic progress is also expected to be larger with stringent than with moderate selection. Equation 9–3 is therefore seen to evaluate, relative to one another, the three items that determine the rate of genetic advance expected under directional selection.

Experiments to Analyze Genotypic and Environmental Variability

The utilization of this theory in making predictions in practical situations requires numerical estimates of the various parameters influencing selection. These estimates can be made from data obtained in replicated field trials. The form of the analysis of variance for data on families tested at two or more locations in two or more years is shown in Table 9–1. The genotypic variance among families and

TABLE 9–1. Form of the Analysis of Variance for Data on Families Compared in Replicated Trials at Two or More Years and Locations*

(Adapted from Comstock and Robinson, 1952)

Source of Variance	d.f.	Expectation of Mean Squares
Families	$(F-1)$	$\sigma_e^2 + R\sigma_{aly}^2 + RY\sigma_{al}^2 + RL\sigma_{ay}^2 + RLY\sigma_a^2$
Families × years	$(F-1)(Y-1)$	$\sigma_e^2 + R\sigma_{aly}^2 + RL\sigma_{ay}^2$
Families × locations	$(F-1)(L-1)$	$\sigma_e^2 + R\sigma_{aly}^2 + RY\sigma_{al}^2$
Families × years × locations	$(F-1)(Y-1)(L-1)$	$\sigma_e^2 + R\sigma_{aly}^2$
Error	$(R-1)(FYL-1)$	σ_e^2

* *F, R, L,* and *Y* refer to the number of families, replications, locations, and years, respectively.

the various interaction components of variance can be obtained by algebraic manipulation of the mean-square expectations shown in the right-hand column of this table. Thus it is possible to obtain estimates of the following pertinent components of variance:

$\sigma_a{}^2$ A genetic component arising from genetic differences among families.

$\sigma_{al}{}^2$ A component arising from interaction of families and locations. It measures whether families behave consistently over locations.

$\sigma_{ay}{}^2$ The family-by-year component of variance.

σ_{aly}^2 The family-by-location-by-year component.

$\sigma_e{}^2$ The error variance.

An example of this type of analysis is provided by an experiment with Korean lespedeza reported by Hanson, Robinson, and Comstock in 1956. In this experiment a total of 284 families from 3 different hybrids were studied in the F_3 and F_4 generations. There were 136 families in population 1, 65 families in population 2, and 83 families in population 3. Each family traced back to the seed of a single randomly chosen F_2 plant. Individuals within an F_3 line were bulked for F_4 tests. The tests were conducted with two replications at each of two locations in each of two years. The two locations differed environmentally, particularly in soil type.

TABLE 9–2.   Estimates of Components of Variance for Two Traits in Korean Lespedeza

(After Hanson, Robinson, and Comstock, 1956)

Character	Population	s_a^2	s_{al}^2	s_{ay}^2	s_{aly}^2	s_e^2
Total yield	1	2,480	0000*	2,057	3,099	16,578
	2	3,354	0000*	858	2,070	17,432
	3	2,084	1,777	1,745	1,461	15,071
Seed yield	1	317	27	101	259	1,061
	2	206	9	123	322	933
	3	122	34	59	117	403

* Negative estimates for which the most reasonable value is zero.

The components of variance shown in Table 9–2 were obtained from the combined analysis of the data obtained at the two locations in the two years. The most conspicuous feature of these data is the small size of the genetic variances compared with the error variances. It is of interest to note, however, that the genetic variances were about the same size as the interaction variances involving years but were larger than the interaction variances involving locations. Despite the different soil types at the two locations, seasonal differences turned out to be a more important source of variation than location differences. Although there were differences among the estimates for the two traits (seed yield and total yield) and for the different populations, the data were consistent in indicating that the lines differed genetically. The experiment thus makes it clear that there are opportunities for increasing yield by selecting in each of the populations. The next step in the analysis is to estimate the heritability, since it provides a measure of the effectiveness with which selection can be expected to exploit the genetic variability.

The method of estimating the heritability will be illustrated using the data for total yield for population 1. Since heritability is defined as the ratio of the genotypic variance divided by the total variance (equation 9–4), we must obtain estimates of both σ_a^2 and σ_A^2. The genotypic variance, that is, the variance due to differences among the means of families, can be taken directly from Table 9–2 and has the value 2480. The total variance among the means of lines compared in R replications, L locations, and Y years is given by

$$\sigma_A^2 = \sigma_a^2 + \frac{\sigma_{al}^2}{L} + \frac{\sigma_{ay}^2}{Y} + \frac{\sigma_{aly}^2}{LY} + \frac{\sigma_e^2}{RLY}$$

which is

$$\sigma_A^2 = 2480 + \frac{0000}{2} + \frac{2057}{2} + \frac{3099}{4} + \frac{16{,}578}{8} = 6355.48.$$

Thus
$$H = \frac{2480}{6355.48} = 0.3902 = 39.0 \text{ per cent.}$$

The expected genetic advance under selection can then be calculated from equation 9–3, assuming the extreme 5 per cent of individuals are saved, as

$$G_s = (k)(\sigma_A) \frac{(\sigma_a^2)}{(\sigma_A^2)} = (2.06)(\sqrt{6{,}355.48})(0.3902) = 64 \text{ gm./plot.}$$

Expressed in per cent of the population mean, this is $64/829 = 7.7$ per cent. Similar values for the other populations appear in Table 9–3. The meaning of these estimates can be made clear by studying Figure 9–2, in which the results for total yield for population 1 are presented in graphical form.

This experiment with Korean lespedeza shows that estimates of the variances that are important to the plant breeder can vary greatly, depending on the environmental unit for which the variances are considered. When the genotypes are evaluated at only one time in one location, then only the *sum* $\sigma_a^2 + \sigma_{al}^2 + \sigma_{ay}^2 + \sigma_{aly}^2$ can be estimated as the genetic variance. When comparisons are made at succes-

TABLE 9–3. Heritability and Estimated Genetic Advance under Selection for Total Yield and Seed Yield in Three Populations of Korean Lespedeza

(After Hanson, Robinson, and Comstock, 1956)

Character	Population	Heritability	G_s in Grams per Plot	G_s in Per Cent of the Mean	Mean* in Grams per Plot
Total yield	1	39.0	64	7.7	829
	2	52.8	86	11.6	741
	3	34.2	55	7.3	751
Seed yield	1	59.8	27	18.9	144
	2	43.2	19	14.4	135
	3	49.2	16	20.6	77

* Total yield and seed yield in gms./plot. To convert to pounds per acre, multiply by 5.1514.

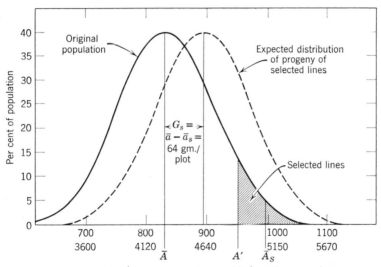

Yield in gm./plot (upper number) and lb./acre (lower number)

FIGURE 9-2. Diagrammatic representation of the effect of selection for total yield in Korean lespedeza (population 1). The original population was one with mean and standard deviation estimated to be 829 and 79.7 gm./plot, respectively. The expected mean of the 5 per cent of highest-yielding lines is $\overline{A} + k\sigma_A = 829 + (2.06)(79.7) = 993$ gm./plot (5115 lb./acre). The progeny of these lines are expected to have a mean of $\overline{A} + [(k)(\sigma_A)(H)] = 829 + 64 = 893$ gm./plot (4600 lb./acre). Thus, the genetic advance under selection is expected to be $G_s = 64$ gm./plot ($= 0.80\sigma$). If the heritability had been 100 per cent, the expected G_s would have been $k\sigma = (2.06)(79.7) = 164$ gm./plot, and the expected mean of the progeny of the selected lines would have been 993 gm./plot (5115 lb./acre).

sive times in one location, the estimable quantities are $(\sigma_a{}^2 + \sigma_{al}{}^2)$ and $(\sigma_{ay}{}^2 + \sigma_{aly}^2)$. A similar situation exists when comparisons are made at two locations in one year, the estimable variances then being $(\sigma_a{}^2 + \sigma_{ay}{}^2)$ and $(\sigma_{al}{}^2 + \sigma_{aly}^2)$. Thus the progeny variance estimated from a single test contains interaction variances in addition to the genetic variance, whereas progressively more of the interaction variances can be identified and removed as more and more locations and times are sampled. Therefore, the data provide a better estimate of genetic variance, according to its definition in equations 9–1 and 9–2, and the estimate of $\sigma_a{}^2$ tends to be closer and closer to its true value as more and more interaction effects are removed. The estimates of heritability, defined as a ratio of genotypic to phenotypic variance, also approximate more closely to their definition value as more of the interaction variances are disentangled from the genotypic variance.

Since only the genetic portion of the total variability contributes to gain under selection, the importance to breeders of information about the parameters of the genotype-environment complex is clear. As better estimates of these parameters are obtained for a variety of different plant materials, the breeder will be better able to anticipate the gain he can expect from different types and intensities of selection.

REFERENCES

Comstock, R. E., and H. F. Robinson. 1952. Genetic parameters, their estimation and significance. *Proc. Sixth Intern. Grasslands Congr.*, 284–291.
Hanson, C. H., H. F. Robinson, and R. E. Comstock. 1956. Biometrical studies of yield in segregating populations of Korean Lespedeza. *Agron. Jour.* 48: 268–272.
Lush, J. L. 1945. *Animal breeding plans.* 3rd ed. Collegiate Press, Ames, Iowa.
Smith, H. F. 1936. A discriminate function for plant selection. *Ann. Eugenics* 7: 240–250.

10

Genetic Components
of Continuous Variation

When families are selected from a base population on the basis of superior performance in field trials and used to form a new population, the net improvement of the new population over the old can be expressed as KG_s. In this expression G_s is the expected genotypic superiority of the selected families over the families making up the base population, and K is a coefficient relating the expected difference in merit between new and base populations to G_s. If we are dealing with pure lines or clones, K is 1, and the expected gain is represented simply by G_s. This is because such materials reproduce themselves precisely, aside from any deviations that might result from mutation. In this situation genetic advance will clearly depend only on the factors discussed in Chapter 9.

Another factor must be taken into consideration when materials subjected to selection do not reproduce themselves as precisely as clones or homozygous lines. This factor relates to the new combinations of genes that can arise from segregation and recombination if the new populations are reconstituted from sexually produced progeny of heterozygous materials. If the superiority of the selected materials stems partly from specific combinations of genes, these combinations will not necessarily remain intact in the population derived from the selected materials, and it follows that K can be other than unity, that is, the genotypic value of the new population can be different from that of the selected parents from which it was descended. This problem exists in experiments such as the one with Korean lespedeza considered in Chapter 9. There the materials under test were F_4 families which might not have the same genotypic values as their F_5 descendants. If we are to make accurate predictions of rates and amounts of genetic advance under selection, it is clear that we must understand the significance of hereditary parameters and be able to estimate them.

The Components of Genetic Variance

The first attempt to partition the genotypic variance into its component parts was made by Fisher in 1918. He recognized three components of *hereditary* variance: (1) an additive portion describing the difference between homozygotes at any single locus, (2) a dominance component arising from interactions of alleles (intra-allelic interaction), and (3) an epistatic part associated with interactions of nonalleles (interallelic interaction or epistasis). This classification has proved useful both in describing types of gene action and in estimation of the magnitude of various types of gene action.

TABLE 10–1. Gene Models for Two Loci Each with Two Alleles

Model I, Additive				Model II, Dominance			
AABB	*AABb*	*AAbb*	*AA - -*	*AABB*	*AABb*	*AAbb*	*AA - -*
7	6	5	6	4	4	2	$3\frac{1}{2}$
AaBB	*AaBb*	*Aabb*	*Aa - -*	*AaBB*	*AaBb*	*Aabb*	*Aa - -*
5	4	3	4	4	4	2	$3\frac{1}{2}$
aaBB	*aaBb*	*aabb*	*aa - -*	*aaBB*	*aaBb*	*aabb*	*aa - -*
3	2	1	2	3	3	1	$2\frac{1}{2}$
- - BB	*- - Bb*	*- - bb*		*- - BB*	*- - Bb*	*- - bb*	
5	4	3		$3\frac{3}{4}$	$3\frac{3}{4}$	$1\frac{3}{4}$	

Model III, Complementary				Model IV, Complex			
AABB	*AABb*	*AAbb*	*AA - -*	*AABB*	*AABb*	*AAbb*	*AA - -*
3	3	1	$2\frac{1}{2}$	4	2	3	$2\frac{3}{4}$
AaBB	*AaBb*	*Aabb*	*Aa - -*	*AaBB*	*AaBb*	*Aabb*	*Aa - -*
3	3	1	$2\frac{1}{2}$	4	3	1	$2\frac{3}{4}$
aaBB	*aaBb*	*aabb*	*aa - -*	*aaBB*	*aaBb*	*aabb*	*aa - -*
1	1	1	1	3	2	1	2
- -BB	*- -Bb*	*- -bb*		*- -BB*	*- -Bb*	*- -bb*	
$2\frac{1}{2}$	$2\frac{1}{2}$	1		$3\frac{3}{4}$	$3\frac{1}{2}$	$1\frac{1}{2}$	

Numbers indicate the genotypic value, $\mu + a$, for each genotype. The border rows and columns represent the mean genotypic values for the three phases possible at each locus, assuming gene frequency is $\frac{1}{2}$ at each locus. See text.

Four gene models are presented in Table 10–1 to illustrate the effect of the three types of gene action on the statistics which we shall use in describing quantitative characters.

The simplest conception of the way genes act assumes that the substitution of one allele for another produces a plus or minus shift on the scale of measurement and that this shift is the same regardless of what other genes are present. This situation is illustrated in the additive model of Table 10–1. The effect of replacing a by A is two units and that of replacing b by B is one unit. Since the effect is the same whether the replacement occurs in a homozygote or heterozygote, dominance is absent. The effect is also the same regardless of phase at the other locus; hence there is no epistasis. If all genes acted in this way, all the hereditary variance would be additive and our coefficient, K, would have a value of 1.

Complete dominance without epistasis is illustrated in the dominance model of Table 10–1. The alleles at a single locus interact in that the substitution of A for a in the genotype aa is not the same as in the genotype Aa. If all genes acted in this way, hereditary variance would be of two types: (1) additive, and (2) deviations from the additive scheme caused by dominance.

One of the classical types of epistasis, complementary gene action, is also illustrated in Table 10–1. The loci interact in that genotypic values associated with genetic phase at one locus are affected by genetic phase at the other locus. The hereditary variance here includes components resulting from all three types of gene action.

The last model of Table 10–1 illustrates a complex type and shows some of the difficulties that can be anticipated in real situations. The three phases of the A locus illustrate full dominance, overdominance, and full recessiveness when with BB, Bb, and bb, respectively. The three phases of the B locus illustrate underdominance, partial dominance, and no dominance when with AA, Aa, and aa, respectively. In the dominance, complementary, and complex models, K is obviously not equal to unity.

Definition of Hereditary Parameters

It is well beyond the scope of this book to cover the specification of genetic variability in detail, and it must suffice to illustrate the most simple concepts and procedures and to provide references for further study.

In the description of genetic variability, the significant hereditary parameters are defined precisely in terms of genetic models. In a model with two gene pairs with two alleles each, the simplest upon which interallelic interaction can be represented, 9 genotypes are possible. With gene frequency ½ at each locus, the Mendelian variation in the system can be specified by eight parameters representing the eight independent comparisons possible among the nine genotypes. One of these parameters, d_a in Hayman and Mather's notation, represents the difference between the homozygotes at the Aa locus (see diagram on page 104). Another parameter, d_b, represents the corresponding additive effect of the Bb locus. A third parameter, h_a, represents the deviation of the heterozygote, Aa, from the AA–aa midparent—that is, the dominance effect at the Aa locus; h_b represents the similar deviation of the Bb heterozygote. Four of the remaining degrees of freedom represent interactions between these "main effects." Thus the interaction between d_a and d_b is an additive $\times$ additive interaction, and the interaction between h_a and h_b is a dominance $\times$ dominance interaction. There are two additive-by-dominance interactions, those between d_a and h_b and between d_b and h_a.

Let us now consider the significance of these hereditary parameters by examining their contributions to genetic variability in the four models of Table 10–1. The total genetic variance in an F_2 with completely additive gene action, as in model I, can be calculated

$$\tfrac{1}{16}(7)^2 + \tfrac{1}{8}(6)^2 + \ldots \tfrac{1}{16}(1)^2 - (4)^2 = 2\tfrac{1}{2}.$$

TABLE 10–2.   Genetic Variances in F_2 Populations

Hereditary Parameter	Additive Model	Dominance Model	Complementary Model	Complex Model
d_a	2	$\frac{2}{16}$	$\frac{18}{64}$	$\frac{18}{256}$
d_b	$\frac{1}{2}$	$\frac{8}{16}$	$\frac{18}{64}$	$\frac{162}{256}$
h_a	0	$\frac{1}{16}$	$\frac{9}{64}$	$\frac{9}{256}$
h_b	0	$\frac{4}{16}$	$\frac{9}{64}$	$\frac{1}{256}$
$d_a d_b$	0	0	$\frac{4}{64}$	$\frac{4}{256}$
$d_a h_b$	0	0	$\frac{2}{64}$	$\frac{18}{256}$
$d_b h_a$	0	0	$\frac{2}{64}$	$\frac{18}{256}$
$h_a h_b$	0	0	$\frac{1}{64}$	$\frac{25}{256}$
Total variance	$2\frac{1}{2}$	$\frac{15}{16}$	$\frac{63}{64}$	$\frac{255}{256}$

Genetic models are the ones given in Table 10–1.   See text.

As shown in Table 10–2, two units of the total are associated with d_a and one half unit with d_b, the additive effects of the Aa and Bb loci, respectively. Thus all the variance in this model is additive, being attributable to the difference between the AA–aa and BB–bb homozygotes.

The total genetic variance for the dominance model is $^{15}/_{16}$, of which $^{10}/_{16}$, or $\frac{2}{3}$, is attributable to the difference between the AA–aa and BB–bb homozygotes. The remainder, $^{5}/_{16}$, arises from the failure of the heterozygotes Aa and Bb to fall on the AA–aa and BB–bb mid-parents, respectively. For the complementary model the total genetic variance is $^{63}/_{64}$, of which approximately 57 per cent can be attributed to additive, 29 per cent to dominance, and 14 per cent to epistatic effects of genes. This epistatic portion of the variance arises from the failure of the Aa and aa phases at the first locus to behave consistently over the Bb and bb phases at the second locus. Numerically this epistasis can be represented as $AaBb - Aabb - aaBb + aabb = 3 - 1 - 1 + 1 = 2$. This value would have been zero in the absence of interaction (epistasis). For the complex model the variance due to additive, dominance, and epistatic effects of genes are in the ratio of 70 : 4 : 26 per cent, respectively.

These gene models were introduced to clarify what is meant by the three types of gene action and to illustrate that in an interacting system involving dominance and epistasis, the genes may still be described in terms of their average effect and a portion of the variance ascribed to them, even though they have lost some of their singularity of effect.

Experiments to Estimate Hereditary Parameters

In real populations it is to be expected that the genetic situation will be much more complex than in the digenic models we have used to define the hereditary parameters. It is true that the models can be expanded to represent the Mendelian variation in more complex cases. For example, with 3 genes, 27 genotypes are possible; and of the 26 degrees of freedom, 3 represent additive, 3 dominance, and 20 epistatic effects of genes. However, the partitions of the hereditary variance relating to the higher-order interaction effects are not sufficiently understood at present so that any distinctions among them are relevant to decisions among plant-breeding procedures. Moreover, estimation of all the multiplicity of parameters that are undoubtedly required for complete specification of quantitative genetic systems is completely beyond the power of existing experimental methods. Practical experiments to evaluate the genetic components of variation therefore make a series of simplifying assumptions. The most frequent assumption is that parameters of the $d_a d_b$, $d_a h_b$, $d_b h_a$, and $h_a h_b$ type, and their higher order analogs, have values of zero. The effect of nonvalidity of this assumption is to inflate the estimate of the additive and/or dominance components of variance. With this difficulty in mind we can turn to a simple experiment designed to estimate the additive and dominance components of variance.

Considering only the Aa locus of our model, the values associated with the genotype are as follows:

$$
\begin{array}{ll}
AA & d_a \\
Aa & h_a \\
aa & -d_a
\end{array}
$$

Since all deviations are measured from the AA–aa midparent as the origin, the difference between the AA and aa homozygotes has the value of $2d_a$.

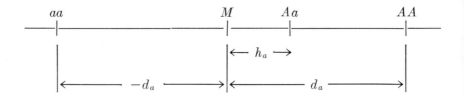

The capital-letter symbol refers to the allele with a plus effect on the character and does not necessarily denote either dominance or recessiveness. When the character is governed by a number of genes the phenotype of a true-breeding strain will be $\Sigma(+d) + \Sigma(-d) + c$, where c represents a constant base level dependent on genes not segregating and nonheritable influences. Henceforth, for simplicity, we will not write c in our equations.

The F_1 of the hybrid between AA and aa will be phenotypically h_a. If $h_a = 0$, no dominance is indicated, whereas $h_a = d_a$ represents full dominance and $h_a > d_a$ represents overdominance. The F_1 between two homozygous strains differing in a number of genes will therefore be $\Sigma(h)$, where h can take sign. If the F_1 deviates from the midparent, some dominance obviously must exist at one or more loci. But the observation that $\Sigma(h) = 0$ does not necessarily indicate that all individual values of h are zero because individual dominance effects may have different signs and exactly counterbalance one another. When all $(+)$ alleles by which two strains differ are concentrated in one parent and all $(-)$ alleles in the other, the difference between the two strains must be $2\Sigma d$. In practice, however, the difference will usually have a lesser value than $2\Sigma d$ because isodirectional distribution of $(+)$ and $(-)$ alleles is unlikely.

The constitution of the F_2 generation as regards the gene Aa is expected to be

Genotype:	$\frac{1}{4}AA$	:	$\frac{1}{2}Aa$	:	$\frac{1}{4}aa$
Phenotype:	d_a		h_a		$-d_a$

The mean of the F_2 generation is expected to be

$$\tfrac{1}{4}(+d_a) + \tfrac{1}{4}(-d_a) + \tfrac{1}{2}(h_a) = \tfrac{1}{2}h_a. \tag{10-1}$$

The contribution of this gene to the total sum of squares is

$$\tfrac{1}{4}(d_a)^2 + \tfrac{1}{4}(-d_a)^2 + \tfrac{1}{2}(h_a)^2 = \tfrac{1}{2}(d_a)^2 + \tfrac{1}{2}(h_a)^2. \tag{10-2}$$

Since the generation mean is $\tfrac{1}{2}h_a$, the variance of the F_2 is

$$\tfrac{1}{2}(d_a)^2 + \tfrac{1}{2}(h_a)^2 - \tfrac{1}{4}(h_a)^2 = \tfrac{1}{2}(d_a)^2 + \tfrac{1}{4}(h_a)^2. \tag{10-3}$$

We are considering a unit population, so that this is also the mean of the squared deviations from the mean, or the variance of the F_2. This is only the genetic portion of the variance, and an environment portion will occur in actual experiments. Letting $D = \Sigma(d)^2$, $H = \Sigma(h)^2$, and E the environmental influences, the variance of an F_2 segregating for many genes can be represented as

$$V_{F2} = \tfrac{1}{2}D + \tfrac{1}{4}H + E. \tag{10-4}$$

By similar methods the summed variance of the backcrosses to the parents is found to be

$$V_{B1} + V_{B2} = \tfrac{1}{2}D + \tfrac{1}{2}H + 2E. \tag{10-5}$$

Many other relations involving other generations can be worked out in similar fashion.

An experiment will serve to illustrate the general features of the estimation procedure. In this experiment heading date was determined for the wheat varieties Ramona (P_1) and Baart (P_2) and for their F_1, F_2, and first backcross generations. The means and variances for these generations are given in Table 10–3. The theory we have discussed above permits us to set up the following simultaneous equations:

$$V_{F2} = \tfrac{1}{2}D + \tfrac{1}{4}H + E = 40.350$$

$$V_{B1} + V_{B2} = \tfrac{1}{2}D + \tfrac{1}{2}H + 2E = 51.640$$

Hence

$$2V_{F2} = D + \tfrac{1}{2}H + 2E = 80.700$$

$$\underline{V_{B1} + V_{B2} = \tfrac{1}{2}D + \tfrac{1}{2}H + 2E = 51.640}$$

and

$$\tfrac{1}{2}D \qquad\qquad = 29.060$$

TABLE 10–3. Dates of Heading of the Homozygous Wheat Varieties Ramona (P₁) and Baart 46(P₂) and Hybrid Generations Derived from the Cross of $P_1 \times P_2$

Genera-tion	Days to Heading from an Arbitrary Date													n	$\bar{x}$	s^2
	4–6	7–9	10–12	13–15	16–18	19–21	22–24	25–27	28–30	31–33	34–36	37–39	40–42			
P₁	4	21	60	48	20	4	2							159	12.99	11.036
B₁	1	12	88	77	85	50	6	4	1	1			1		326	15.63	17.352
F₁		1	2	20	83	51	12	2	1					171	18.45	5.237
F₂		4	25	66	156	115	50	41	38	34	16	4	3	552	21.20	40.350
B₂			4	34	49	47	45	61	41	26	6	1		314	23.38	34.288
P₂							33	56	35	19	5			148	27.61	10.320

Data of 1953 season.

Using the mean variance of the nonsegregating generations,

$$\frac{V_{P1} + V_{P2} + V_{F1}}{3} = 8.894,$$

as the best available estimate of E and substituting in either equation 10-4 or 10-5, we estimate $H = 9.584$.

In the F_2 generation, 22 per cent of the total variance,

$$\left(\frac{8.894}{40.350} = 0.22\right),$$

is environmental in origin, and the remainder, or **78** per cent, is due to genetic causes. Thus heritability in the broad sense is **78** per cent. We are, however, more interested in the heritability computed on the basis of the additive genetic variance, because it better indicates the degree to which the progeny of these F_2 plants will resemble their parents. Calculated on this basis the heritability is

$$\frac{29.060}{40.350} = 72.0 \text{ per cent.}$$

Separating the genetic variance into two parts has therefore somewhat improved our estimate of the degree to which selection is likely to be effective.

Assuming that the earliest 5 per cent of the F_2 plants were selected for further propagation, the expected genetic advance can be calculated

$$G_s = (k)(\sigma_A)\left(\frac{\sigma_a^2}{\sigma_A^2}\right) = (2.06)(\sqrt{40.350})(0.720) = 9.42 \text{ days.}$$

The mean of the F_2 population was 21.20 days; hence the mean of the F_3 progeny of the selected individuals is expected to be $21.20 - 9.42 = 11.87$ days. It was possible to check this prediction by comparing the expected advance with the actual advance made in F_3 progenies grown in an adjacent plot. These progenies were derived from the earliest 5 per cent of the plants of an F_2 population grown in a previous year. The actual gain, as measured by the mean of these progenies and the F_2 population mean, was 8.45 days. Hence the procedure followed in this analysis predicted quite accurately the advance that was achieved in an actual selection experiment.

Value to the Plant Breeder

For a long time it has been apparent, at least in a general way, that progress in plant breeding depends on the nature, magnitude, and interrelations of genetic and nonheritable variation in agriculturally significant plant characteristics. However, the ways in which genotype and environment and their interactions affect progress under selection have been poorly defined. Perhaps one of the most important contributions of the procedures presented above has been to indicate the type of information that is required and the forms in which it may be most useful. By defining clearly the forces with which the plant breeder must deal in manipulating quantitative characters, these procedures have served a useful purpose.

As yet there have been comparatively few experiments performed that have permitted unbiased estimates of the hereditary parameters. The breeder bases his hopes of improvement on evidence of genetic variation. But estimates have rarely been made in such a way as to exclude bias from genotype-environment interactions, and it is only from unbiased estimates that realistic appraisals of the rate and magnitude of the improvement will be possible. Similarly, few estimates of genotypic covariances between characters have been made from adequately designed experiments. Obtaining such estimates is presently one of the most important problems in plant breeding.

Nevertheless, progress is being made in both theory and experiment, and plant breeders can look forward to the gradual accumulation of information on the relative magnitude of additive, dominance, and epistatic variances and their interactions with nonheritable agencies. With this information will come the ability to foresee the consequences of various plant-breeding procedures and the ability to make accurate decisions about the most effective breeding practices.

REFERENCES

Cockerham, C. C. 1956. Analysis of quantitative gene action. *Brookhaven Symposia in Biology,* No. 9: 53–68.

Fisher, R. A. 1918. The correlation between relatives on the supposition of Mendelian inheritance. *Trans. Roy. Soc. Edinburgh* 52: 399–433.

Fisher, R. A., F. R. Immer, and O. Tedin. 1932. The genetical interpretation of statistics of the third degree in the study of quantitative inheritance. *Genetics* 17: 107–124.

Hayman, B. I., and K. Mather. 1955. The description of genic interactions in continuous variation. *Biometrics* 11: 69–82.

Mather, K. 1949. *Biometrical Genetics,* Dover Publications, New York.

11

Pure-Line Breeding and Mass Selection

We have seen that selection in self-pollinated crops is an ancient method of breeding (Chapter 6), but that understanding of the scientific basis of the method was a much more recent development. The genetic diversity providing the basis for selection has its origin in spontaneous heritable changes which occur at a slow rate and in a haphazard manner. The dispersal of the mutant alleles through populations depends upon natural hybridization and recombination compounded over many generations. As a consequence of the mating system of these species, the plant breeder can reasonably expect that any plant he selects will be homozygous, and hence will give rise to a pure line that reproduces itself with great precision.

As a result of the establishment of these principles, modern programs of selection in variable populations of self-pollinated crops generally follow one or the other of two distinct but related patterns which differ from each other only in the number of pure lines saved to form the new variety. These alternative procedures have come to be called *Pure-Line Breeding* and *Mass Selection*. In pure-line breeding, the new variety is made up of the progeny of a single pure line. In mass selection, the progeny of many pure lines are composited to form the new variety.

Pure-Line Breeding

Pure-line breeding generally involves three distinct steps. In the first step a large number of selections are made from the genetically variable original population. These initial single-plant selections are clearly of the greatest importance with this method of breeding because nearly all the genetic diversity exists between lines and little within

109

lines. Selection within lines is therefore not worth while, and, if superior types are not included among the original selections, subsequent manipulations are incapable of compensating for their absence. No set number of individual selections can be designated as appropriate for all circumstances. In general, however, the number of initial selections should be as great as considerations of time, expense, space, and competitive plant-breeding projects will permit. With the small grains several hundreds or even a few thousands of individual selections can be made with moderate effort by choosing individual heads or panicles more or less at random. If some specific goal is sought, for example, earlier types or types with larger seeds, special methods of selection can be very useful. For other crops, particularly the row crops, the population to be selected can be grown in such a way that each individual can be examined for variations and more careful selection practiced. Careful selection initially is of course particularly important where space or other limitations restrict the number of lines it is possible to grow in subsequent generations.

The second step consists of growing progeny rows from the individual plant selections for observational purposes. This evaluation by "eye" may extend over several years. Lines with obvious defects are eliminated promptly. Frequently, artificially produced disease epiphytotics or other aids to selection permit the elimination of additional unsuitable types. After such obvious eliminations, the selections are grown over a shorter or longer period of years to permit the observation of performance under different environmental conditions, and still further eliminations are made. A drastic reduction in the number of lines must be accomplished during this observational period because the last step in the program is demanding and costly, thereby limiting the number of lines that can be accommodated.

The third and final stage starts when the breeder can no longer decide among lines on the basis of observation alone and he must turn to replicated trials to compare the remaining selections with each other and with established commercial varieties in relative yielding ability and other aspects of performance. The length of time required for evaluation clearly depends on circumstances, but ordinarily this stage of the program lasts at least three years.

In its earliest use it is probable that the method had the essential features of mass selection, but toward the end of the pre-Mendelian era it became more and more characteristically pure-line breeding in the form described above. Some of the varieties developed by this method in the period of its greatest use, that is, the late nineteenth and early twentieth centuries, are still represented among present-day commercial varieties. Much of the success of the method during that period appears to be related to the existence of land varieties that were waiting to be exploited. These varieties had been grown for long periods and frequently on a large scale. Seed sources were not protected from natural crossing as in present practice; nor is there evidence that any except the most obvious variants were purged. There was thus ample opportunity for diversity to accumulate, and plant breeders, both amateur and professional, made the most of their opportunities. The result was a large number of types offered to the trade as wonderful new varieties. Many of these types were inferior and soon disappeared, but, all and all, this period was a very fruitful one that resulted in substantial improvement in the varieties of most self-pollinated species.

Most of the land varieties characteristic of the more advanced agricultural countries have now been well exploited, and, although good types were undoubtedly missed, the method is certainly less valuable now than in the past. There seems to be little likelihood that it will regain its former importance because the new pure-line varieties are kept true to type by seed associations and are not favorable populations for selection. In some of the less advanced agricultural areas, however, land varieties still exist, and the situation is similar to that of Western Europe and North America of half a century ago. Since these resources still exist, the pure-line selection method has an important role to play in delaying the day predicted by Malthus.

Varieties Developed by Pure-Line Selection

The successful use of pure-line selection could be illustrated with any of a very large number of examples. The present purpose will be adequately served, however, by consideration of the development of a few varieties representing the more common variations of the method.

The complex of wheat types originating in the Crimea and grown under the designation of "Turkey" or "Crimean" wheat has all the features of a land variety. This complex was the foundation of the vast hard red winter-wheat industry of the United States, and it was understandably the object of selection by many wheat breeders, both private and public. Almost without exception the approach was the one described above, namely, selection of large numbers of individual lines for observation and final evaluation. Turkey wheat is quite homogeneous for gross plant type, and most of the selections made from it and grown as agricultural varieties have much the same appearance as Turkey. Also, the more important named varieties selected from the Turkey complex, such as Kanred, Blackhull, Nebred, and Cheyenne, all have the same general adaptation and productivity of the original variety. However, they have each represented a distinct improvement in one or more important characteristics such as earlier maturity, improved stiffness of straw, or resistance to diseases that improved their performance relative to Turkey. Only Blackhull has departed in general appearance sufficiently to be readily distinguished from Turkey or the other selections.

Another common variation of pure-line selection depends upon the discovery of an obvious "off-type" in the mother variety. Fulghum oats is a good example of a variety developed in this manner. J. A. Fulghum, a farmer-breeder of Warrenton, Georgia, noticed a plant that was taller and earlier than other plants in his field of Red Rustproof Oats. He harvested the plant and increased seed from it, thus giving rise to the Fulghum variety. Earlier maturity, more vigorous plant development, better grain quality, and fewer awns, compared with Red Rustproof, were some of the characters that have made Fulghum the most important varietal type of red oats of the United States. Fulghum has subsequently been reselected and has in turn given rise to important pure-line strains such as Kanota, Frazier, and Franklin oats.

Occasionally an opportunity arises for improvement by pure-line selection in some specific character, frequently one that was previously unimportant. In 1926 a new root rot, caused by *Periconia circinati*, was discovered on sorghum in Kansas. This disease, called sorghum disease or milo disease and first ascribed to *Pythium arrhenomanes*, soon threatened the milo crop of Kansas. Its effect was catastrophic. In 1930, F. A. Wagner of the Kansas Agricultural Experiment Station found two plants of Dwarf Yellow Milo that were apparently normal in every way although they were growing on infested soil where all other milo plants had died. In the following year, one of these plants produced progeny free of disease and the other produced both healthy and diseased offspring. From the resistant plants there was developed a resistant strain of Dwarf Yellow Milo that was very similar to its commercially important parent strain, but could be grown in *Periconia*-infested soil without any evidence of injury.

This sorghum disease soon appeared in Oklahoma, Texas, New Mexico, and California. Plant breeders in those states found the same low incidence of resistant plants in the local varieties, and these plants provided the basis for a quick and easy solution to the control of this very destructive disease. In California the incidence of resistant plants in commercial fields was of the order of one plant per ten acres in Yellow Double Dwarf Milo, the most widely grown variety, and the opportunity to select resistant individuals from these fields was seized. Besides saving at least one year of time, selecting from the commercial fields avoided the expense and effort of growing large acreage expressly to locate resistant individuals. In addition to illustrating an important use of pure-line selection, this example shows the importance that opportunistic timing can assume in plant breeding.

Mass Selection

The main difference between pure-line selection and mass selection in self-pollinated crops relates to the number of lines retained. With pure-line selection the derived type consists of a single pure line. With mass selection the majority of the lines selected are likely to be retained. Retention of many lines has an important effect on step two (mentioned previously under Pure-Line Breeding), tends to eliminate step three, and influences the purposes the program can serve effectively. Although mass selection in self-pollinated crops has had more limited use than pure-line selection, at least in the twentieth century, it does have two important functions in plant breeding.

The first of these functions derives from the safety and rapidity with which mass selection can effect improvement in land varieties. These land varieties are the basis of agriculture in some of the less developed agricultural areas, and such varieties are likely to include lines that are too early or too late in maturity, disease susceptible, or for other reasons do not contribute their full share to the total yield. There is evidence that the residue remaining after discarding the obviously unproductive or defective lines retains the best features of the original variety in general adaptation and yield and can be released to growers without the extensive testing that pure-line varieties should require.

There is no experimental evidence to indicate the size of population to be selected or the proportion of the selections to be retained. Populations of several hundred or preferably two or three thousand individuals appear to be desirable whenever circumstances permit. If too few lines are eliminated, the possibility of improvement is reduced. On the other hand, drastic elimination may place the adaptation of the variety in jeopardy. As long as large populations are selected initially and perhaps no more than 25 per cent of the lines are discarded, adequate numbers should remain to minimize the danger of upsetting the essential agronomic or horticultural features of the original variety. Such a conservative approach appears to have two great advantages. First, the plant breeder is relieved of the effort required to test the new type, allowing him to place more emphasis on other breeding programs. Second, the elimination of the testing should allow the selected type to be released for commercial production in the shortest possible time.

There is no reason why the selection should not be repeated every few years until it no longer seems to produce results. The termination of a repetitive mass-selection program would no doubt be influenced by the success of single pure lines that would almost certainly have been selected from the original population as well as by the progress of parallel breeding programs involving hybridization methods.

The second function of mass selection is the purification of existing varieties in the production of pure seed by seed associations. Usually a few hundred plants are selected from fields known to be representative of the variety. Progenies are grown from these individual plants the following year and observed at critical stages of development to permit progenies including mutants, natural hybrids, varietal mixtures, or other off-types to be eliminated. The remaining progenies are generally harvested in bulk to constitute the pure-seed source. This procedure can be modified to suit crops with special requirements, and it can be repeated as often as necessary to maintain the desired purity of the variety.

One of the most important features of pure lines is the great precision with which they reproduce themselves. Many plant breeders believe that the inclusion of numerous closely related pure lines imparts useful flexibility to varieties and are therefore careful during seed purification to eliminate only those progenies that obviously do not conform to type.

REFERENCES

Akerman, Ake, et al. 1938. *Swedish contributions to the development of plant breeding.* Bonniers Bokytryckeri. (Discusses pure-line and mass selection as well as other plant-breeding methods.)
Yearbook of Agriculture, U. S. Dept. Agric. 1936, 1937. (Gives numerous examples of pure-line selections from self-pollinated crops.)

12

Pedigree Method of Plant Breeding

The object of hybridization in breeding self-pollinated species is to combine, in a single genotype, desirable genes that are found in two or more different genotypes. In deciding which method of handling the segregating generations is likely to be most successful in achieving this goal, the plant breeder must take a number of factors into consideration. For example, the yielding ability, adaptation, and disease reaction of the parents available to him, knowledge of the genetic control of these characters, and technical considerations such as the ease with which hybrids can be made and the space required to grow segregating generations in field and greenhouse will all influence the decision. Of the three basic methods of handling segregating materials in self-pollinated species (Chapter 5), the pedigree procedure has been the most widely used. We shall now consider the circumstances where the pedigree procedure is appropriate and describe in some detail the steps to be followed.

In the pedigree method superior types are selected in successive segregating generations, and a record is maintained of all parent-progeny relationships. Selection begins in the F_2 generation, when individuals are selected which in the judgment of the breeder will produce the best progeny. Most hybrids can be expected to segregate for a large number of genes, and it is certain that every F_2 individual will differ from every other individual. In the F_3 and F_4 generations, many loci will have become homozygous and family characteristics begin to appear. Much heterozygosity persists in these generations, however, so that plants within families are still likely to differ from one another genetically. In these generations, then, selection is practiced for the best plants in the best families. By the F_5 or F_6 generation most families can be expected to be homozygous at most loci. Hence selection within families is no longer very effective, and emphasis shifts almost entirely to selection among families. In these generations

115

families with a common ancestor one or two generations removed are likely to be very similar to one another. The pedigree record at this point makes possible the elimination of all except one member of such closely related families.

The pedigree record consists of a series of notes which give the relationships among the families that have been grown. The distinguishing features of families are usually also included in the record. These records, properly taken, can be useful in deciding the families to be continued and the ones to be discarded. Unless handled with discrimination, however, the taking of notes can be so demanding of time and effort as to constitute a major weakness of this method of breeding. There is usually no reason to keep detailed records of large numbers of characters and certainly no reason to waste time on families destined to be discarded because of observable weaknesses. Often designation for continuation of a family or reselection within a family will suffice. It should be recognized that the pedigree does not establish a retraceable route that can be followed in developing the same variety by repetition of the same cross. The pedigree does, however, provide a record of the precise relationship of all plants of each kindred. This information is useful primarily in avoiding the selection of closely related individuals whose probable worth is nearly identical.

Choice of Parents

It is usually helpful in planning a pedigree program to regard the variety to be produced as a replacement for some well-established variety. The new variety usually cannot be much poorer in yield, adaptation, or dependability than the variety it is intended to replace, irrespective of improvements in specific features. For this reason, almost without exception, one parent is selected for its proven performance in the areas of intended use. The other parent is usually chosen because it complements specific weaknesses of the first parent. Ordinarily the strain singled out as the second parent exhibits these specific characteristics in an intense form, even if intensity in the second parent is greater than that desired in the new variety. This is because some intensity is usually lost in the different genetic environment of the new variety. Recombination occasionally leads to the production of desirable features not found in either parent. However, the best chance of success lies in selection of parents which between them include the features desired in the new variety.

Many times all the characters for which improvement is necessary or desirable do not occur in any two parents. Then a third parent may be included by crossing it with the F_1 of the first two parents. If a fourth parent is necessary the hybrid of two F_1's can be made the starting point in the selection program. An alternative procedure is to attempt the improvement in two or even more steps instead of attempting to solve a large number of problems in a single step.

Although the pedigree-method breeder probably always hopes to increase yield, sometimes this may be his sole objective. It is true that increases in yield often attend the adjustment of reaction to diseases, maturity dates, plant size, or other specific characteristics, but the opportunity exists for increasing yield as a result of enhancing basic metabolic rates or other features of plant growth about which little is known at present (Watson, 1952).

Before a breeder makes a hybrid, he should have clearly defined his objectives, selected the parents to be hybridized, and decided on the method of breeding to be followed. Rarely is the hybridization procedure itself troublesome with self-pollinated species. The actual making of the hybrid is therefore a minor part of the total breeding program. It does, however, have an importance beyond the time and effort expended, because making the hybrid symbolizes the beginning of a process that the breeder hopes will end in a significant contribution to human welfare.

Aids to Selection

If selection of the parents determines the potential of a pedigree breeding program, the skill with which selection is practiced in the segregating generations determines whether the potential of the hybrid will be realized. Accurate judgment of the worth of the plants selected in segregating generations requires that the breeder be a careful student of his crop. Since, in the segregating generations, decisions must be made among hundreds of individuals, most judgments must be based on quick visual evaluation rather than precise measurements. Hence the breeder must know the morphological as well as the physiological characteristics that characterize acceptable varieties. He should have a feel for the effect that adverse as well as favorable environmental conditions have on his selections and also perception concerning the probable market quality of the plants he saves. Without the perceptivity born of thorough knowledge of the crop, it is doubtful whether the visual evaluation necessary in the early generations of a pedigree program can be successful in identifying the infrequent valuable types that most hybrid combinations are capable of producing by segregation.

Although keen discernment is necessary to identify plants of high basic worth, no amount of acumen will allow the breeder to select for specific characteristics when the conditions necessary for the expression of these characteristics do not occur. In order to practice effective selection for specific characters the breeder is often forced to devise special techniques that allow him to separate desirable from undesirable plants.

Special techniques are frequently necessary when selecting for disease resistance, because natural epiphytotics adequate to distinguish between resistant and susceptible plants may not occur regularly. When they do not, the solution is usually found in artificial inoculation that enhances the probability of a heavy epiphytotic in the breeding nursery. Sometimes, however, the breeder requires control over races of the pathogenic organism or over other variables that cannot be achieved in regular nurseries. Greenhouses frequently allow the necessary controls, but they may impose severe limitations on population sizes, especially if plants must be grown to maturity. Another solution is off-season nurseries, as, for example, have been used extensively in California to determine the reaction of small grains to stem rust. When planted in early July the development of the plants proceeds under conditions of temperature and humidity favorable to the rusts. Most spring varieties mature a good crop of seed by October in time for normal planting in November or December. These off-season nurseries have several advantages: (1) Uniformly severe epiphytotics of rust develop regularly, permitting positive identification of susceptible plants. (2) Artificial inoculation of susceptible spreader varieties is required to start the epiphytotic; hence the desired race or races can be used with little chance of contamination. (3) Selection is especially effective for spring types because winter or semiwinter types fail to head. (4) The off-season planting serves the added purpose of advancing the generation.

This fourth factor is a matter of importance, because the time required to produce a new variety is often a source of worry to plant breeders and has sometimes led to the release of new varieties before they were proved and ready for release. Where there is urgent need, considerable expense is justified in growing off-season breeding nurseries. Greenhouses can be useful in reducing the time necessary for the breeding of a variety, particularly for the making and growing of F_1 hybrids or other operations with a small space requirement. Recently there has been a great increase in the growing of extra generations in the field in subtropical or tropical areas or even in the opposite hemisphere.

Selection for characteristics such as winter hardiness and resistance to postharvest sprouting, drought, and lodging present problems similar to those encountered in selecting for disease resistance. Lodging, for example, is a difficult character to evaluate under field conditions because in some seasons little or no lodging occurs, whereas in other seasons severe conditions may cause all varieties to lodge. The severe losses in yield and quality that attend lodging have led to many attempts to develop aids to selection. Some of the more successful methods that have been developed depend on machines to measure factors such as the breaking strength of the straw or the force required to pull plants from the soil, as these characteristics tend to be correlated with lodging. Winter hardiness is another character difficult to evaluate under field conditions. Only in certain years are the conditions favorable for differential killing, and then the situation is likely to be complicated by killing in patches, caused by slight variations in environment in the breeding nursery. Laboratory tests for winter hardiness usually depend on differential response to artificially produced low temperatures as an index of ability to survive in the field. The results of laboratory tests have often been reliable as a measure of resistance to cold, but they have been less successful in measuring winter survival, because it depends not on low temperature alone but other factors as well.

With still other characters special techniques are more than merely aids to selection, they are essential if progress is to be made. For example, the diastatic power of malting barley is an important quality factor and one which no plant breeder can determine visually except in the grossest way. Effective selection in cases such as this is clearly dependent on the development of suitable tests. To be of greatest value to the breeder such tests should be rapid and inexpensive, and the amount of material required for the test should be small enough to permit its application to single plants. Such a test has been developed to measure the prussic-acid content of sudan grass. Individual plants vary widely in content of a cyanogenic glucoside, and this rapid and inexpensive test has played an important role in the development of nontoxic varieties of this valuable forage grass.

The basis of quality is complicated in many cases. The desirability of wheat strains depends on whether they are to be used in making bread, pastries, breakfast foods, crackers, macaroni, or other products. Wheat excellent for one use may be virtually useless for other purposes. Some wheats give excellent results for a particular purpose but only after special handling in milling and baking. Thus complete evaluation of a new strain may require a series of tests in which a number of variables are examined at several levels each. Although several rapid methods have been developed for evaluating certain properties of wheat, and these short-cut methods have been useful to breeders, none can replace actual milling and baking trials. Complete trials are expensive and require substantial quantities of wheat and are consequently one of the last steps in wheat-breeding programs.

These examples of the role of special techniques as aids to selection are a small sample of the hundreds described in the literature of plant breeding, genetics, phytopathology, and food sciences. Though the literature is a rich source of ideas, modifications are frequently necessary to suit local circumstances, and much may depend on the ingenuity of the plant breeder if efficient procedures for selection are to be developed for particular situations.

Handling the Hybrid Materials

Procedures that may be followed in the various stages of a pedigree breeding program will now be described in some detail. These procedures should not be regarded as fixed or exclusive but rather as a general guide that can be modified to suit the singular requirements of individual pedigree breeding programs. (See, e.g., Culbertson, 1954; Harrington, 1952; Quinby and Martin, 1954; Weiss, 1949.)

In *the F_1 generation* enough F_1 hybrid plants should be grown to produce the seed necessary for an F_2 population of the desired size, leaving enough reserve seed to sow a similar F_2 in case of crop failure.

In *the F_2 generation* the size of the population to be grown will be influenced by the number of F_3 families it is possible to handle. Although no hard and fast rules can be laid down, few plant breeders are satisfied with fewer than 50 F_3 families and usually grow many more. The ratio of F_2 individuals to F_3 families ordinarily varies from about 10 to 1 up to 100 to 1; occasionally the ratio may be still higher, particularly if a large proportion of F_2 plants can be eliminated on the basis of easily observable deficiencies. Usually the ratio is higher for wide crosses than for crosses between closely similar parents.

The F_2 generation affords the first opportunity for selection in pedigree programs. An obvious first criterion of selection is the elimination of all plants carrying undesirable major genes. Next, those plants having high intensity of the visible characteristics desired in the new variety are selected. Further selection is usually necessary to reduce the plants to a number that is possible to accommodate in the F_3 generation. At this point selection may depend on differences among plants, and keen judgment may be necessary to select the individuals that will produce outstanding progeny. The vigor of many F_2 plants may depend on heterozygosity and consequently lead to the selection of the more heterozygous individuals. However, F_3 progeny tests give an indication of those F_2 plants that depend excessively on heterozygosity for their desirability.

The beginning plant breeder must develop an attitude of ruthlessness in selection. He must dispel the feeling that among the plants he discards may be the one that will lead to the variety he has in mind. Unless he keeps this natural inclination under control he will shortly find himself overwhelmed with materials that he should have discarded, and his effectiveness as a plant breeder may be impaired or lost.

Since, in the F_2 generation, selection is on a single-plant basis, planting rates in this generation must be adjusted to allow individual plants to be handled conveniently. This is a severe disadvantage in some crops because growth habit under thin experimental rates of planting may differ from that obtained with thicker commercial rates of planting.

In *the F_3 generation* family differences make their appearance. Each family should therefore be represented by enough plants in the F_3 to give an indication of the general features of the family and also of the heterozygosity of the family. Again, rigid rules cannot be laid down, but usually not less than ten individuals and preferably thirty or more individuals are grown per family. Selection in this generation continues on a single-plant basis, but the emphasis is on selection in the superior families. Outstanding individuals in otherwise undistinguished families need not be neglected entirely.

The number of selections to be made in the F_3 generation will be influenced by several factors. By this generation the potentiality of the hybrid will sometimes have become apparent. If there is a dearth of promising families the entire cross might be discarded. If, on the other hand, the parents have nicked well, there may be a wealth of superior types among which to select. Rarely, however, should the total number of selected plants exceed the number of families grown in the F_3 generation.

The F_4 generation is usually handled in much the same way as the F_3 generation but with further shift in emphasis toward selection among families. Many families will have become quite homozygous, but usually enough genetic diversity remains to make selection within families intriguing. Single plants thus remain the basis for selection in the F_4 because homozygosity has usually not reached the point where most plant breeders are willing to propagate families in mass.

Selection in the F_2 and F_3 generations will have eliminated many families with obvious deficiencies by the F_4 generation. Furthermore, by this generation genetic differences among families are much greater than differences within families. In this generation, therefore, the first good opportunity occurs to reduce the number of families drastically. This is done in part by visual comparison of families and in part on the basis of the pedigree record. Many times two or more plants will have been selected in a family in the F_3 generation. If, in the F_4, there is little to choose among such related families, one line will usually suffice to perpetuate the lineage.

In *the F_5 generation,* commercial planting rates are usually in order. The potentialities of individual families have usually become fixed by this generation. Consequently many plant breeders harvest F_5 rows in mass to obtain the larger amounts of seed they require for yield and quality tests in the F_6 generation. If, after thick seeding, it is decided to select again on a single-plant basis, there may be some difficulty in separating single plants from one another. At this stage little harm is done if two or more plants are inadvertently used to perpetuate a family, instead of only one plant.

By the F_5 generation the emphasis in selection will have shifted almost entirely from easily observable differences to a visual estimation of characteristics that are often subjective and difficult to judge. These characteristics are sometimes more easily compared in multi-row plots than in single-row plots. Consequently, in the F_5 generation, the basic planting plan is sometimes altered from that of earlier generations. In cereals, for example, three-row blocks 5 feet long often permit better comparisons among families than longer single-row plots. Plots of this size are consequently sometimes used in the F_5 and later generations. In other crops, such as tomatoes, a single row may still be as valuable as multi-row plots.

Some plant breeders initiate preliminary-yield trials in the F_5 generation, usually using two or three replications of plots of appropriate size for the materials under selection. The yield data are then used as an additional criterion of selection.

In *the F_6 and F_7 generations,* operations are usually a continuation of those of the F_5 generation. In the F_6 and F_7 generations the primary objective is identification of the few best families or, stated conversely, drastic elimination of mediocre families. In so far as possible this reduction will, for reasons of economy, be based on continued visual observation. When the number of families has been reduced to manageable proportions, precise quality determinations are initiated. Quality tests may start as early as the F_5 or as late as the F_8 generation, depending on the expense and complexity of the test. Ordinarily, preliminary-yield trials begin in the F_6 or F_7 generations, relative yielding ability being used as an additional criterion in reducing the families to a number that can be handled in the precise yield and quality trials upon which final evaluation is based.

Final Evaluation

The final evaluation of promising strains involves (1) further observation for obvious weaknesses that may not have appeared in previous seasons, (2) quality tests, and (3) precise-yield tests. The same plantings will usually fulfill all these purposes. In the evaluation of a new variety, the plant breeder must always compromise between temerity and overcaution. If the evaluation is not sufficiently extensive, agriculture can be misserved by the release of an inferior variety. On the other hand, delays in putting valuable varieties into production due to overcaution are also no service to agriculture. The compromise reached by many plant breeders is to release when a variety proves to be superior in 5 years of trials at each of five representative locations in the intended area of use. Regional trials conducted cooperatively by plant breeders in different counties, states, or provinces, or even in different countries, are often useful for this purpose. The United States Department of Agriculture has been especially active in organizing regional uniform nurseries. These regional trials usually contain about fifteen to thirty entries, allowing each cooperating breeder to enter one or two of his most promising strains. Each cooperating breeder grows the uniform set of materials and submits the results to a central office. This office summarizes the data for release to all interested persons.

Early-Generation Testing

Earlier in this chapter we discussed some general principles useful in choosing the parents that should be hybridized to produce particular results. Despite the care with which parents are chosen, it is a common experience to find that certain combinations "nick" to produce many superior offspring, and other hybrids between apparently equally promising parents produce disappointing progeny. Thus the appearance, yield, or adaptation of the parents is not always a good indicator of superior "combining ability." This is presumably because combining ability often depends on complex interaction systems among genes.

In the pedigree system of breeding and also in the bulk system to be considered in the next chapter, any appraisal of the breeding materials permitting early elimination of materials of low potential is clearly advantageous, because all improvement programs have size limitations, and elimination of poor materials enhances the probability of finding superior segregates in the remaining materials. More precisely, the problems involved are, first, identification of the hybrids most likely to give the highest proportion of superior segregates and, second, early evaluation of the segregates from the promising crosses.

Selection among Crosses

One of the criteria that has been used for selection among hybrids is performance of early-generation bulk-hybrid populations. As early as 1932 Harrington concluded that yield tests of a composite of F_2 plants gave an indication of the yield potentialities of wheat crosses. Harlan, Martini, and Stevens in 1940 found that high-yielding F_5 bulk populations of barley tended to produce more high-yielding familial isolates than low-yielding bulk populations. Immer, in 1941, reached this same conclusion from his studies of bulk populations in barley, finding that some hybrids were consistently higher in yield in the F_2, F_3, and F_4 than other hybrids. He found, however, that the necessity for space planting of F_1 hybrids in barley resulted in a hybrid $\times$ planting-rate interaction that greatly reduced the value of F_1 yields for purposes of prediction. In oats, on the other hand, Atkins and Murphy found that F_7 bulk progeny tests were limited in their values for predicting the number of high-yielding segregates to be expected from any hybrid. In cereals, therefore, the value of yield trials of bulk hybrid materials for selecting among crosses is conflicting.

A number of similar studies have been conducted with soybeans. These studies generally indicated no close agreement of F_1 yield with yields in subsequent generations. Similar nonsignificant correlations between the yields of F_2 plants and their F_3 progenies were usually noted. When F_3 lines were carried to the F_4 as progenies of individual F_3 plants, associations between yields of F_3 lines and yields of their progenies were also generally weak, and it was concluded that little worth-while information on the potential yielding ability of subsequent selections can be obtained before the F_4 generation.

In the foregoing studies, the seed for each succeeding generation of cereals and soybeans was produced in different years. Thus the populations studied may not have been comparable, owing to genetic shift resulting from differing pressures of natural selection in different seasons. Weiss (1949) believes this factor to be particularly important in soybeans, stating that "the unreliability of bulk population tests in predicting yield of segregates from soybean crosses would seem to be largely attributable to the fact that soybeans are a full season crop and therefore subject to varying natural selection pressures." In other studies the advanced generations were tested in succeeding years. Here interactions among hybrids, generations, and seasons were all confounded. Mahmud and Kramer in 1951 attempted to avoid these difficulties by testing F_4 populations made up of bulks of equal quantities of seed from individual F_3 plants; in addition, F_3 and F_4 lines were all tested in the same year to avoid interaction with season. These workers concluded that F_3 lines provide good estimates of the yielding potential of later generation segregates when genetic shift and interaction with environmental factors such as differences among seasons and spacing were taken into account.

Further refinements in technique have been made in more recent studies. For example, Johnson, Robinson, and Comstock (1955) designed an experiment in which it was possible to estimate for two segregating populations of soybeans: (1) genotypic variance among F_3 lines, (2) variance due to genotypic-environmental interactions, and (3) progress to be expected from selection. They found that serious misjudgments are likely to result from estimates of genotypic variance based on single trials. This is because variance due to genotypic-environmental interactions cannot be separated from the genotypic variance in experiments conducted at a single location in a single year, and if the interaction components of variance are important, predictions based on the inflated estimate of the genotypic variance will not be realized in future years or at other locations (Chapter 9). The magnitude of this effect is illustrated by one of their hybrids in which the genotypic variance for yield indicated by a single test was reduced by 71 per cent when average performances over locations and years were considered. They concluded that much of the failure of yields of soybean lines to predict the yields of progenies resulted from genotypic-environmental interactions and that earlier studies may have failed because interactions between genotype and environment had not been taken into account.

The type of information that is necessary for accurate decisions regarding the prepotency of different hybrids is gradually being recognized as a result of experiments such as the ones discussed above and those considered in Chapter 9. Thus the hybrids most likely to produce high-yielding progeny will be the ones that combine high mean-yielding ability in the early generations with large genetic variance for yielding ability.

Calculation of the progress to be expected from selection clearly depends on accurate estimates of yielding ability and of the pertinent genetic, environmental, and interaction components of variance in the early generations. It should be remembered from Chapter 10 that genetic advance under selection is defined in terms of the genetic values of the actual generations tested. These predictions may therefore not be descriptive of the effect of selection on genotypes derived by sexual reproduction from the materials tested. For example, the superiority of lines selected on the basis of F_3 data may not be retained, because of segregation, in succeeding generations. What is to be anticipated in this connection depends on the nature of the gene action involved. As we saw in Chapter 10, the nature of gene action in quantitative inheritance is being studied, but critical evidence has not yet been obtained. As more of this information becomes available, we can look forward to increasing precision in predicting the prepotency of hybrids.

Attempts to classify varieties of self-pollinated crops into groups depending on ability to transmit high yield in crosses has hardly begun. The way in which the estimates we are now capable of making might be used in selecting among crosses can be illustrated with the data of Hanson, Comstock, and Robinson, discussed in Chapter 9. In that study (Table 9–3) the mean yields and the expected genetic advance for seed yield for three hybrids were population 1, 144 and 27, population 2, 135 and 19, and population 3, 77 and 16 grams per plot. Of these three hybrids, number 1 is the most promising because of its high mean yield and high genetic variance. Hence, in an actual pedigree program with limited facilities, the greatest advance might well result if hybrids 2 and 3 were eliminated and the subsequent effort concentrated on hybrid 1.

Selection within Crosses

When making selections in the F_2 and later segregating generations the breeder attempts to save the plants that in his judgment will produce the best progeny. Two questions might be posed: (1) How accurately can the performance of particular progenies be predicted from the characteristics of single plants or single rows? (2) How early in the inbreeding process are various agricultural characters sufficiently fixed to permit accurate selection? Many studies have been conducted in attempts to answer these questions.

The results of these studies can be summarized as follows: The effect of environment on single-plant yields is so large that selection for heritable high yield in the F_2 is virtually futile. On the other hand, effective selection among spaced F_2 plants for disease resistance and other characters of high heritability is frequently possible. Since selection in the F_2 generation must be based on performance in a single season, effectiveness of selection in that generation for characters moderately subject to seasonal fluctuations (e.g., plant height, maturity date) is often small. The effectiveness of selection among individual plants is therefore seen to be highly sensitive to the magnitude of the heritable variability *relative to* environmental variability.

Thus the question arises whether F_3 or F_4 lines, in which larger populations can be studied, provide a suitable basis for predicting the agricultural value of subsequent generations. The studies reported to date seem to indicate that F_3 line tests permit reasonably effective selection among lines for the moderately heritable characters. This is particularly true if tests are replicated in time and space. F_3 or F_4 performance, as measured in single trials, has generally been a poor basis for predicting the yields of subsequent selections; trials conducted in more than one environment have been moderately good for purposes of prediction. Hence attempts to select for high yield in early generations should probably be limited to truncated selection in which only the poorest lines are eliminated.

These studies thus appear to support the preference of many plant breeders for the methods of selecting within hybrids that were outlined earlier in this chapter. These methods, it will be recalled, call for visual inspection of individual plants and unreplicated progenies of selected individuals as the basis for selection in the segregating generations. On the other hand, selection for yielding ability and other characters that are much influenced by environment is generally postponed until later generations.

General Conclusions

One of the chief advantages of the pedigree method of handling segregating generations in hybridization breeding may not have become apparent in the discussion above. The pedigree method permits the plant breeder to exercise his skill in selection to a greater degree than any of the other prominent methods used with self-pollinated species; hence it is the most satisfying of the several possible procedures open to the plant breeder following hybridization.

The main disadvantage of the method is the limitation it places on the amount of material one plant breeder can handle. These limitations can be overcome to some extent by starting different hybrids in different years, thus staggering the periods when peak work loads occur. Nevertheless, a conflict frequently develops between the necessity for exploring the potentialities of large numbers of crosses and the need to handle large numbers of selections within each cross.

The pedigree method has been used to develop many of the important varieties of self-pollinated crop species. Plant-breeding literature describes the origin of many of these varieties in detail, so examples will not be given here. For particularly extensive lists of varieties developed by this method, the reader is referred to the 1936 and 1937 *Yearbooks of Agriculture.* Other examples are given in review articles on particular crops in the series *Advances in Agronomy.*

REFERENCES

Akerman, Ake, et al. 1938. *Swedish contributions to the development of plant breeding.* Alb. Bonniers Boktrycheri, Stockholm. (Includes a discussion of the pedigree method of plant breeding.)

Culbertson, J. O. 1954. Seed-flax improvement. *Adv. in Agron.* 6: 143–182. (Describes in detail pedigree systems used with flax.)

Harlan, H. V., M. L. Martini, and H. Stevens. 1940. A study of methods in barley breeding. *U. S. Dept. Agric. Tech. Bull.* No. 720, 26 pages.

Harrington, J. B. 1932. Predicting the value of a cross from an F_2 analysis. *Canadian Jour. Res.* 6: 21–37.

Harrington, J. B. 1952. Cereal breeding procedures. *Food and Agric. Organ., United Nations,* Paper 28, Rome, Italy. (Includes key literature citations to pedigree breeding.)

Immer, F. R. 1941. Relation between yielding ability and homozygosis in barley crosses. *Jour. Amer. Soc. Agron.* 33: 200–206.

Johnson, H. W., H. F. Robinson, and R. E. Comstock. 1955. Estimates of genetic and environmental variability in soybeans. *Agron. Jour.* 47: 314–318. (Includes key literature citations on early-generation testing.)

Quinby, J. R., and J. H. Martin. 1954. Sorghum improvement. *Adv. in Agron.* 6: 305–359.

Salmon, S. C., O. R. Mathews, and R. W. Leukel. 1953. A half-century of wheat improvement in the United States. *Adv. in Agron.* 5: 3–141.

Watson, D. J. 1952. The physiological basis of variation in yield. *Adv. in Agron.* 4: 101–145. (A general discussion of the factors affecting crop yield.)

Weiss, M. G. 1949. Soybeans. *Adv. in Agron.* 1: 78–158.

13

Bulk-Population Breeding

In the bulk-population method of breeding, the F_2 generation is planted in a plot large enough to accommodate several hundreds or even several thousands of plants. Planting rates and cultural practices are usually the normal ones of commercial practice. At maturity the plot is harvested in bulk and the seeds used to plant a similar plot the following season. This process is repeated as many times as desired by the plant breeder. The method is operationally best suited for seed crops such as small grains and beans. It is totally unsuited for fruit crops and for most vegetable crops.

During the period of bulk propagation natural selection presumably plays a role in shifting gene frequencies in the bulk population. This role is probably an important one, particularly if bulk handling is continued over a large number of generations. There is, of course, no need to rely exclusively on natural selection in bulk-population breeding since artificial selection can be practiced at any time during the period of bulk propagation. When the plant breeder decides that the bulking has accomplished the purpose he has assigned to it, single-plant selections are made and evaluated in the same way as in the pedigree method of breeding.

Nilsson-Ehle of Sweden was apparently the first to use the bulk-population method. He adopted it because of the advantages it offered in handling the segregating generations of a hybrid made to combine the winter hardiness of the Squarehead variety with the high yield of the Stand-up variety of winter wheat. Nilsson-Ehle recognized the important features of the bulk-population method as they are known today. He assisted natural selection by discarding plants that had suffered winter damage but would nevertheless not have been completely prevented from reproducing. In this way he increased the rate of shift toward hardy types. He recognized that growing large populations would increase the chance of high-yielding types appearing among the winter-hardy selections, and that the bulk-population method was well suited for handling the large numbers he needed because he would be relieved of keeping pedigree records. He also realized that homozygosity would increase during the period of bulk handling so that selections made after a few generations of bulking could be expected to breed true.

Natural Selection in Bulk Populations

Natural selection is such an important feature of the bulk method that it will be well to consider its effects in some detail before considering the applications of the method. Natural selection has been an important factor in the evolution of cultivated species, prior to domestication, certainly, and probably afterwards as well. In nature, of course, the selective pressures are exclusively those associated with survival. It does not follow that the characteristics that make a domesticated species useful to man are necessarily associated with survival; in fact, quite the opposite may be true for some characters. Nevertheless, many of the characteristics that contribute to survival in wild species must also be associated with the basic adaptation and productiveness of these same species under cultivation. The central issue in bulk-population breeding is the nature of this correlation— that is, whether ability to survive in competition is related to agricultural worth and, if so, the extent of the relationship. We shall return to this question after considering briefly some theoretical aspects of survival in competition.

Theory of Survival in Competition

When two or more types compete against one another in a bulk plot large enough to avoid drift, survival depends on two factors: (1) the number (not the weight) of seeds each type produces, and (2) the proportion of the seeds of each type that reach maturity and produce offspring. Both of these factors are in themselves summaries of a large number of genetic and environmental factors (and their interactions), some of which may be known but most of which escape our powers of observation or measurement. The unknown factors may in fact be our primary interest for the very reason that they escape detection and hence are immune from artificial selection.

When only two types compete, the theoretical survival curve of the poorer competitor is

$$A_n = as^{n-1}, \tag{13-1}$$

where A_n is the proportion of the poorer competitor in the nth generation, a is the initial proportion of the poorer competitor, and s is the
selective index. The selective index takes into account the two factors
mentioned above. As an example, consider that two types have been
mixed in equal proportion $(A_1 = 0.5)$, and survival values for the
better and poorer competitors are 1 and 0.9 respectively $(s = 0.9)$.
The expected proportion of the poorer type in the fifth generation is

$$A_5 = (0.5) (0.9)^4 = 0.3645 \quad \text{or} \quad 36.45 \text{ per cent.}$$

The proportion of the better competitor is obviously $1 - 0.3645 = 0.6355$ or 63.55 per cent.

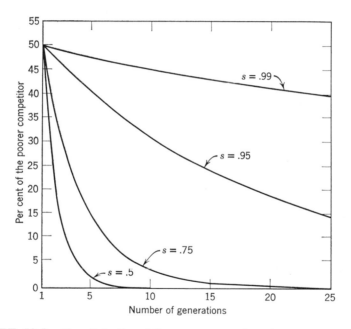

FIGURE 13-1. The elimination of the poorer competitor when two types compete in a bulk plot. Initial proportions are assumed to be equal in all cases.

Theoretical curves for hypothetical two-genotype mixtures are given
in Figure 13–1 for several different selective indexes. For simplicity,
only the curve of the poorer competitor is given. If there is a striking
difference in the competitive abilities of the two genotypes, the elimination of the poorer competitor (and the increase of the better competitor) is rapid for the first several generations. However, the last
individuals of the poorer competitor are slow to disappear. When the

difference in competitive abilities is small, changes are slow at first and still slower later. For example, if the difference between the two types is 5 per cent ($s = 0.95$), fourteen generations are required to reduce the poorer competitor from 50 to 25 per cent, but after fifty generations it will still make up nearly 4 per cent of the total population.

When three types, each making up one third of the initial population, compete against one another, the curves for the best and the poorest competitors resemble the curves for a two-type mixture. The intermediate type decreases slowly as long as part of its competition is provided by the poorest type, but it is eliminated more rapidly once it competes principally against the best type. The middle curve of Figure 13–2 illustrates the elimination of a type of intermediate competitive ability in a three-genotype mixture.

In a typical bulk plot, many types are expected to be in competition. When we consider the theoretical curves expected in multi-type mixtures (Figure 13–2), we see that the best type increases rapidly for several generations, and the poorest types are eliminated rapidly. However, a third type of curve appears, a humped curve. This type of curve represents good competitors that increase in prevalence so long as part of the competition is against the poorer competitors. Even moderately poor competitors do not decrease so long as the least productive types are represented by any considerable number of individuals. The second-best type does not start to decrease until half of the total population is made up of plants of the best competitor. Eventually only the single-best type will remain, but the last plants of the next-best sorts are slow to disappear.

These theoretical curves ignore at least two important factors. First, it is likely that the number of seeds produced by good competitors will be greater in the beginning, when the poorer types still constitute a considerable proportion of the population, than later when competition will be against more aggressive types. This factor affects only the pitch, not the type of the curve. Second, seasonal variations are such that first one type and then another might be expected to be favored. This should lead to an undulating curve unlike any of the theoretical curves depicted.

The next question to consider is the correspondence of these theoretical curves to actual survival curves. Survival in populations is undoubtedly affected by many factors such as mean temperature, diurnal temperature range, soil moisture at various stages of growth, light intensity, density of population, prevalence and kind of weeds, and a host of others. Few of these factors can be isolated and tested one at a time under field conditions because their actions will vary from season to season. However, their mean aggregate effect can be measured by means of a census figure indicating the survival of various genotypes in the population.

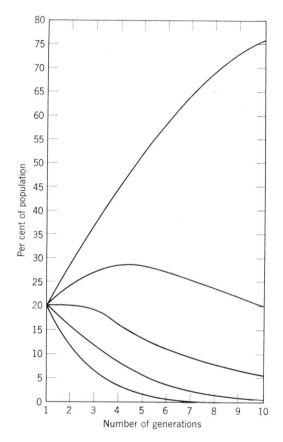

FIGURE 13-2. Survival in multi-type mixtures. In this five-type mixture the selective values for the two better and two poorer types are assumed to be 40 and 20 per cent greater or less than the intermediate type.

Perhaps the simplest type of experiment designed to measure changes in populations is one in which changes in the proportions of a limited number of pure lines are plotted against time. The competing types are homozygous and, in the absence of natural crossing, remain constant from season to season. Genetic variability is eliminated as a complicating factor because no new genotypes arise. The census data permit an estimate of the effectiveness of natural selection in eliminating unsuited genotypes. At least three such experiments have been conducted that can be compared with the theoretical curves, allowing us to test whether good agricultural types are also good competitors.

Natural Survival in Varietal Mixtures

Perhaps the most significant and comprehensive experiment on natural selection in domestic plants was conducted by Harlan and Martini. Seeds of eleven barley varieties were mixed, and the mixture was grown for a number of years at ten experiment stations in the northern and western United States. Each year the plots were threshed and sufficient seed saved to plant two plots the following year. One plot was again harvested in bulk and the other used to take a census of the number of plants of each of the eleven varieties remaining in the population. Because the Coast and Trebi varieties could not be distinguished readily, these varieties were reported together. The final census given in Table 13–1 represents the end effects of 4 to 12 years of natural selection.

The eleven varieties included in the mixture represent a wide range of types, some with wide adaptation, some with more specific adaptation, and a few poorly adapted at all locations. Trebi is probably grown on more acres than any other barley variety in the United States; it is the major variety of the nothern intermountain region, but it is also well adapted to the eastward. Manchuria-type barleys constitute the major commercial types of the northern plains, north central, and northeastern regions of the United States. On the Pacific Coast, the commercial acreage is dominated by coast-type barleys. Hannchen and White Smyrna are good-yielding two-row barleys, Hannchen being well adapted to cool, moist areas, while White Smyrna does well under dry conditions. None of the other varieties included in the mixture can be regarded as good types. Deficiens and Meloy are particularly wanting when all locations are considered.

A striking feature of the experiment is the rapidity with which one or two varieties became dominant at certain locations. This change was accompanied by equally rapid elimination of certain other varieties. At some places, for example, Pullman, Moro, Davis, and Arlington, one variety dominated in as few as 4 years of competition. At other locations, for example, Moccasin, North Platte, and Aberdeen, changes occurred more slowly, and a greater number of varieties maintained a moderate percentage of plants. These results indicate that natural selection in barley is a significant force at all locations and a force of great magnitude at some locations. Two questions are raised

TABLE 13–1. Final Census Showing Effect of Natural Selection in a Mixture of Barley Varieties Grown at Ten Locations for 4 to 12 years

(After Harlan and Martini, 1938)

Variety	Arling-ton Va.	Ithaca N.Y.	St. Paul Minn.	Fargo N. Dak.	North Platte Neb.	Moc-casin Mont.	Aber-deen Idaho	Pull-man Wash.	Moro Ore.	Davis Calif.
Coast and Trebi	446	57	83	156	224	87	210	150	6	362
Gatami	13	9	15	20	7	58	10	1	0	1
Smooth Awn	6	52	14	23	12	25	0	5	1	0
Lion	11	3	27	14	13	37	2	3	0	8
Meloy	4	0	0	0	7	4	8	6	0	27
White Smyrna	4	0	4	17	194	241	157	276	489	65
Hannchen	4	34	305	152	13	19	90	30	4	34
Svanhals	11	2	50	80	26	8	18	23	0	2
Deficiens	0	0	0	1	3	0	2	5	0	1
Manchuria	1	343	2	37	1	21	3	1	0	0

Numbers are the recorded number of plants of each of 11 varieties found in a sample of 500 plants.

by the data. First, do the trends approach the theoretical curves? Second, is agronomic merit related to competitive ability?

Theoretically, the best variety at any place will eventually dominate the population. With the exception of Aberdeen, Fargo, and North Platte, where the situation is confused by inability to separate Coast and Trebi there is no question which variety is dominant at any location. The data for White Smyrna at Moro are plotted in Figure 13–3 as an illustration of the seven curves of this type which were found.

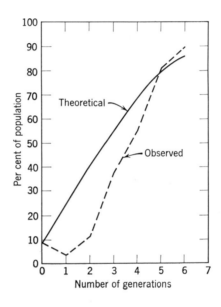

FIGURE 13-3. Curve of White Smyrna at Moro, Oregon, compared with the theoretical curve of a dominant variety. Seven curves of this type were observed. (After Harlan and Martini, 1938.)

The curves of the poorer varieties are expected to be similar to each other in shape, differing mostly in the time required for complete elimination. In the actual data Harlan and Martini found many instances of curves of rapid elimination. Such trends were shown by Deficiens at all ten stations. At least one other variety was eliminated rapidly at each of the other stations, and at some locations, for example, Davis, nine of the eleven possible curves were of this type.

The intermediate curves representing mediocre types are expected to fall into two classes when applied to the actual data. The poorer of the intermediate types should decrease at a rate represented by a nearly straight line and the better types according to a definitely humped curve. A number of examples of both types of curve were found. The behavior of Svanhals at Moccasin (Figure 13–4) is an example of the latter type.

It is perhaps remarkable that the curve of Hannchen at Ithaca was the only definitely undulating curve. At the beginning of the experiment conditions were apparently favorable to Hannchen. Then came a series of years in which the percentage of Hannchen decreased rapidly. When Hannchen had been reduced to about 2 per cent of

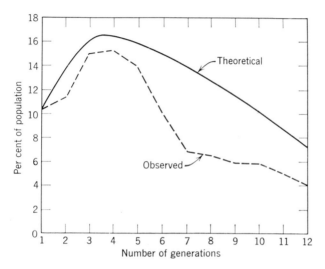

FIGURE 13-4. Curve of a good competitor (Manchuria at Mocassin, Montana) compared with the theoretical curve of a better than average competitor. (After Harlan and Martini, 1938.)

the population, another series of more favorable years seems to have ensued, and this variety slowly increased, reaching more than 12 per cent before the onset of another series of unfavorable years brought about another decline. The fluctuations of Hannchen did not, however, seriously affect the trend of the dominant variety. Several other less definite curves of this type occurred.

This experiment provides considerable information about the relation between competitive ability and agricultural value. Coast and Trebi are unquestionably the outstanding commercial varieties included in the mixture. These varieties made a poor showing only at Moro, and then because of the competition of White Smyrna, a variety noted for good yielding ability and excellent adaptation to arid conditions. Hannchen, another commercial variety of good yielding potential and wide adaptation, also made a very good showing when all locations are considered.

The dominant variety differed from that commonly grown by farmers in the vicinity at only two locations. In New York, Manchuria, a six-rowed type, became dominant, whereas most of the commercial acreage at the time of this experiment was in Alpha, a two-rowed variety similar to Hannchen. On the other hand, in Minnesota, where Manchuria and Manchuria hybrids constitute the greater part of the commercial acreage, Hannchen, a two-rowed type, was predominant. Although two-rowed spring barleys are grown in New York, the preference for such varieties depends partly on considerations other than yield. Similarly, published accounts of productivity indicate that Hannchen could be successfully grown in Minnesota. In general, then, the agreement was good between the list of superior varieties and the list of good competitors.

The agreement between the list of mediocre or inferior types and the list of varieties rapidly eliminated in competitive mixtures was more striking. Deficiens and Meloy are unquestionably inferior agricultural types, and their elimination was rapid and complete or nearly complete at all locations. Svanhals, Lion, Smooth Awn, and Gatami lack distinction as agricultural types. They were either eliminated or declined sharply at virtually all locations.

A similar experiment was conducted by Suneson and Wiebe (1942) with four widely cultivated varieties of California barley. Starting with equal portions of the four varieties, they measured by means of an annual census, the proportion of each variety over a 9-year period. On plots adjacent to the bulk population they also measured the yield of each variety in pure stands. Survival and yield for each of the varieties are given in Tables 13–2 and 13–3.

The most conspicuous feature of this experiment was the rapidity with which Atlas dominated the bulk plot and, conversely, the rapidity with which Hero and Vaughn were virtually eliminated. The differences in yield among the four varieties were small. Hence this experiment is not conclusive in establishing a relation between yielding ability and survival. Since Vaughn, the highest-yielding variety, was the poorest competitor, it seems reasonable to conclude that there was no very strong positive association between these characteristics. The marked differences in competitive ability that were observed must therefore have depended on characteristics other than the weight of seed produced. Competitive ability also does not seem to be related to the standard criteria of agronomic desirability, such as heading date, plant height, and level of resistance to diseases, since Vaughn is the "best" variety of the four when judged according to such standards.

The history of the competition among these varieties for the favor of California farmers is, however, more enlightening. Seed of Vaughn

TABLE 13–2. Proportion of Each Variety in Varietal Mixtures of Barley Grown at Davis, California, from Successive Seeding of a Mixture Originally Containing Equal Numbers of Seed

(After Suneson and Wiebe, 1942)

Percentage of Plants of Each Variety in Mixture*

Variety	1933	1934	1935	1936	1937	1938	1939	1940	1941	1948†
Atlas	25.4	38.1	47.4	42.8	49.2	54.4	47.7	63.2	65.5	88.0
Club Mariout	24.7	23.4	18.6	22.7	24.3	20.1	27.6	17.3	18.8	10.5
Hero	24.7	20.5	15.9	12.5	12.2	9.2	13.7	8.3	7.7	0.7
Vaughn	25.2	18.0	18.1	19.9	14.3	16.2	11.1	11.3	7.5	0.4

* Disparity from 100 per cent total represents other varieties or natural hybrids.
† Suneson (1949) continued the mixture for 7 more years with the final results indicated.

TABLE 13–3. Yields of Varieties of Barley When Grown Separately in One Fiftieth of an Acre Plots Replicated Five Times, at Davis, California, 1933–1940

(After Suneson and Wiebe, 1942)

	Annual Yield in Bushels per Acre*								Percentage of Total Yield (8 Year Average)
Variety	1933	1934	1935	1936	1937	1938	1939	1940	
Vaughn	87.5	88.5	121.5	101.5	66.9	69.4	40.2	56.6	27.0
Atlas	92.2	81.7	91.1	92.9	69.8	66.2	36.6	55.6	25.1
Hero	81.0	81.7	95.0	83.6	66.4	61.0	39.7	53.5	24.0
Club Mariout	87.5	80.6	103.5	82.5	66.2	54.3	38.7	46.1	23.9

* Estimates of error for the mean yields ranged from 1.8 to 3.7 bushels per acre.

has been readily available since 1932, and its "superiority" over the other varieties in yield and other agronomic characteristics in many parts of California has been known and publicized. Nevertheless, it has never occupied more than 5 per cent of the 1½ million acres planted to barley in California annually. Atlas has ranked first in popularity, with Club Mariout a close second. Hero, like Vaughn, has never been widely grown. Except for this census experiment, there are no data to support the apathy of California farmers for Vaughn. Conservatism among farmers in changing varieties can also be ruled out because substantial shifts in the popularity of barley varieties have occurred in the period in question. Suneson has concluded: "This suggests that the bulk population method of breeding will not necessarily perpetuate the highest yielding or the most disease resistant progenies, but that the otherwise intangible character of competitive ability may measure other very important characteristics." Apparently the standard criteria applied by plant breeders to judge the "superiority" of their selections are not all inclusive or unerring. Experiments such as this one emphasize the importance of learning more about the nature of survival in competitive mixtures.

Survival in Hybrid Populations

Competition among homozygous genotypes, such as was measured in the experiments reviewed above, indicate that adapted varieties are highly competitive against unadapted types. Although this information is suggestive, it does not provide critical evidence about the effects of natural selection on survival in hybrid populations. The population structure of hybrid bulks can be expected to differ from the variety mixtures that have been studied in at least two particulars. First, segregation will occur in the hybrid bulks with the result that the competing genotypes are not expected to be constant from generation to generation; rather, new types will occur each generation until selfing produces homozygosity. Second, only close to the point where homozygosity has been achieved (perhaps by the F_6 to F_8 generation) will the population become comparable to the variety mixtures. Even then, such populations will differ from the varietal mixtures in that a vastly larger number of genotypes will be competing against each other.

In bulk-hybrid populations three categories of observations can be made which bear on the effects of natural selection. First, changes in gene frequency at particular segregating loci can be studied by recording phenotypic frequencies in succeeding generations. Second, the appearance and performance of entire populations can be observed and compared in various generations; these comparisons can be made among generations and/or against some standard genotype. Third, individual lines can be isolated from the bulk population periodically and their performance used as an index of change in the population.

Survival of Genes in Populations

When the survival of alternative alleles of a particular gene is studied, it is not the effect of the gene alone that is observed, but the combined effect of the gene and closely linked loci. Data on temporal changes in gene frequencies are meager in plants. Perhaps the most comprehensive study is that of Suneson and Stevens (1953), who measured the survival of alternative alleles at 5 loci in a barley-bulk population derived from one of Harlan's composite crosses. This stock, called Composite Cross II, originated from a mixture of equal amounts of hybrid seed of 28 varieties crossed in all possible combinations but one—a total of 378 hybrids. Among the parents were types differing in the following allelic alternatives: rough versus smooth awn, two- versus six-rowed, hooded versus awned, deficiens versus lateral florets present, and black versus white lemma.

The alleles for hooded and deficiens, on the basis of experience by both farmers and breeders, can be regarded as inferior to their alternatives. At Davis, California, two-rowed types have not proved satisfactory. The allele for six-rowed spikes might therefore be expected to be superior in competition. A priori, there is no reason to suspect either of the alternative alleles to be superior at the smooth awn and black loci. However, since the black parents involved in these crosses were all distinctly inferior types at Davis, unfavorable linkages might be expected to play a role. Conversely, it is known that favorable yield genes are linked with the smooth-awn allele in at least one of the parents involved in these crosses, and increases in the proportion of that allele at the expense of its alternative might thus be expected.

Changes in the proportions of these various allelic pairs during the 12-year period from 1925 to 1936 are shown in Figure 13–5. As expected, the obvious morphological defectives, hooded and deficiens, were promptly eliminated. The black allele also did not survive, and two-rowed types decreased in prevalence. On the other hand, the smooth-awn allele gradually increased in the population. Thus the three types known to be agronomically inferior were eliminated or drastically reduced in the population, and the two presumably neutral allelic pairs behaved as anticipated from known or inferred linkages.

These authors obtained similar information about these five allelic pairs at seven other locations in the United States. The hooded and deficiens alleles became progressively less numerous at all seven locations. Two-row types, however, increased at six of the seven locations, a trend previously reported by Harlan and Martini for the two-row

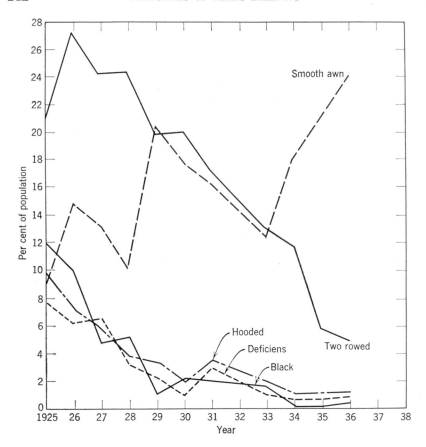

FIGURE 13-5. Survival of genes in a hybrid barley population. (After Suneson and Stevens, 1953.)

parent varieties grown at these same stations. Black and smooth-awned alleles in general showed the trend typical of moderately well adapted types—types well enough adapted to remain in competition for many years. This study, like the experiment of Harlan and Martini, suggests that agronomically poor types are also poor competitors.

Yield and Appearance of Bulk Hybrid Populations

During the 25 years in which Composite Cross II, a bulk hybrid composited from 378 hybrids among 28 varieties, has been grown at Davis, California, by Suneson without selection, its yielding ability has been compared periodically with Atlas barley as a standard (Table 13–4). Yield was determined in trials involving from 10 to 20 paired comparisons between Atlas and Composite Cross II; hence considerable confidence can be placed in each comparison. The composite bulk was conspicuously inferior to Atlas in yielding ability and in general agronomic appearance in the early generations. There was a gradual improvement, however, in both yielding ability and agronomic type until, by the F_{15} generation, the bulk consisted almost entirely of coast types and it equalled Atlas in yielding ability. This bulk continued to improve in yielding ability over 10 more generations of bulk handling without artificial selection. Similar results have also been reported with several other long-term bulk-hybrid populations (Suneson and Stevens, 1953; Suneson, 1956).

TABLE 13–4. Progressive Changes in Yield of Composite Cross II, Compared with Atlas Barley, Grown at Davis, California, 1924–1950

(After Suneson and Stevens, 1953)

Test Years	Genera- tions	Yield Bu. Per Acre		Per Cent of Atlas
		Bulk	Atlas	
1937–1938	F_3–F_4	58.3	86.2	67.6
1933–1934	F_7–F_8	71.3	83.8	85.0
1937–1940	F_{11}–F_{14}	72.6	81.7	88.8
1941–1946	F_{15}–F_{20}	75.1	70.8	106.0
1947–1949	F_{21}–F_{23}	56.8	49.9	113.8
1950	F_{24}	51.3	37.8	135.7

Among the 28 parents of Composite Cross II, 3 were resistant to scald disease, considered to be the most destructive disease of barley in the environment in which the composite bulk was grown. Yet only 17 resistant progenies were found in a random sample of 356 random progenies isolated from the bulk in the F_{12} generation. The recovery of resistant progenies was therefore less than expected, assuming scald resistance to be neutral in survival. In the F_{23} generation, about 7 per cent of highly resistant progenies remained in the population. Resistance to scald is therefore not a potent factor in survival.

The studies of Adair and Jones (1946) to determine the effect of natural selection on three bulk hybrid populations of rice grown under three different environments (Arkansas, California, Texas), with special attention to the survival of desirable plant characteristics, are also enlightening. They found that average heading dates were different for the same bulks grown in the different locations and were "in accord with differences in length of growing season at the three stations, suggesting that this was an important factor in survival." The relation between the other characteristics and survival was less marked. Adair and Jones concluded "that bulk hybrid populations can be used to advantage in breeding rice, for in each of the lots grown at three locations for eight generations plants representing all degrees of maturity, height, grain type and awn development survived. Selections made from hybrid lots should be reasonably well adapted in the respective locations for otherwise they would have been eliminated by natural selection."

Progenies Isolated from Bulk Populations

Suneson (1956) has used the characteristics of individual lines, selected in different generations of bulking, as a gauge of the changes which have occurred in barley Composite Cross II in response to natural selection. Among 356 selections made in the F_{12} generation, none combined generally good agronomic type with ability to outyield Atlas barley in replicated-yield trials. In the F_{20} generation, when the average performance of the bulk population was superior to that of Atlas, another attempt was made to extract superior lines from the mass population. Two among 50 single-plant selections were considered to be outstanding in agronomic characteristics and yielding ability; in yield trials conducted over a 7-year period, the better of the two outproduced Atlas by 37 per cent and had an advantage of 3.5 pounds in test weight. This selection had moderate to good resistance to all five of the principal barley diseases occurring in California. Sixty-six selections were made in the F_{24} generation. Like the selections made in the F_{20} generation all were good in yielding ability and at least moderately good in agronomic type. A few lines appeared to be exceptional; for example, the top three selections exceeded Atlas in yield by an average of 56 per cent in 4 years of replicated trials. It seems clear that superior selections have made up a greater and greater proportion of Composite Cross II as the number of generations has advanced.

The effects of natural selection on survival in bulk hybrid populations can be summarized as follows: (1) survival of competing alleles is nonrandom in that inferior competitors are usually also inferior agriculturally; (2) morphological uniformity of bulk populations increases with advance in generation but much variability remains even after twenty or more generations of natural selection; (3) steady improvement occurs in yielding ability with continued bulk propagation; (4) characteristics such as maturity, plant height, and adaptation are rather rapidly adjusted by natural selection to fit the environment in which the bulk is grown. However, certain characters which might logically be supposed to have survival value, for instance, resistance to some diseases, are apparently neutral in survival; and (5) the proportion of superior selections appears to increase steadily over long periods of bulk handling.

Careful and precise studies of the effects of natural selection on bulk hybrid populations have been limited in number and restricted to a few species. Until a greater variety of evidence is available, the foregoing conclusions must be regarded as tentative.

Artificial Selection in Bulk Populations

It seems reasonable to conclude from the evidence now available that natural selection exerts strong selective pressures in bulk hybrid populations, particularly in respect to intangible characteristics that provide for good adaptation. Since, however, other important characters (notably disease resistance) appear to be neutral in competition, opportunities exist for practicing artificial selection to shift the population toward agriculturally desirable types. Obviously the population should be purged of unwanted genotypes, especially when the unwanted types are highly competitive. For example, determinate (bush) types of beans do not compete well against types with indeterminate (vine) habit of growth. If, as is often true, interest is only in the determinate types, artificial selection against the indeterminate types is clearly called for. There is also good reason to apply artificial selection against seed colors, pubescence, awns, barbs, or other characters that are undesirable in commercial varieties but may be more or less neutral in survival. When a character has low survival value, as, for example, nonhardiness in Nilsson-Ehle's wheat, the reason for artificial selection may not be as compelling as when an undesirable character is neutral or has high survival value. Artificial selection may nevertheless aid natural selection in more rapid elimination of unwanted types.

Bulk-population handling often provides opportunity to make use of artificial aids to selection. For example, bulk harvesting when only part of the plants are mature is an effective and inexpensive way to select for earliness. Or if particular sizes or shapes of seeds are required, screens or other mechanical devices provide a rapid way of eliminating inferior types. Naturally occurring or artificially induced disease and insect attack, of course, provide opportunities to rogue susceptible individuals from the bulk plot.

In an attempt to determine the effect of continued artificial selection in barley for vigorous plants with large, well-filled, disease-free spikes, Atkins (1953) selected in seven different barley bulks for three generations beginning in the F_3. Comparisons of the yields of the selected populations and those of comparable unselected populations showed that the selected populations consistently outyielded the unselected lots, although the differences were not always significant.

Duration of the Bulking Period

In 1937 Harrington proposed the "mass-pedigree" method as a solution to one of the deficiencies in the pedigree method of breeding. Harrington found that this method was much less effective with wheat in some years than in others. For example, critical selection for straw strength and resistance to some diseases, and to a large extent for plant height, earliness, and resistance to shattering, was not possible in dry seasons; yet such seasons occurred frequently. Pedigree breeding is clearly wasteful of labor and expense in such circumstances. Harrington therefore proposed to avoid this wasted effort by carrying the segregating generations in bulk until a favorable season provided the conditions necessary for efficient selection for some major character under consideration. Then single-plant selections could be made and subsequently handled as in the pedigree scheme. With this method, bulk handling is terminated whenever suitable conditions occur for selection for some primary character. At one extreme, the bulking might end as early as the F_2 generation; here the scheme is no different from typical pedigree procedure. On the other hand the bulking might be continued for many generations if the advent of conditions suitable for selection is long delayed.

At present the most common use of the bulk method is to obtain homozygous lines with a minimum of effort and expense. When the primary objective of the bulk handling is to achieve homozygosity, the duration of the bulking depends on genetic considerations and can thus be specified quite precisely. It will be recalled that the average percentage of homozygosity under selfing is very high by the F_6 generation and approaches 100 per cent by the F_{10} generation, even when a very large number of genes are segregating. Should heterozygotes be more productive than homozygotes, the rate of return to homozygosity is not accurately represented by a descending Mendelian series, but will be delayed. Selfing, however, is an exceptionally powerful inbreeding mechanism, and striking heterosis is required to retard the advent of homozygosity to an important extent. Hence, in species where little cross-fertilization occurs, selections made in the F_6 to F_8 generations are almost certain to have reached a state of homozygosity where their subsequent evaluation is not likely to be seriously complicated by segregation.

In short-time bulks, natural selection operates principally on individuals that are heterozygous at many loci. Competition is therefore primarily among heterozygotes. Since the ultimate fate of the population is complete or near complete homozygosity, it appears that natural selection is halted at precisely the time when it might be expected to become effective in selecting among homozygous lines. This being true, the effectiveness of natural selection is not likely to have been very great, and it is probably desirable to select more plants in F_6 for testing in plant rows than would be necessary with the pedigree method.

A few plant breeders have chosen to emphasize natural selection in their bulk-breeding programs by continuing the bulking for an extended number of generations, in some instances twenty-five generations or even more. This procedure is based on the following reasoning: (1) Upon attainment of near homozygosity by the F_6 to F_8 generation, bulk hybrid populations are composed of a vast number of mostly homozygous types, nearly all of poor or indifferent agricultural value. This will be true whether or not rigorous artificial selection was practiced in the segregating generations. (2) The proportion of homozygotes with obvious agricultural defects will be reduced by both natural and artificial selection. (3) Natural selection may be more discerning than the plant breeder in selecting for subtle differences related to adaptation and certain other desirable characteristics; hence natural selection may be essential if the very best types are to be increased to the point where they make up a significant proportion of the population. (4) Selection differentials are likely to be small and not constant from generation to generation. Thus long periods of bulk handling are desirable.

Evidence is inadequate at present to permit a critical evaluation of long-continued bulk handling as a method of handling hybrid populations in self-pollinated species. Undoubtedly, the appeal of the method is drastically reduced for most plant breeders by the long-time requirement. The time requirement is not an adequate reason to reject the method if indeed bulk handling is the only effective way of dealing with certain intangibles that are important agriculturally. Some breeders have incorporated long-term bulks as a part of their programs on the basis that they require little effort and do not detract seriously from other types of programs that are more suited to solving pressing problems.

Varieties Developed by the Bulk Method

It is difficult to determine from the plant-breeding literature the extent to which the bulk method, in its various modifications, is used in plant breeding and the number and importance of the varieties that have been developed by the method. In the United States V. H. Florell appears to have been one of the first breeders to use bulk-population breeding. He was particularly successful with the cross of Atlas × Vaughn barley made in 1927. This population was carried in bulk until 1934, when a large number of single-plant selections were made and the forty-five most promising distributed to barley breeders in the western United States. Ultimately four varieties tracing to this bulk population were released for commercial production. These varieties are Arivat, released by the Arizona Station in 1940, Beecher, released by the Colorado Station also in 1940, Glacier by Montana in 1943, and Gem by Idaho in 1947. Nearly twenty other varieties of barley which can be traced directly or by parentage to composite-cross populations are listed by Harlan (1956).

REFERENCES

Adair, C. R., and J. W. Jones. 1946. Effect of environment on the characteristics of plants surviving in bulk hybrid populations of rice. *Jour. Amer. Soc. Agron.* 38: 708–716.

Atkins, A. E. 1953. Effect of selection upon bulk barley populations. *Agron. Jour.* 45: 311–314.

Harlan, J. R. 1956. Distribution and utilization of natural variability in cultivated plants. *Brookhaven Symposia in Biology*, No. 9: 191–208.

Harlan, H. V., and M. L. Martini. 1938. The effect of natural selection in a mixture of barley varieties. *Jour. Agric. Res.* 57: 189–199.

Harlan, H. V., M. L. Martini, and H. Stevens. 1940. A study of methods in barley breeding. *U. S. Dept. Agric. Tech. Bull.* No. 720, 26 pp.

Harrington, J. B. 1937. The mass-pedigree method in the hybridization improvement of cereals. *Jour. Amer. Soc. Agron.* 29: 379–384.

Laude, H. H., and A. F. Swanson. 1943. Natural selection in varietal mixtures of winter wheat. *Jour. Amer. Soc. Agron.* 34: 270–274.

Newman, L. H. 1912. *Plant breeding in Scandinavia.* Canadian Seed Growers Association, Ottawa.

Suneson, C. A. 1949. Survival of four barley varieties in a mixture. *Agron. Jour.* 41: 459–461.

Suneson, C. A. 1956. An evolutionary plant breeding method. *Agron. Jour.* 48: 188–190.

Suneson, C. A., and H. Stevens. 1953. Studies with bulked hybrid populations of barley. *U. S. Dept. Agric. Tech. Bull.* No. 1067, 14 pp.

Suneson, C. A., and G. A. Weibe. 1942. Survival of barley and wheat varieties in mixtures. *Jour. Amer. Soc. Agron.* 34: 1052–1056.

14

Backcross Breeding

The backcross method of breeding provides a precise way of improving varieties that excel in a large number of attributes but are deficient in a few characteristics. As the name implies, the method makes use of a series of backcrosses to the variety to be improved during which the character (or characters) in which improvement is sought is maintained by selection. At the end of the backcrossing the gene (or genes) being transferred, unlike all other genes, will be heterozygous. Selfing after the last backcross produces homozygosity for this gene pair and, coupled with selection, will result in a variety with exactly the adaptation, yielding ability, and quality characteristics of the recurrent parent, but superior to that parent in the particular characteristic for which the improvement program was undertaken. It is apparent that this method, in contrast to the pedigree and bulk population methods of breeding, provides the plant breeder with a high degree of genetic control of his populations.

Although line breeding (a type of backcrossing) has been used by animal breeders for more than a century and was an important method used to fix breed characteristics in domestic animals, the potential of backcross breeding in plants seems to have escaped attention until 1922, when Harlan and Pope pointed out the possibilities of the method in small-grain improvement. They observed that the backcross method had been "largely if not entirely neglected in any definite programs to produce progeny of specific types." They suggested the probability that there were many instances in which backcrossing would be of greater value than the more common procedure of selecting during a program of selfing following hybridization. It was also in 1922 that Briggs started an extensive backcrossing program to develop bunt-resistant varieties of wheat. He emphasized that, within certain limits, the method was scientifically exact because the morphological and agricultural features of the improved variety could be

150

described accurately in advance, and because the same variety could, if it were desired, be bred a second time by retracing the same steps.

If a backcross program is to produce a successful variety, the following three requirements must be satisfied: (1) a satisfactory recurrent parent must exist; (2) it must be possible to retain a worthwhile intensity of the character under transfer through several backcrosses; and (3) sufficient backcrosses must be used to reconstitute the recurrent parent to a high degree. These and other features of this method of breeding will be considered in this chapter.

The Genetic Basis of Backcross Breeding

In segregating generations obtained by selfing, one half of homozygous individuals are expected to be of the desired genotype at any particular locus. For example, the F_2 generation of the cross $AA \times aa$ is expected to consist of $\frac{1}{4}AA : \frac{1}{2}Aa : \frac{1}{4}aa$. Although half of the progeny are homozygous, only half of these homozygotes are of the desired genotype, say AA. If, instead of selfing the F_1, it is backcrossed to the superior parent, say AA, the expectation is $\frac{1}{2}AA$ and $\frac{1}{2}Aa$. In the backcross then, one half of the total progeny are expected to be homozygous for AA. This same expectation exists, of course, for each gene pair by which the two parents differ. If additional backcrosses are made to the same parent, the hybrid population progressively becomes more like the recurrent parent, that is, the population converges on a single genotype instead of breaking up into 2^n homozygous genotypes as with selfing. In the backcrossing program homozygosity is attained at the same rate as with selfing and in accord with the familiar formula

$$\text{Proportion of homozygosity} = \left(\frac{2^m - 1}{2^m}\right)^n,$$

where m is either the number of generations of selfing or backcrossing, and n is the number of gene pairs. Thus Figure 6–1 can be used to find the expected percentage, after m generations of backcrossing, of plants homozygous for the n alleles entering the cross only from the recurrent parent. If, for example, the parents differ in ten gene pairs, and no selection is practiced, six backcrosses will produce a population in which 85 per cent of the individuals will be homozygous and identical with the recurrent parent at all ten loci.

The rate at which genes entering a hybrid from the nonrecurrent parent are eliminated during the backcrossing will, of course, be influenced by linkage. Suppose, in a backcross program undertaken to transfer the desirable allele A from a generally undesirable variety (the donor parent) to a superior variety (the recurrent parent), that an undesirable allele b is linked to A. The genotype of the F_1 hybrid will be Ab/aB, and selection for A in the first backcross generation will tend to pull along b, making it difficult to obtain the desired recombination AB. However, since B is reintroduced with each backcross, there will be a number of opportunities for the crossover to occur. If we assume that no selection is practiced, except for A, the gene being transferred, it can be shown that the probability of eliminating b is

$$1 - (1 - p)^{m+1}, \qquad (14\text{--}1)$$

where p is the recombination fraction and m is the number of backcrosses. Hence, if b is located 50 or more crossover units from A, or is located on another chromosome, the probability that it will be eliminated is $1 - (.5)^{m+1}$. Substituting in this formula gives the values for $n = 1$ and $m + 1$ found in Figure 6–1; for example, after 5 backcrosses the probability that b will have been eliminated is $1 - (.5)^6 = .984$. In a selfing series, with selection only for A, the corresponding probability is .50. These probabilities will, of course, become progressively smaller as p becomes smaller—that is, as the linkage between A and b becomes tighter. For example, if the recombination fraction is .01, the linkage will be broken in only 1 line in 100 with a selfing series, but the probability is 74 per cent that the desired crossover will have occurred during a series of 5 backcrosses.

TABLE 14–1. Effect of Linkage on the Probability of Eliminating an Undesirable Gene (b) Linked to a Desirable Gene (A)

Recombination Fraction	Probability That the Undesirable Gene Will Be Eliminated	
	With Five Backcrosses	With Selfing
.50	.98	.50
.20	.74	.20
.10	.47	.10
.02	.11	.02
.01	.06	.01
.001	.006	.001

It is assumed that selection is practiced for A but is not practiced against b.

Table 14–1 gives the expectancies for several additional recombination fractions. The chances of obtaining the desirable recombination is clearly greater with backcross mating than with selfing, assuming that no selection is practiced against the undesirable allele. When effective selection can be practiced against undesirable alleles which are linked to desirable ones, both selfing and backcrossing afford opportunities for obtaining the recombinations required. In fact, a selfing series is here more efficient, because effective crossing over is possible in both male and female whereas with backcrossing effective crossing over cannot take place in the recurrent parent. Effective selection is not always possible, however. In dealing with genes governing characters with low heritability, for instance, selection is not effective, and therefore the backcross procedure is advantageous.

It should now be clear that recurrent backcrossing, even in the absence of selection, is a powerful mechanism for achieving homozygosity and that any population obtained by backcrossing must rapidly converge on the genotype of the recurrent parent. When recurrent backcrossing is made the basis of a plant-breeding program, the genotype of the recurrent parent will be modified only with regard to genes being transferred, which must of course be maintained in the population by selection. Properly executed, backcross breeding programs thus allow all the desirable characteristics of the recurrent parent to be recovered, except for the possibility that characters governed by genes tightly linked with the gene (or genes) being transferred will be modified inadvertently. This is at the same time both the strength and the weakness of the method. Recurrent backcrossing provides a certain and precise way of making gains of predictable value with little possibility that uncontrolled segregation will produce subtle weaknesses which may be difficult to discover in a necessarily finite period of evaluation. At the same time, the method sets an upper limit on the amount of advance, a limit that will often be lower than that possible when segregation is not rigidly controlled.

Selecting the Recurrent Parent

Since a series of backcrosses reproduces the genotype of the recurring parent with little modification except for the character (or characters) being transferred from the donor parent, it is obvious that backcross programs must be based on suitable recurrent parents. In well-established crops, the successful varieties grown in any area represent the end products of long periods of evolution in which natural selection, selection by farmers, and more recently the efforts of plant breeders have each played a role. In common wheat, for example, there are several varieties, such as Pawnee, Comanche, Thorne, and Ramona, with excellent performance records suggesting that they may be difficult to replace. Other crop species, as well, tend to be dominated by a few varieties that have maintained their importance for many years despite the release of numerous "improved" varieties intended to replace them. These dominant varieties are manifestly the ones to be considered most seriously for selection as recurrent parents in backcross breeding programs.

Maintenance of the Character under Transfer

In backcross breeding, level of heritability is not of any particular consequence to the progress of the program except for the character under transfer. But for that character high heritability is important. This is because selection must be practiced for the character being transferred through each of several rounds of backcrossing. At the same time all other characters are taken care of automatically by the backcrossing procedure, with the result that the breeder need pay them little or no attention. Backcross breeding therefore has its greatest ease of application when the character being transferred can be identified readily in hybrid populations by visual inspection or by simple tests. As with all other methods of breeding, characters of high heritability governed by single genes are the easiest to handle. However, backcross breeding is by no means restricted to characters governed by one or a few genes. In fact, the number of genes is probably less important in determining the speed and precision with which a character transfer can be effected than is the certainty with which the character can be identified in segregating populations. Said in another way, a character of high heritability governed by several genes might well be more easily transferred by backcrossing than a character of low heritability governed by fewer genes.

Regardless of the number of genes governing the character to be transferred, it is necessary that a worth-while intensity of the character be maintained throughout the series of backcrosses. Some of the intensity of the character may, of course, be lost even when the genetic control is predominantly monogenic. This was true in the breeding of bunt-resistant wheats in California. Martin, the donor parent, is almost completely free of bunt when inoculated artificially, whereas the several backcross-derived commercial varieties carrying the "single" gene governing the resistance of Martin often have as much as 3 or 4 per cent of bunted plants when artificially inoculated. There is substantial evidence that retention of the full resistance of Martin was not essential in this breeding program. First, Wiebe and Briggs (1937) showed that, even under favorable conditions for this pathogen, the bunt organism was unable to maintain itself on the derived varieties. Second, even though bunt was a very serious disease of wheat in California prior to the release of these "resistant" varieties, causing losses in yield and quality every year and severe losses in many years, the disease has not been a factor in wheat production in California in the 20 years subsequent to their release.

There may, however, be circumstances in which the plant breeder desires a particularly high intensity of the character being transferred. The general agricultural worth of the donor parent, for obvious reasons, need not be an important criterion in selecting the parents for backcross programs. Hence the breeder is free to select the donor parent almost exclusively on the basis that it exhibits the character in particularly intense form. This is often done, even when the genetic control is known to be predominantly monogenic, because modifier genes in the new genetic background often cause some intensity to be lost even though the most stringent selection has been practiced throughout the backcrossing program. Several cases of this type and their implications in backcross breeding have been discussed by Briggs and Allard (1953).

In dealing with quantitative characters, the donor parent may often be selected with the view of sacrificing some of the intensity of the character for which it was chosen. J. W. Jones and L. L. Davis selected Lady Wright, a long-grain variety, and Coloro, a short-grain variety, as parents in breeding a medium-grain variety. Calady, which was produced by backcrossing four times to Coloro, has the high yield and good adaptation of Coloro coupled with a good medium-grain type. This and other successes in dealing with quantitative characters, as well as promising results with still others, indicate that the backcross method is not, as commonly supposed, limited as a method to simply inherited qualitative characters. Like all other methods it depends on the ability of the breeder to distinguish between genetic and environmental variability and to select those individuals that are desirable for genetic reasons.

The Number of Backcrosses

If backcross breeding is to succeed with certainty, the genotype of the recurrent parent must be recovered in its essential features. This is primarily a function of the number of backcrosses, although selection for the type of the recurrent parent in the early backcross generations is effective in shifting the population toward the characteristics of that parent. Additionally, it should be recognized that the recurrent parent is not composed of a single pure line but is likely to be made up of many closely related pure lines. Hence enough plants of the recurrent parent must be used to recapture the variability of the recurrent parent if there is to be real assurance that the improved type will have the same basic agricultural characteristics as its predecessor.

Six backcrosses coupled with rigid selection in early generations have proved satisfactory in a large number of backcross-breeding programs completed at the California station. It is believed that selection for the type of the recurrent parent, if based on moderate-sized populations, is equivalent to one or two additional backcrosses without selection. After the third backcross, however, the population usually resembles the recurrent parent so closely that selection on an individual-plant basis is largely ineffective except for the character being transferred.

At this point it should be emphasized that backcrosses can also serve a useful purpose in pedigree breeding programs. When, in a pedigree program, one of the parents is superior to the other in a majority of its characters, one or even several backcrosses to the better parent may be advantageous. Sufficient heterozygosity remains after two or even three or four backcrosses to furnish a satisfactory basis for selection for yield (or other characters) in families derived from backcrossed plants. Stopping the backcrossing early preserves the possibility that transgressive segregation will provide additional increments of yield, but the price is relinquishment of control over the genetic forces that the backcross method regulates so precisely. As a consequence the evaluation tests which the backcross method circumvents are a necessary feature of pedigree programs which employ some backcrossing. Some breeders refer to programs of this type as backcross programs; this terminology is misleading, however, and such programs would be more accurately described by the term "backcross pedigree."

Backcross Breeding Procedures

The general plan of a backcross breeding program can be outlined as follows:

1. Selection of parents. The recurrent parent (A) is generally a dominant variety in the area. The donor parent (B) is selected because it possesses in high degree some character in which A is deficient.

2. The F_1 of $A \times B$ is backcrossed to A. Selection is practiced for the desirable character of parent B. Selfed seed from the selected

plants are used to produce a large F_2 in which intensive selection is practiced for the desirable character of parent B and for the general features of parent A. The selected plants are used to produce F_3 lines, among and within which further selection is practiced for the type of the recurrent parent.

3. The selected plants are backcrossed to parent A to produce second-backcross seeds $(A^3 \times B)$. The second-backcross plants are again backcrossed to parent A to produce third backcross seeds $(A^4 \times B)$. The procedure of (2) is repeated.

4. The fourth, fifth, and sixth backcrosses are made in succession, with an F_2 and F_3 being grown after the sixth backcross and intensive selection being practiced for the character being transferred and for the plant type of the recurrent parent.

5. A number of lines, homozygous for the character from parent B and as similar as possible to the recurrent parent, are bulked, increased, and released for commercial production.

If minimum numbers of plants are grown at each step in the foregoing program, the transfer of a single dominant gene can theoretically be accomplished (through six backcrosses) with 53 plants from backcrossed seeds, 96 F_2 plants, and 68 F_3 rows of 24 plants each. These numbers are based on a probability of .999 of having at least one Aa plant after each backcross and at least one homozygous AA progeny in the F_3. In species where artificial hybridization is difficult, the number of hybrid seeds necessary can be greatly reduced by growing F_2 and F_3 populations after each backcross and crossing on homozygotes. Somewhat smaller total populations are required to transfer incompletely dominant or recessive genes because homozygotes can be recognized in the F_2 generation, making F_3 progenies unnecessary.

It will have been noted that the program outlined above does not call for successive backcrosses but for the growing of F_2 and F_3 populations after the first, third, and sixth backcrosses. It will also have been noted that substantially larger numbers of plants are called for than the theoretical minimum. The reasons for this recommendation are grounded in two considerations believed to outweigh the advantages of successive backcrossing and small populations. First, there is good evidence that selection for the type of the recurrent parent, if based on F_2 and F_3 populations of moderate size, is equivalent to at least one or two, and perhaps even three additional backcrosses in a continuous series without rigorous selection. Second, selfing after the first, third, and sixth backcrosses provides the breeder, with little expenditure of effort, the large populations required if effective selection is to be practiced for a worth-while intensity of expression of the character under transfer.

The Breeding of Baart 38

The procedures used in breeding Baart 38, a stem-rust-resistant and bunt-resistant variety of wheat, will illustrate the main features of backcross breeding (Briggs and Allard, 1953). The steps taken in transferring stem-rust resistance from Hope wheat to Baart, and merging this type with bunt-resistant Baart to form Baart 38, are shown in Table 14–2. One modification that was especially useful in testing for resistance to rust, and also in reducing the time required to complete the program, was the growing of a summer nursery, seeded in July and harvested in October (the normal nursery is seeded in November and harvested in June). Because of the urgent need for rust-resistant varieties when this program was started, the first three backcrosses were made in succession; thus the program was atypical in that respect. Rigid selection was practiced for the characteristics of Baart, especially in the period 1931–1934. Baart 35, developed independently by backcross transfer of the Martin resistance to bunt, was used as the recurrent parent for the fifth backcross, thus merging resistance to the two diseases.

In the summer of 1936, 3-row blocks of 1056 F_3 families were subjected to a rust epiphytotic. In the spring of 1937, remnant seed of the 340 families homozygous for resistance to rust were tested for resistance to bunt, again in 3-row blocks. A total of 157 lines resistant to both diseases were bulked to form Baart 38. Final testing for resistance to the two diseases was thus combined with the first seed increase. The entire program, including the first seed increase, required the growing of slightly more than 5000 nursery rows over a period of 8 years. In all, 1867 hybrid seeds were made, requiring the emasculation and pollination of about 125 spikes. Baart 38, it should be noted, has one less backcross than the minimum suggested earlier. Subsequently, two additional backcrosses were added by Suneson, and the product was released as Baart 46.

The breeding of Baart 38 required about 8 years from the time of the original cross to the release of seed for commercial production. This is a substantially shorter period than is ordinarily required to produce a new variety by pedigree or bulk breeding.

Improvement by Steps

Although the backcross method has its most straightforward application in the transfer of monogenic characters, it offers a number of

TABLE 14–2. The Breeding of Baart 38 Wheat

(From Briggs and Allard, 1953)

Year	Nursery	Generation	Remarks	No. of Nursery Rows*	No. of Plants Used for Crossing†	No. of Hybrid Seeds Made
1930	Spring	Hope × Baart	Original cross		1	20
1931	Spring	Hope × Baart[2]	First backcross	1	4	49
1931	Summer	Hope × Baart[3]	Rust test; 3 heterozygous resistant plants used for second backcross	4	3	66
1932	Spring	Hope × Baart[4]	All 60 plants were used for third backcross	3	60	904
1932	Summer	F_1	Rust test; 31 heterozygous resistant plants selected	60		
1933	Spring	F_2	Without benefit of rust test, selected 405 plants on basis of agronomic characters	31		
1933	Summer	F_3	Rust test; selected 180 plants from homozygous resistant rows	405		
1934	Summer	Hope × Baart[5]	Subjected F_4 to rust, selected 14 plants for fourth backcross	180	14	240
1935	Spring	(Hope × Baart[5]) × Baart 35‡	Selected 28 plants for fifth backcross to Baart 35	14	28	588
1935	Summer	F_1	Selected 290 plants	28		
1936	Spring	F_2	Inoculated with bunt, saved 1056 bunt-free plants	290		
1936	Summer	F_3	Rust test; grew 1056 lines in 3-row blocks, 340 were homozygous for resistance	3168		
1937	Spring	F_3	Remnant seed of 340 stem-rust-resistant lines inoculated with bunt. 157 were homozygous for bunt resistance. These were bulked as Baart 38.	1020		

* Rows of Baart grown for checks and parents excluded. Rows were 16 feet long.

† Excludes Baart parent.

‡ Baart 35 is a bunt-resistant strain of Baart† developed independently from Martin × Baart[7].

procedures by which a variety can be improved in two or more characters, or by which additional genes governing any one character can be accumulated in one variety. Once a variety has been improved in some characteristic, its use as the recurrent parent in subsequent backcross programs will, of course, automatically preserve that improvement. This can be illustrated with the breeding of wheats for bunt resistance in California, a program in which the resistance of Martin, governed by a single gene, was transferred to twelve commercial varieties. The Hope resistance to stem rust was subsequently added to five of these varieties in independent projects and the end products merged as illustrated by Baart 38. In one variety, Big Club, there was urgent need for resistance to Hessian fly. This resistance was added by further backcrossing to the bunt-resistant and stem-rust-resistant lines to produce Big Club 53, resistant to all three organisms.

One more example of step breeding may be cited, that of Onas 53 wheat. Bunt resistance, awns, and stem-rust resistance were incorporated into Onas in three steps. Bunt resistance was added to produce Onas 41. Interest in awns stems from the fact that, as mentioned in an earlier chapter, they improve the variety in both yield and bushel weight. The awned type was developed independently and merged with Onas 41 to produce bunt-resistant, awned Onas 49. Meanwhile a project was started to add stem-rust resistance from a Kenya variety of wheat. For this Onas 41, Awned Onas, and Onas 45 served in turn as recurrent parents, bringing together in Onas 53 the three characteristics mentioned above.

Another approach that may be used to improve a variety in two or more characteristics is to transfer the necessary genes simultaneously in the same program. This procedure has certain disadvantages. In the first place, somewhat larger populations are required to transfer two genes together than to transfer them independently. This disadvantage is even greater with three or more genes because genetic complexity increases exponentially with the number of genes. Second, experience has shown that conditions allowing expression of one of the characters may not always occur, thus delaying both transfers. There is an added advantage with separate transfers, when, because of favorable circumstances, one transfer can be completed in time to serve as the recurrent parent for the other.

Applications to Cross-Pollinated Crops

Backcross improvement of cross-pollinated species differs in no fundamental way from that of self-pollinated species. Particular care is necessary with cross-pollinated crops, however, to assure that the sample of gametes taken from the recurrent parent represent the gene frequencies characteristic of that variety. Caliverde Alfalfa, developed by E. H. Stanford (1952) and B. R. Houston, is resistant to bacterial wilt, mildew, and leaf spot. It is a good example of the application of backcross breeding to a cross-pollinated species. Mildew-resistant and leaf-spot-resistant plants found in California Common, the progenitor variety, were used as a source of resistance to those diseases. Wilt resistance was transferred from Turkestan, a type wholly unsuited for production in California. About two hundred plants of California Common were used to represent the recurrent parent in each of four backcrosses. Although this backcross transfer may have been complicated by tetrasomic inheritance, this necessitated only minor modifications in procedures and did not adversely affect the end result. The breeding of this variety, including seed increase for release to growers, took 7 years. Caliverde, being indistinguishable from California Common except for its resistance to three important diseases, rapidly replaced its progenitor in commercial production of alfalfa in California.

Influence of Environmental Conditions on Backcross Programs

One of the advantages of the backcross method is that the breeding program can be carried out in almost any environment that will allow the development of the character being transferred. With pedigree programs the use of greenhouses or "off-season" nurseries is limited to tests for disease resistance or to adding generations, because attempts to evaluate agricultural performance are for the most part valueless under such environmental conditions. Backcross programs do not suffer from this restriction, and can be greatly accelerated, if necessary, by growing several generations per year. It is also possible for the breeding to be done outside the area where the variety will be used, if this is convenient for the breeder.

General Remarks on Backcross Breeding

Most of the examples of backcross programs considered in this chapter concern breeding for disease resistance. This is appropriate because the backcross method has probably been used more widely for this purpose than for all other purposes combined. Nevertheless, it should be emphasized that the backcross technique is also suitable for the adjustment of morphological characters, color characteristics, and simply inherited quantitative characters such as earliness, plant height, and seed size and shape. In fact, the method can be used to adjust any character that is moderately to highly heritable. A large number of examples of backcross programs dealing with crops other than cereals have been given by Thomas (1952). Additional examples are given by Briggs and Allard (1953).

The reason the backcross method has come to be particularly associated with breeding for resistance is undoubtedly related to the dramatic effect that diseases and insects can have on production and on quality. To the farmer it matters little whether the variety he grows is the ultimate in respect to general agronomic or horticultural features if either production or quality are drastically reduced by some malady. Under such circumstances it is only reasonable for plant breeders to come to grips with the problem at hand and to relegate secondary matters to their appropriate places. The value of the backcross method for handling the urgent problems of plant breeding has been amply demonstrated, and the use of the method has increased markedly, particularly in recent years. The trend among wheat breeders is shown by a survey conducted by R. F. Peterson (1957). He found that:

> In the 1920's only a few wheat breeding institutions were using the backcrossing method. . . . During the forties, when new races of stem and leaf rust were being discovered, an increasing number of wheat breeders made use of the backcross method to transfer rust resistance to adapted varieties. A considerable number of wheat breeders adopted backcrossing for the first time, and some who had used it as a secondary method now adopted it as their main method. . . . The tremendous outbreak of 15B stem rust in North America in 1950 and the more recent appearance of still more virulent rust races have given new impetus to the use of the backcross method.

Peterson reported that the backcross method was used extensively in Australia and was rapidly increasing in use in South America. However, in Europe, Asia, and Africa, where the cereal rusts are in general less destructive, "the backcross method of plant breeding seems to be used considerably less than in North America, Australia, and South America. It appears that the danger of losses due to new physiologic races of rusts has been one of the main factors in inducing wheat breeders to adopt this method."

In this survey Peterson also found that many wheat breeders "did not use the full backcross technique to reconstitute the recurrent parent, but used only one, two, or three backcrosses so as to retain the benefits of transgressive segregation for agronomic characters such as yield and adaptation." The possible benefits of a compromise of this sort are obvious, but against these benefits must be weighed some disadvantages. Campbell (1957), discussing the requirements of hard red spring wheat varieties in Canada, has pointed out one of these disadvantages. The place of Canadian hard red spring wheat in world markets depends on its excellent quality, and the established standards of quality must be met by any new Canadian variety. The goal of breeding programs then is

... a rust resistant variety with excellent quality which is high yielding, and which is suitable for combine harvesting. Time has been a particularly important factor in our production of new varieties. New rust races have been appearing with disturbing regularity. This means that new varieties with resistance to these races must be produced quickly. Testing of the milling and baking quality of new varieties has been rigorous and lengthy. We have reached the point where it takes at least as long to test a new variety as it does to produce it. The backcross method of breeding appears to be the most efficient answer to all of these problems and an intensive program was begun in 1951.

This program employs full-course backcrossing to avoid the necessity for lengthy evaluation trials.

Another aspect of the problems associated with the release of new varieties has been discussed by Borlaug (1957):

... there are many undesirable aspects associated with precipitous changes in varieties. Farmers are reluctant to rapidly shift their production from an old proven variety to an unknown new one. Their reluctance is based on their familarity with the old variety which permits them to exploit to the maximum its potential yielding ability. They know the best rates and dates of seeding of the old variety for their local conditions. They are familiar with the amount of fertilizer, the number and timing of irrigations which can be safely applied without unnecessary danger of lodging. When a new variety is introduced many of these considerations must again be worked out by the grower for his own local conditions before he is able to utilize a new variety in such a way as to approach its potential optimum productivity. Similarly the milling industry is often opposed to variety changes, except when absolutely necessary, since it requires modifications in the blending of varieties going into their flours, and thereby complicates their industrial operations. ... When the appearance of new races threatens the commercial crop the breeder sometimes has no choice but to begin multiplication of a new variety which possesses the necessary resistance but may be inferior to the old varieties in one or more agronomic characteristics. The sudden appearance of the new races of the rust organism thereby often leaves the plant breeder in a dilemma. On the one hand the growers and millers are reluctant to change varieties, and on the other hand failure to insist on such varietal changes, as soon as a suitable new variety is available, may result in severe economic losses throughout a large area.

The conventional backcross method of plant breeding comes closest to overcoming the dilemma. It provides, if properly carried out, new varieties which are phenotypically similar to the recurrent parent and thereby readily received by both farmer and miller.

In emergency situations, therefore, the value of the backcross method appears to be clear cut. This leaves open the question whether the backcross method should be restricted to such situations or whether it has a place when the need for improvement of dominant varieties in a few of their characteristics is less compelling. An issue such as this clearly hinges on whether a well-adapted variety which has been improved in some specific way by backcrossing can serve a useful purpose before it is superseded by some basically superior type. This in turn depends on the rate at which advances can be made by breeding methods which take advantage of transgressive segregation. Evidence available at present does not appear to allow a clear-cut answer to this question. Indeed, an unambiguous answer may not exist because backcross breeding might well serve a useful purpose in some situations and not in others.

REFERENCES

Borlaug, N. E. 1957. The development and use of composite varieties based upon the mechanical mixing of phenotypically similar lines developed through backcrossing. *Rept. Third Intern. Wheat Conf.,* 12–18.

Briggs, F. N. 1930. Breeding wheats resistant to bunt by the backcross method. *Jour. Amer. Soc. Agron.* 22: 239–244.

Briggs, F. N. 1938. The use of the backcross in crop improvement. *Amer. Nat.* 72: 285–292.

Briggs, F. N., and R. W. Allard. 1953. The current status of the backcross method of plant breeding. *Agron. Jour.* 45: 131–138.

Campbell, A. B. 1957. Plant breeding in the rust area of Canada. *Rept. Third Intern. Wheat Rust Conf.,* 35–37.

Harlan, H. V., and M. N. Pope. 1922. The use and value of backcrosses in small grain breeding. *Jour. Hered.* 13: 319–322.

Patterson, F. L., J. F. Schafer, R. M. Caldwell, and L. E. Compton. 1957. Results of different methods of breeding in small grain improvement at Purdue University. *Rept. Third Intern. Wheat Rust Conf.,* 26–29.

Peterson, R. F. 1957. Accomplishments of the pedigree and backcross methods in wheat breeding. *Rept. Third Intern. Wheat Rust Conf.,* 5–9.

Pugsley, T. A. 1949. Backcrossing for resistance to stem rust of wheat in South Australia. *Emp. Jour. Exp. Agric.* 17: 193–198.

Schaller, C. W. 1951. The effect of mildew and scald infection on yield and quality of barley. *Agron. Jour.* 43: 183–188.

Stanford, E. H. 1952. Transfer of resistance to standard varieties. *Proc. Sixth Intern. Grasslands Congr.:* 1585–1589.

Suneson, C. A. 1947. An evaluation of nine backcross-derived wheats. *Hilgardia* 17: 501–510.

Thomas, M. 1952. Backcrossing, the theory and practice of the backcross method of breeding some non-cereal crops. *Commonwealth Bur. Plant Breeding and Gen. Tech. Communication* No. 16, 139 pages.

Weibe, G. A., and F. N. Briggs. 1937. The degree of bunt resistance necessary in commercial wheat. *Phytopath.* 27: 313–314.

15

Theory of Selection in Populations of Cross-Pollinated Crops

The properties of populations of self-pollinated crops and the ways these properties are modified by selection have occupied our attention to this point. We shall now shift our attention to populations with a different kind of organization—populations which are made up of freely interbreeding organisms. As a consequence of its reproductive system and its previous evolutionary history, each population of this type is believed to possess an integrated genetic structure which is at least partly specifiable in terms of systems of gene frequencies. This kind of population was defined by Dobzhansky (1951) as "a reproductive community of sexual and cross-fertilizing organisms which share in a common gene pool."

In seeking to understand the responses of Mendelian populations to selection we shall consider, in the present chapter, the simplest postulates of population genetics and the predictions they allow us to make about genetic progress under selection. In Chapter 16 we shall compare predicted responses to selection with the responses actually observed in some representative experiments. Then we shall discuss the different ways in which selected individuals can be mated together, the influence of various systems of mating on population structure, and the effectiveness of selection (Chapter 17). Chapters 18 and 19 will be devoted to the effects of inbreeding and outcrossing on genotypes and phenotypes of populations, and as a final topic in this section we shall consider, in Chapter 20, some of the ways in which the genetic mechanisms governing mating systems can be manipulated to produce the types of mating that are effective in the improvement of cross-pollinated crops. Against this background we should then

166

be prepared to discuss methods of breeding cross-pollinated crops in the chapters of Section Five.

In our discussion of the genetic organization of populations, it is necessary to make certain mathematical representations in which the biological attributes of populations will be reduced to a mere skeleton of their probable real complexity. It is important to understand these simplifications for what they are—working models to help us grasp the essential points among the complex biological relationships that in summation give Mendelian populations characteristic organizations and patterns of response to selection.

The Hardy-Weinberg Law

Modern population genetics is founded on a proposition deduced independently by Hardy in England in 1908 and Weinberg in Germany in 1909. In its modern form this proposition can be stated as follows: If a gene is represented in an infinitely large random-mating population by the adaptively neutral alleles, A and a, in the ratio of $qA : (1 - q)a$, the frequencies of these alleles will remain constant in the genotypic proportions

$$q^2AA + 2q(1 - q)Aa + (1 - q)^2aa = 1, \qquad (15\text{--}1)$$

unless alteration is brought about by (1) selection, (2) nonrandom mating, (3) differential migration, or (4) differential mutation of $A \to a$ or $a \to A$. Moreover, this constant state (equilibrium) with respect to this allelic pair is reached in one generation of random mating regardless of the initial composition of the population. This proposition is so important in population genetics that its derivation must be thoroughly understood before proceeding to the effects of selection.

Let us suppose that there are N individuals in a population and that any individual is equally likely to mate with any other individual (or with any individual of opposite mating type if sexuality or incompatibility is involved). Suppose, furthermore, that in this *random-mating* population D individuals are homozygous dominants, H are heterozygotes, and R homozygous recessives. Since each AA individual has two A alleles and each Aa individual one A allele, the total number of A alleles in this population is $2D + H$. The proportion of A alleles is therefore

$$\frac{2D + H}{2N} \quad \text{or} \quad \frac{D + \frac{1}{2}H}{N}.$$

This proportion is called the *gene frequency* of A. Similarly, the gene frequency of a is

$$\frac{H + 2R}{2N} \quad \text{or} \quad \frac{\frac{1}{2}H + R}{N}.$$

The sum of the frequencies of A and a is unity. The frequencies of A and a are usually denoted either by q and $1 - q$ or by p and q, respectively, where

$$q + (1 - q) = 1$$

or

$$p + q = 1.$$

As an example, suppose a sample of 100 individuals drawn from a population is made up of $25AA$, $10Aa$, and $65aa$ individuals, that is, $D = 25$, $H = 10$, and $R = 65$. The gene frequencies are

$$A \qquad q = \frac{25 + 5}{100} = .30,$$

and

$$a(1 - q) = \frac{5 + 65}{100} = .70.$$

Frequently genotypes are expressed in proportions rather than actual numbers. Then, since $D + H + R = 1$,

$$q = D + \tfrac{1}{2}H,$$

and

$$(1 - q) = R + \tfrac{1}{2}H.$$

In this notation $AA = .25$, $Aa = .10$, and $aa = .65$. Thus

$$q = .25 + \tfrac{1}{2}(.10) = .30,$$

and

$$(1 - q) = .65 + \tfrac{1}{2}(.10) = .70,$$

exactly as before.

Let us now consider a large population in which the genotypic proportions are $AA = D$, $Aa = H$, and $aa = R$, where $D + H + R = 1$. If mating is at random, the frequencies of various types of matings are

		Females		
		AA	Aa	aa
		D	H	R
	AA D	D^2	DH	DR
Males	Aa H	HD	H^2	HR
	aa R	RD	RH	R^2

For example, the mating $AA \times AA$ occurs in the proportion D^2 of the total matings, and all offspring produced by this mating will be AA. The results of all matings will be as given in Table 15–1.

TABLE 15–1.   Frequency of Matings in a Random-Mating Population in Which Zygotic Proportions Are D $AA:H$ $Aa:R$ aa

Type of Mating	Frequency of Mating	Offspring		
		AA	Aa	aa
AA × AA	D^2	D^2		
AA × Aa	2DH	DH	DH	
Aa × Aa	H^2	$\frac{1}{4}H^2$	$\frac{1}{2}H^2$	$\frac{1}{4}H^2$
AA × aa	2DR		2DR	
Aa × aa	2HR		HR	HR
aa × aa	R^2			R^2
Total	1.00	$(D + \frac{1}{2}H)^2$	$2(D + \frac{1}{2}H)(R + \frac{1}{2}H)$	$(R + \frac{1}{2}H)^2$

The sum of the matings is unity since this sum is the product of two unities. Hence

$$(D + \tfrac{1}{2}H)^2 AA + 2(D + \tfrac{1}{2}H)(R + \tfrac{1}{2}H)Aa + (R + \tfrac{1}{2}H)^2 aa = 1.$$

Since $D + \frac{1}{2}H = q$ and $R + \frac{1}{2}H = (1 - q)$, the foregoing expression can also be written

$$q^2 AA + 2q(1 - q)Aa + q^2 aa = 1,$$

which is equivalent to 15–1, our original statement of the Hardy-Weinberg law.

This population is said to be in *equilibrium*. This means that the genetic composition of the population will remain constant from generation to generation, with respect to this locus, in the absence of the disturbances mentioned earlier. Furthermore, for populations that are not in equilibrium it can be shown that equilibrium is reached in one generation of random mating, regardless of the original proportions of AA, Aa, and aa.*

As a numerical illustration, consider again the sample $AA = 25$, $Aa = 10$, $aa = 65$, in which $q = .30$ and $1 - q = .70$. Substituting

* With two gene pairs, there are nine genotypes in the population, and zygotic equilibrium is not reached in a single generation of random mating. The approach to equilibrium is very rapid, however, in the absence of close linkage. With many gene pairs there is a constant change of nonequilibrium populations toward an equilibrium point which is approached only after long-continued random mating.

into Table 15–1, we find that the composition of the next generation will be $.09AA + .42Aa + .49aa$. The original population was thus not in equilibrium. We could have arrived at the same result by considering that random mating of individuals in a population is equal to the random combination of the pool of gametes

	A .30	a .70
A .30	.09	.21
a .70	.21	.49

which is the identical result obtained by the longer method above. Using either of these methods, we find that in all subsequent generations the genotypic frequencies remain

$$AA \quad Aa \quad aa$$
$$.09 \quad .42 \quad .49.$$

Selection in Mendelian Populations: One Pair of Genes

In the previous section we established the important principle that, in a random-mating population, gene frequencies remain constant from generation to generation, and that zygotic frequencies are in the proportion of $q^2AA : 2q(1 - q)Aa : (1 - q)^2aa$ unless there are disturbances due to selection, nonrandom mating, differential mutation rates, or gene migration. These disturbances, however, particularly selection and nonrandom mating, are of primary interest to breeders. In fact, the single most important reason for introducing the Hardy-Weinberg law was to provide a basis for discussion of the disturbances. The effect of these four forces on the genetic changes in a population under the control of a breeder will be examined in this and the next two chapters. From the standpoint of applied breeding, selection is the foremost among the factors and will consequently be considered first.

Let us suppose that in a Mendelian population comprised of aa, Aa, and AA genotypes the gene frequency is $(1 - q)a$ and qA. Furthermore, suppose that selection causes the three genotypes to reproduce in the proportion of $(1 - s)aa : (1 - ks)Aa : 1AA$. In this proportion the ratio of the number of dominants to recessives $(1 : 1 - s)$ is a measure of the reproductive rates of the A and a alleles, and k is a measure of the degree of dominance. For example, if the reproductive rates of aa, Aa, and AA are 90, 95, and 100, respectively, $s = 0.10$,

$ks = 0.05$, and $k = 0.05/0.10 = 0.5$. This represents a situation where there is no dominance, and the selective disadvantage against a is 10 per cent. If there is full dominance so that selection does not distinguish between AA and Aa, $k = 0$, and the reproductive rate becomes $(1 — s):1:1$. Thus, when the reproductive rate of AA is taken as 1, the superiority of the AA form to the others leads to positive values of s and k, whereas the superiority of the heterozygote to both homozygous forms causes k to assume a negative value. For characters with an all-or-none expression, the two parameters s and k provide a convenient method of calculating expected rates of genetic change with selection.

In using the preceding relationship to investigate the theoretical effects of selection, consider first the case where only the AA individuals are selected. (What values do k and s have here?) The next generation will obviously be homozygous for A, and the frequency of the A allele will be 1.0. Similarly if only aa individuals are allowed to reproduce, the frequency of A will drop to zero. In either of these instances selection will have produced the total effect of which it is capable in a single generation, and further attempts at adjustment of the character by continued selection will be futile, excluding the possibility that mutation can create additional alleles at the locus. Selection of this accuracy is, of course, rarely possible, and in actual practice it is to be expected that gene frequencies will be changed by selection much more slowly than in the extreme examples we have just considered.

Turning to situations where selection is not fully efficient, let us represent the frequencies of the two alleles by $[(1 — q)a + qA]$ and the reproductive rates of the aa, Aa, and AA genotypes by $(1 — s):(1 — ks):1$. It can be shown* that the change in gene frequency (Δq) is closely approximated by

$$\Delta q = sq(1 — q)[(1 — q) + k(2q — 1)]. \qquad (15\text{--}2)$$

From this equation it is seen that Δq is a function not only of the coefficient of selection (s) and the degree of dominance (k), but also of q, the gene frequency. The values taken by Δq at different values of q and for three conditions of dominance ($k = 0$, 0.5, and 1.0) are plotted for a constant selection pressure in Figure 15–1. The magnitude of Δq is appreciable, that is, selection has a substantial effect in changing gene frequency, only when the value of q is intermediate. It follows that selection of a given intensity (s fixed) is most

* See Li (1955), Lerner (1950), and other books on population genetics.

effective for characters that are common and ineffective for characters that are rare in populations.

The statement is often heard that progress under selection is rapid at first but declines in later generations. Figure 15–1 shows that this is not necessarily the case. If a trait for which a constant rate of selection is practiced is rare in the original population, progress will be slow at first. As selection causes the frequency of the desired allele to increase in the population, the tempo of progress increases, reaching

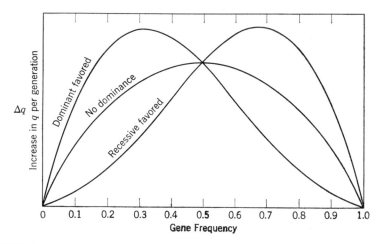

FIGURE 15-1. Rate of change of gene frequency under constant selection (s). The reproductive rates of AA:Aa:aa are dominant-favored, 1:1:1 − s; no dominance, 1:1 − ½s:1 − s; recessive-favored 1 − s:1 − s:1. The height of the curved lines indicates the rates at which selection would change the frequency of the desired allele under the three conditions of dominance that are considered. When selection favors a dominant allele, Δq is greatest at q = .33; for a recessive, Δq is a maximum at q = .67.

a maximum, after which it decreases again. This applies regardless of the degree of dominance; the main effect of dominance is to alter the value of q at which progress (Δq) reaches a maximum. In real selection experiments it is improbable that s would be maintained at a fixed value. It is likely that the breeder would alter the intensity of selection as the desired allele became more and more frequent in his populations.

The change in Δq as q changes is associated with altered amounts of free variability at different gene frequencies. In a Mendelian population mating at random, $(1 - q)^2$ of the a alleles will be in aa genotypes. The remainder, $2q(1 - q)$, will be in Aa genotypes and, if dominance is complete ($k = 0$), these a alleles will be sheltered from selection. Thus if selection is practiced for A in a population in which a is abundant and $k = 0$, rapid progress can be made at first by preventing aa individuals from reproducing. As the frequency of a decreases, however, a progressively smaller proportion of a alleles occur in aa genotypes and a larger proportion in Aa genotypes. It follows that selection will become less and less effective and Δq progressively smaller.

This explains why selection is a powerful tool in reducing undesirable recessives to low levels in Mendelian populations but is ineffective in eliminating them completely. Changing the mating system to one of the nonrandom systems does not alter this situation much, as we shall see in Chapter 17, with the exception of schemes of mating that involve inbreeding. With inbreeding, only $q(1 - F)$ of the undesirable alleles are sheltered in heterozygotes (F is Wright's coefficient of inbreeding; see Chapter 17). Since F varies from 0 with random mating to 1.0 with complete inbreeding, it is seen that the more severe forms of inbreeding are powerful tools in eliminating recessive defectives from populations.

Selection in Favor of the Heterozygote

It is not yet entirely clear whether or not there are instances in which the heterozygote at a locus is superior to both of the homozygotes. If there is indeed such superiority, the result of selection will differ from the situations previously considered in that gene frequency will ultimately reach a stable equilibrium, rather than approach zero or unity as a limit. If we represent the reproductive rates of the genotypes $aa : Aa : AA$ as $(1 - sa) : 1 : (1 - sA)$, the change in q in successive generations can be shown to be

$$\Delta q = \frac{q(1 - q)[sa(1 - q) - sAq]}{1 - sAq^2 - sa(1 - q)^2} . \qquad (15\text{--}3)*$$

The condition of equilibrium is, of course, the point where $\Delta q = 0$ (constant gene frequency). It can be found by setting equation 15–3 equal to zero and solving for q. It turns out that the equilibrium frequency of A is

$$qA = \frac{sA}{sa + sA} . \qquad (15\text{--}4)$$

The equilibrium frequency obviously depends only on the selection

* See Li (1955), Lerner (1950), and other books on population genetics for the derivation of this equation.

coefficients against the homozygotes and is independent of the initial gene frequency of the population.*

It follows that fixation of type is not possible in a random mating population when effective selection is practiced in favor of the heterozygote. The maximum proportion of heterozygotes will occur when the gene frequency is .5, that is, when equally effective selection is practiced against both homozygotes, but even then half of the population will continue to exhibit the less desirable homozygous phenotypes. It should be recognized that a gene frequency of .5, even though it produces the maximum proportion of the best phenotypic class, does not necessarily produce the best average phenotype for the population. It is at the equilibrium point that the ratio of homozygotes to heterozygotes will be such as to place the mean phenotypic expression of the character for which selection is practiced at an optimum in the population. Thus, if reproductive rates are $AA = 1.0$, $Aa = 1.2$, and $aa = 0.8$, the point at which the population has maximum fitness occurs at $qA = \frac{2}{3}$. At $qA = \frac{2}{3}$, the population is made up of 44.44

---

* Many different notations have been used to represent this relationship. In the notation we used earlier in this chapter, the respective reproductive rates of the three genotypes were represented as $(1 - s)aa : 1 - ks(Aa) : 1AA$ and the change in q in successive generations was found to be $\Delta q = sq(1 - q)[(1 - q) + k(2q - 1)]$. Again putting $\Delta q = 0$ and solving for q, we obtain

$$qA = \frac{1 - k}{1 - 2k}.$$

In the notation used in Chapter 10, the relations among the three genotypes were described as

$$AA - aa = 2d$$

$$Aa - \frac{AA - aa}{2} = h.$$

In the d and h notation equilibrium for qA is

$$qA = \frac{d + h}{2h}.$$

Suppose that the respective reproductive rates of the three genotypes are

$$AA = 1.0$$

$$Aa = 1.2$$

$$aa = 0.8.$$

In the first notation, $sA = 0.2$ and $sa = 0.4$; in the second notation $s = 0.2$ and $k = -1.0$; in the third $d = 0.1$ and $h = 0.3$. At the equilibrium value A and a will have frequencies of two thirds and one third respectively, according to any of the foregoing three methods of calculation.

per cent AA, 44.44 per cent Aa, and 11.11 per cent aa genotypes, and the mean fitness of the population is 1.07, that is, at a point intermediate between the fitness of the AA and Aa genotypes. If the relative reproductive rates of $aa:Aa:AA$ are $0.0:1.2:1.0$ (i.e., aa is sterile), $qA = 0.86$, and at equilibrium the zygotic frequencies will be 73 per cent AA, 25 per cent Aa, and 2 per cent aa, producing a mean fitness of 1.03.

Wright (1949) has shown that the frequencies of multiple alleles will be proportional to the reciprocals of the coefficients of selection of their respective homozygotes, that is,

$$q_1:q_2:q_3 \cdots = \frac{1}{s_1} : \frac{1}{s_2} : \frac{1}{s_3} \cdots . \qquad (15\text{--}5)$$

Equation 15–5 is based on the assumption that all heterozygotes are equally desirable, and the homozygotes are inferior in various degrees. If, however, one heterozygote (say A_1A_2) is substantially superior to all others, then A_1 and A_2 will eventually dominate the population. The condition in which selection favors heterozygotes is the only one which allows many alleles to remain in a population in considerable numbers in stable equilibrium. This results in *balanced polymorphism*, a genetic mechanism that theoretically can provide for substantial genetic plasticity and hence must be regarded as a factor of possible importance in both evolution and breeding. Biological evidence has been presented by many investigators, notably Ford (1949), that balanced polymorphism is indeed a factor of consequence in evolution.

Selection of Characters Governed by Many Genes

The phenotypes of characters governed by many genes are distributed on a continuous scale, and the methods used to gauge selection intensity for all-or-none characters are impracticable for describing and analyzing these quantitative traits. Selection intensity can, however, be usefully assessed as the difference between the mean of the plants selected to be the parents of the next generation and the mean of their own generation. For example, if among a group of plants whose mean phenotype is 100 units, only a superior group with mean phenotype of 110 units is taken for breeding, the selection differential is 10 units. This measure refers to the differential between the selected group and the parent population expressed in terms of the actual units in which the measurements were taken. It is usually more convenient, however, to express selection differentials in standard deviation units because (1) the selection differential can then easily be transformed into percentages of individuals saved for breeding, as we have already seen in Chapter 9, and (2) comparisons of selection intensities for different characters or different populations are more easily made than with the actual units of measurement. This measure of selection intensity is the statistic denoted by k in Chapter 9. It will be recalled that the value of k depends on the ordinate (z) at the point of truncation and the area under the normal curve to the right of this ordinate (v), so that

$$k = \frac{z}{v}. \tag{15-6}$$

Some of the values assumed by k at different selection intensities were given in Chapter 9, for instance, when the extreme 5 per cent of the population is selected, $k = 2.06$.

Since, in dealing with quantitative characters, it is convenient to express selection intensities in standard deviation units, assessment of the probable effects of selection requires some idea of the variance for such characters. It was shown by Wright in 1921 that for n genes of equal additive effect the standard deviation is $\sqrt{2nq(1-q)}$ times the effect of one gene, and the potential range of expression of the character is $\sqrt{2n/q(1-q)}$ times its standard deviation. Plants which deviate from the mean of a normally distributed population by 2 standard deviations are expected once in 21 individuals, and plants varying as much as 3 standard deviations from the mean are expected once in about 370 individuals. The larger the number of genes, the less it is likely that the most extreme type possible will be represented in the population. With even so few as four or five genes governing a character under selection, it is very unlikely that the extreme type will occur, and the best the breeder can do is select the most desirable plants available. With additional cycles of selection the frequency of the desired alleles is expected to increase. The result will be not only a shift of the mean of the population but also the eventual appearance of individuals with more extreme phenotypes than occurred in the original population.

If all the genetic variability of a population were due to additive gene action, improvement theoretically should continue until all desirable alleles have been fixed in the population (ignoring opposing mutation). The rate at which selection leads to fixation, and hence to exhaustion of the genetic variability, depends on both gene frequencies and numbers. If a character is governed by a single gene, and q is .5, the limit of selection is the standard deviation multiplied by $\sqrt{2n/q(1-q)}$, or 2.83σ. If, however, 10 genes govern the character the limit of selection is 8.95σ, assuming all gene frequencies to be .5. With 20 genes the limit of selection increases to nearly 13σ. Other things being equal, Δq will decrease for each gene as n increases. With a single gene, selection should therefore cause q to approach 1.0 comparatively rapidly; it should also reduce the variance of the population to a low level at a comparatively rapid rate because the variance, $2q(1 - q)$, will be very small when q is near 1.0 (or zero). Even though selection may take a long time to fix a monogenic character because Δq is small when q is large (or small), it can rather rapidly reduce the variance of monogenic characters to the vanishing point, giving the appearance of complete exhaustion of the genetic variance. When n is large, Δq will be small for each gene even when q is near .5, and Δq becomes appallingly small when values of q differ much from .5. Hence the limit of selection is not likely to be achieved when n is at all large, regardless of the intensity of selection that can be practiced. In other words slow and steady progress continued over many generations is expected when selection is practiced for a character governed by several genes that act in an additive fashion.

Selection and Nonadditive Gene Action

There are several factors that can upset the progress of selection as depicted under strictly additive gene action, the most serious being various types of nonadditive gene effects. We have already considered the case of intra-allelic interaction where the heterozygote, Aa, is the most desirable genotype. Selection for an intermediate phenotype presents much the same problem in that preference for an intermediate, rather than an extreme, leads to a nonlinear scale of desirability, even though gene action itself may be strictly additive. If the frequency of genes A and B is .5 for each gene and their effects are additive, the genotypic and phenotypic composition of the population with respect to these genes will be as follows:

Genotype	Frequency	Phenotype in Arbitrary Units
aabb	1	1
Aabb, aaBb	4	2
AAbb, aaBB, AaBb	6	3
AABb, AaBB	4	4
AABB	1	5

Saving only individuals with the intermediate phenotype (3 units) increases the proportion of that phenotype from $\frac{3}{8}$ to $\frac{1}{2}$ in the first generation of selection. In the next generation of selection the intermediate phenotype will be increased to 56 per cent and in the third generation to about 58 per cent. The variance of the population will be reduced to about half its original level in the same period. Beyond this point selection is virtually powerless to effect further changes in either the proportion of the desired type or in the variance. Furthermore, if selection is relaxed, the population returns to its original status within three or four generations. It will have been noted that selection in this case was neither for nor against either A or B but was for particular combinations of these two genes. Such selection does not change the frequency of either A or B but merely causes the gametic ratio to be distorted; upon relaxation of selection recombination restores the gametic ratio to its original value of $1:1:1:1$.

In this example the genetic control of the character was by genes acting in an additive fashion, yet the scale of desirability conferred nonadditive properties on the genotypes involved. Nonadditivity can result as well from strictly genetic properties of nonallelic genes. Indeed this is the situation usually brought to mind when selection for epistatic effects is mentioned. For example, it is possible that the substitution of A for a might lead to an increase in merit in the genotype bb but a decrease in merit in the Bb or BB genotypes. If the frequency of B in the population is constant, the net selection pressure for or against A will be zero. Therefore, whenever the number of genotypes in which A is superior to a is exactly counterbalanced by the number of genotypes in which a is superior to A, selection will not change the frequency of A or a ($\Delta q = 0$). If, however, selection changes the frequency of B, or other genes which determine whether A or a is superior, selection for A (or a) can be effective ($\Delta q \neq 0$). In other words the coefficient of selection for A depends on the frequency of all nonallelic genes which influence the relative worth of A versus a. A vast number of nonallelic interactions of this type are possible. The effects of such nonallelic interactions on the progress of selection are not understood at present, and their investigation is one of the most active areas in population genetics.

Potentialities and Limitations of Progress under Selection

The limit to which selection can carry a population will, according to the concepts of population genetics, have been reached when all desirable alleles are fixed in the population. The probability that this limit can be reached for characters governed by many genes or involving various sorts of nonadditiveness is very small for reasons we have already noted. There are many other factors which influence the rate of progress toward this theoretical limit. Some of these factors are purely practical and others are theoretical. Although not enough is known about most of them to form any definite conclusions concerning their relative importance, we can speculate about the role of two of the more obvious factors, namely, (1) heritability and (2) the selection intensity that is permissive under the differing reproductive biologies of various species.

Consider a population in which the frequency of A is .90, and it is desired to fix A, that is, eliminate a by selection.* If no mistakes of classification are made, 12 generations are required to increase the frequency of A to .95 and an additional 32 generations to further change the frequency to .98. Suppose, however, selection is complicated by difficulties of classification such that the coefficient of selection is no longer 1.0 but is only .20. Then the time required to change the frequency of A from .90 to .95 is 54 generations and an additional 155 generations are required to effect the further change from $q = .95$ to $q = .98$. Selection in outcrossing populations is obviously a distressingly feeble means of achieving near fixation of even highly heritable monogenic characters, let alone characters of low heritability. As mentioned earlier, this is one of the reasons why other breeding schemes have tended to supplant mass selection in cross-fertilized species.

Another factor obviously affecting progress under mass selection is the selection intensity that is permissive in particular species. If all of the seeds obtained must be saved merely to maintain the population, the selection pressure is zero. If, however, the species is prolific, selection differentials up to several thousand to one may be possible. Suppose that it is practicable to grow no more than 1000 plants in any one generation, and it is desired to save the extreme 5 per cent, that is, the 50 plants that are most desirable with respect to the character for which selection is practiced. It would then be necessary for these plants to produce 20 offspring on the average if population size is to be maintained. If only the extreme 1 per cent is to be saved as parents, the average number of offspring required increases to 100. For most plant species selection differentials of this magnitude pose no real problem, but in the breeding of some plant species and in the breeding of all large domestic animals they are beyound possibility.

* Several investigators have made calculations of this type. For example, Lush (1943) gives tables to facilitate the computations.

The problem is not so simple as this, however, because the breeder is rarely able to confine his attention to only one character at a time. Suppose it is necessary to select simultaneously for two genetically uncorrelated characters. Furthermore, suppose that in order to obtain enough seed to plant the next generation, seed must be saved from 5 per cent of the plants in each generation. Then, to select the extreme 5 per cent of individuals for character A, it is necessary to ignore character B, or vice versa. Obviously, selection pressure must be relaxed for each character if selection is to be practiced simultaneously for both characters. In this case the best that can be done is to save all plants that are in the best **22** per cent for character A and are also in the best **22** per cent for character B.

In general, when equal attention is paid in selection to n characters and when x per cent of the population must be saved to maintain population size, selection intensity for any one character will be given by $\sqrt[n]{x}$. For example, if equal attention is to be given to 3 characters and the reproductive rate requires that 5 per cent of the total population be saved in each generation, selection for each character must be confined to the best $\sqrt[3]{0.05} = 37$ per cent of individuals. With 5 characters the breeder will be able to save from only the best **55** per cent, and with 10 characters selection must be confined to the best **74** per cent for each character, if 5 per cent of all plants must be saved to maintain the population. It is apparent that increasing the number of characters for which selection is practiced rapidly decreases the intensity of selection possible for any one character to a low level unless enormously large populations can be accommodated.

Migration and Mutation

Although selection is the foremost of the factors that lead to genetic changes in populations under the control of breeders, migration is also of great practical importance. Migration is a term which describes the effects of introducing outside and previously noninterbreeding genotypes into breeders' populations. Migration can be responsible for introducing alleles previously absent from the population, or for changing the frequency, upwards or downwards, of alleles already present. However, the importance of migration is more properly discussed in connection with intervarietal hybridization, and we shall not devote more space to it here.

Point mutation, in contrast to the other three factors enumerated, is ordinarily to be disregarded as unimportant in breeding programs. Mutations of economic value, although not unknown in agricultural species according to Gustafsson's review of the subject, must be regarded as of negligible importance in any particular breeding program. Obviously when favorable mutants make an appearance, advantage should be taken of them. But until methods are developed to increase favorable mutations in a directed way, reliance must be placed on other methods of improving plants genetically. The reverse situation is encountered in connection with the deleterious effects of mutations on the progress of selection. Here the main effect is that some of the effectiveness of selection may be dissipated in overcoming their effects.

REFERENCES

Dobzhansky, Th. 1951. Mendelian populations and their evolution. In *Genetics in the 20th century*, pp. 573–589. The Macmillan Co., New York.
Ford, E. B. 1949. *Mendelism and evolution*. Methuen Co., London.
Gustafsson, A. 1947. Mutations in agricultural plants. *Hereditas* **33**: 1–100.
Lerner, I. M. 1950. *Population genetics and animal improvement*. Cambridge University Press.
Li, C. C. 1955. *Population genetics*. The University of Chicago Press.
Lush, J. L. 1943. *Animal breeding plans*. Iowa State College Press.
Wright, S. 1921. Systems of mating. *Genetics* **6**: 111–178.
Wright, S. 1949. Population structure in evolution. *Proc. Amer. Phil. Soc.* **93**: 471–478.

16

Responses to Selection
and the Genetic Organization
of Populations

Common patterns of response to selection in the multitude of experiments that have been conducted with various characters in numerous plant and animal species are the following:

1. Initial rapid gain followed by a protracted period of slow progress.
2. Slow, steady, long-continued response.
3. Slow response culminating in a plateau.
4. Little or no response.
5. Rapid initial gain, followed by a plateau period during which selection is ineffective, followed by another period of rapid gain, culminating in yet another plateau.

In this chapter one example of each of these types of response to selection will be presented. Then we shall consider the evidence that these selection experiments provide concerning the genetic organization of populations, because, provisional though it must be at present, explanation of this organization is necessary if better control of selective change in populations of domestic plants is to be achieved.

When considering the effects of selection on populations, it is necessary to keep in mind that, although selection can take advantage of any favorable genetic variability arising *de novo* in a population, mutation rates are probably too low for such spontaneous variability to assume much importance, even in long-term experiments in artificial selection. Hence the advances that may accrue from artificial selection are confined, for all practical purposes, to its effects on genetic variability already existing in the population at the start of the experiment. The effects of selection are exerted by changing the frequencies with which various genes or combinations of genes occur or by the assembly of combinations of genes not previously found in the population. Three phenotypically discernible responses are possible: (1) a change in the proportion of previously existing genotypes, accompanied by a shift in the population mean; (2) the appearance of new genotypes; and (3) changes in the variability (variance) of the population.

Rapid Initial Gain in Response to Selection

When selection has been directed at adjusting characters such as color, resistance to certain diseases, and heights and angle of declination of ear (in maize), the usual response has been rapid gain in the first few generations, followed by a protracted period of slow progress. These traits thus behave as if they are governed by a number of genes that have quite different magnitudes of effect on the phenotype. The usual interpretation has been that the rapid progress in the early generations is associated with increases in the frequency of a small number of important genes. According to the principles discussed in the previous chapter, the rate of change (Δq) of these major genes might be expected to be large until selection had carried them to the point of near fixation. The period of rapid response to selection presumably terminates at this point.

The subsequent slow progress is presumably associated with further small changes in the frequency of the major genes and with slow increases in the frequencies of minor genes. For reasons discussed in Chapter 15, it is expected that Δq will be small, even at intermediate gene frequencies, for single members of systems of genes each with individually small effects, and hence progress under selection will also be slow.

In general, then, selection for such quasiqualitative characters causes a shift in the mean of the population in the direction toward which selection is practiced. For example, selection for more intense color often produces a population in which pigmentation is more intense on the average than the unselected parental population. In addition, selection may also produce individuals that are more highly colored than any individuals of the original population. In other words, selection has produced new genotypes by assembling combinations of genes that did not previously exist in the population. Finally, the selected populations tend to be less variable in coloration than unselected populations. This is presumably because selection has increased the frequency of the major genes to near fixation and hence reduced the genetic variability in the population.

Slow Response to Selection

The classical long-time selection experiments for oil and protein content in maize, conducted at the Illinois Agricultural Experiment Station, illustrate a number of important features of response to directional selection.

The mean oil content of the original population, an unselected Burr White stock, was 4.68 per cent. The range of oil contents in this original population varied from a low of about 3.7 to a high of about 6.0 per cent (Figure 16–1). In the first selected generation the mean oil content was greater than in the original population, but there were no plants that exceeded 6.0 per cent in oil content. This can be interpreted as indicating that one generation of selection changed the proportions of different genotypes in the selected population as contrasted to the original population, but that no new genotypes had arisen. In the second and third selected generations, a few plants with slightly more than 6.0 per cent oil content appeared, presumably as a result of new combinations of genes that did not occur in the original population. With further selection, there was a steady increase in the mean oil content. Also more and more plants with oil contents transcending those of the original population appeared. By the tenth generation of selection the original and selected populations no longer overlapped. The effect of ten generations of selection was thus to produce a population in which no genotypes were like those of the original population. Selection for low oil content produced essentially the same result, but in the opposite direction from selection for high oil content.

This experiment, and a parallel one in which selection was practiced for both high and low protein content, has subsequently been continued until fifty generations of selection were completed with the 1949 crop. The results are shown graphically in Figures 16–2 and 16–3, together with either the best fitting straight line or curvilinear trend. Progress toward high content of oil and toward both high and low protein content was remarkably steady throughout the entire course of the experiment. However, little progress toward low oil content was made after twenty to twenty-five generations of selection. Much of the oil in the corn kernel is in the germ so it was not unexpected that selection for low oil content also reduced the germ size. It is apparently impossible to reduce germ size, or perhaps oil content, below a certain point and still retain viability. It may be that genes capable of further reducing the content of oil exist in the population, but that progress was brought to a halt by the physio-

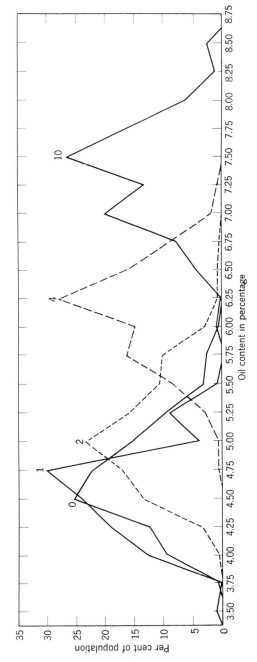

FIGURE 16-1. Selection for high oil content in corn. The frequency distribution of the original population is designated by 0 and the frequency distributions after various numbers of generations of selection by numbers. The mean oil content increased steadily with selection. Transgressive segregants appeared in the second-selected generation, and by the tenth-selected generation all individuals of the selected generations transgressed the original population. (After Smith, 1908.)

logical limitation that germ size, or oil content, cannot be reduced below a threshold value. Presumably there is a similar threshold for

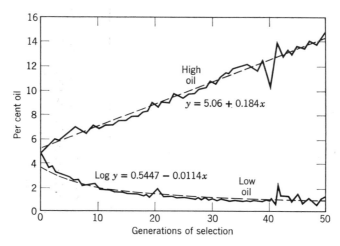

FIGURE 16-2. Effect of fifty generations of selection on oil content of corn. Actual data indicated by solid lines; fitted trend lines are shown as broken lines. (After Woodworth et al., 1952.)

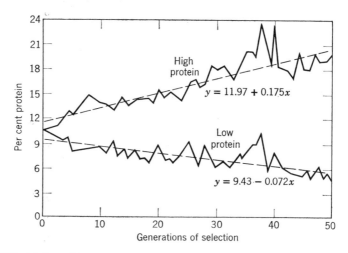

FIGURE 16-3. Effect of fifty generations of selection or protein content of corn. Actual data indicated by solid lines; fitted trend lines are shown as broken lines. (After Woodworth et al., 1952.)

high oil content and also for both high and low protein content, but fifty generations of selection had not carried the populations to these thresholds.

One of the key issues in these experiments is the amount of genetic variability remaining in the four strains after long-continued selection for chemical composition. This was tested in three ways: by comparing the variances of the selected populations in different generations, and by two types of ancillary experiments, one involving relaxation of selection and the other selection in the reverse direction.

Certain technical difficulties arise in comparing variabilities of the selected populations in different generations. In general, however, the evidence seems to indicate that the selection produced no drastic increase or decrease in the variability of either the high or low lines.

In the experiment involving relaxation of selection, part of the seed of the thirty-seventh selected generation was used to start populations in which no selection for chemical composition was made in choosing ears to plant each new generation. In all four populations the nonselected groups showed some tendency to revert to the composition of the original Burr White variety. This tendency was most pronounced in the high-oil line where the nonselected group, after eight generations, had a mean oil content of 1.49 per cent less than that of the selected group. Average differences between the selected and unselected in the other three populations were low oil, 0.22 per cent, high protein, 0.63 per cent, low protein, 0.76 per cent.

The reversed selection populations were started from the six ears of each strain most extreme in the opposite direction from the regular selection. Unfortunately only two generations of the reversed selection have been reported so the results are not conclusive. As was the case in the nonselection experiment, the high-oil population was more affected than the other strains, the two generations of reverse selection reducing the mean oil content by 2.03 per cent. Some progress was made by the reverse selection in the high-protein and low-protein populations but the low oil strain was not discernibly effected in two generations. In general, the experiments in relaxation or reversal of selection suggest that long-continued selection for chemical composition had not resulted in fixation of genes governing high or low content of oil or protein.

Very early in these experiments it was recognized that selection for chemical composition was leading to changes in the characteristics of the ears, kernels, and plants of the selected strains and was also tending to reduce grain yields. These changes became progressively more pronounced as the selection proceeded until after fifty generations each strain had a highly distinctive ear and kernel type; some of these differences are illustrated in Figure 16–4. The four strains also differed in maturity, plant height, ear height, tiller number, and other plant characteristics. The side effect on grain yield was severe, all four strains having been reduced to approximately one half of the yield of adapted hybrids by the end of the fifty generations of selection for chemical composition.

Responses to selection for chemical composition of the maize kernel thus fall into the second and third categories listed at the beginning of this chapter. Selection for high oil and high and low protein

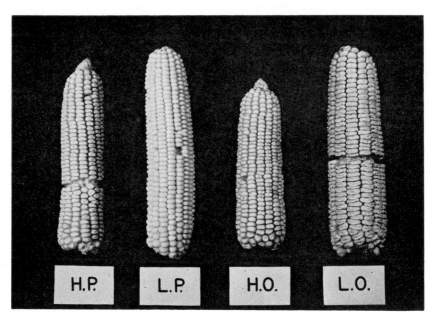

FIGURE 16-4. Ear characteristics of the Illinois high and low protein and high and low oil strains of corn. Why should selection for chemical content also influence ear characteristics? (After Woodworth, Leng, and Jugenheimer, *Agron. Jour.* 44:64. 1952. Photograph courtesy Illinois Agricultural Experiment Station.)

content led to slow, steady change over fifty generations and selection for low oil content produced slow, steady change for about twenty-five generations, after which there was little or no change. Characters governed by complex systems of genes, each with small effects on the phenotype, are expected to respond to selection in this manner. The rate of change in frequency (Δq) for any single gene is expected to be so small that progress toward fixation should be slow. The expected phenotypic manifestations of this genetic situation are exactly those observed: gradual shift of the population mean, the appearance of phenotypes more extreme than any previously found in the population, and little if any decrease in genetic variability. Lack of progress toward lower oil content after twenty-five generations of selection was apparently not due to fixation of genes conditioning low oil content. Rather, it appeared to stem from a physiological threshold that brought progress to a halt in a still genetically variable population. Although the correspondence between expected and observed responses is striking in this experiment, we should recognize that future developments may well lead to modifications of this provisional genetic explanation of these observed responses to selection.

Lack of Response to Selection

The success of selection in modifying the chemical composition of the corn kernel had a great influence in promoting similar attempts to improve yield in maize. The procedure that was followed, as in selection for chemical composition, was a modification of mass selection called ear-to-row selection. The ear-to-row method attempted to determine the relative breeding value of different ears by planting a portion of the seed from them, an ear to a row, and measuring the yielding ability of the resulting plots. Seed from the most productive rows (or remnant seed from the highest yielding mother ears) was planted and the resulting population reselected.

The main difference between simple mass selection and ear-to-row selection is in basing the selection on the performance of progeny, according to the Vilmorin principle, rather than on the characteristics of the parental plant itself. Since the ear-to-row method employs open pollination, the pollen parents of the different kernels on an ear are a more or less random sample from the population, and the method is still essentially one of mass selection. The ear-to-row method was proposed by Hopkins in 1896, and it became popular almost immediately. The early results showed wide differences in the productiveness of the seed from different ears, and some results were also obtained indicating that the yield of stocks made up from high yielding ear rows was larger than the yield of stocks made up from the ear rows of low productivity. As a consequence of the success of the ear-to-row method in modifying chemical composition, together with the foregoing encouraging results in selecting for productivity, the method was taken up by nearly all agricultural experiment stations and seedsmen and by many farmers as well.

Advances in yield resulting from ear-to-row selection were soon found to be much smaller than expected on the basis of the early results. The results obtained at the Nebraska Agricultural Experiment Station were typical. In ear-to-row selection practiced continuously in the period 1911 to 1917 the productivity of the original unselected variety (Hogue's Yellow Dent) and the selected strain were 53.6 and 53.3 bushels per acre, respectively. Moreover, even when small initial responses were reported, no investigator found clear-cut evidence for continued responses to selection. As a result of many experiences of this type it was concluded that ear-to-row selection was virtually powerless to improve yield in adapted varieties, although it was conceded that it might be of some use in improving unadapted varieties. Selection for yield in corn is thus an example of the fourth type of response listed at the beginning of this chapter.

A number of reasons have been suggested for the failure of mass selection to improve yield in maize. Some have suggested that genetic variability at loci affecting yield is limited. Recently (1955), however, Robinson, Comstock, and Harvey obtained evidence that additive genetic variance for yield is substantial in open-pollinated varieties. Experiences with recurrent selection and other methods of breeding, in which great differences have been found in the yielding ability of different plants in open-pollinated populations, also make it difficult to accept the view that open-pollinated varieties lack genetic variability for yield.

Others have suggested that lack of progress can be attributed to the fact that in ear-to-row selection only a small portion of the genotypic variance is subject to selection (the variance of half sibs is one eighth of the additive variance). Although ear-to-row selection is an inefficient breeding scheme, it is equally inefficient genetically regardless of the character under selection, whether color, disease resistance, or chemical content, suggesting that its failure to improve yield is not to be ascribed to the genetic limitations of the method.

It seems likely that much of the failure is to be attributed to low heritability of yield. When ear-to-row selection was in vogue, single plots were used to evaluate the yielding ability of ear rows. As we have seen (Chapters 9 and 12), accurate measurement of yielding ability requires replicated experiments repeated over locations and years. Hence heritability for yield, calculated from single plots grown in one environment, would certainly have been very low in ear-to-row programs. Apparently the failure of mass selection to improve yield in maize can be attributed more to inability to recognize superior genotypes than to lack of genetic variability for yield.

Response to Selection after Many Generations at a Plateau Value

The final type of response to selection that we shall consider is of critical importance to our understanding of the storage and release of variability in Mendelian populations. This is the type of response in which a period of genetic gain is succeeded by a plateau, which in turn is followed by another period of gain from selection. The most comprehensive experiment in which this type of response was obtained was one conducted by Mather and Harrison (1949), in which selection was practiced for more than one hundred generations for abdominal chaetae number in *Drosophila melanogaster*. The essential features of this experiment are presented in schematic form in Figure 16–5. The response to selection was large for approximately twenty generations, but was accompanied by a reduction in reproductive capacity so severe that selection had to be suspended to keep the population in existence. After a few generations of relaxation of selection, fertility restored itself. The restoration of fertility was in turn accompanied by a rapid reduction in number of chaetae, but stabilization of chaetae number occurred at a level substantially above that of the original population.

Upon resumption of selection for increased number of chaetae, it was soon possible to reach a slightly higher-average number than in the initial phase of the experiment, but with the important difference that fertility was not reduced. There were two types of evidence that this new level of bristle number represented a new stable situation: (1) upon suspension of selection, only a slight reduction in bristle number occurred; and (2) continued selection for increased bristle number produced little gain in nearly fifty generations. At that point, however, there was another sudden increase, followed some twenty generations later by still another response. The important features of this experiment are: (1) the large reservoir of hidden variability in the population; (2) the reduction in fecundity that occurred in the period of rapid increase in bristle number, and recovery of reproductive capacity when selection was relaxed; (3) the establishment of new stable levels of chaetae number following relaxation of selection; (4) the responses to selection after many generations at a plateau level; and (5) the effect of selection for one character in changing many morphological and physiological characters, even against the trend of natural selection.

The genetic model envisaged by Mather and Harrison to account for these responses hinges on the assumption that the best adapted individuals in a population are those that exhibit a harmonious com-

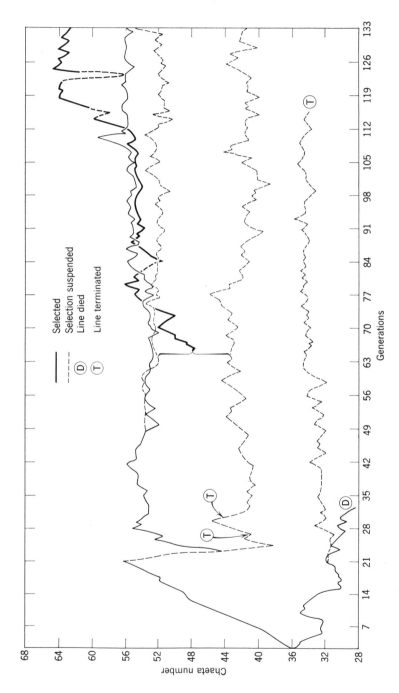

FIGURE 16-5. The results of selection for high number of abdominal chaetae in *Drosophila melanogaster.* (After Mather and Harrison, 1949.)

bination of all characters leading to maximum fitness in the environments in which the population exists. As Wright (1951) has stated, "the best adapted form in a species is usually one that is close to the average in all quantitatively varying characters." According to Fisher (1930), natural selection working on the whole complex of characters tends to favor organisms clustering about mean values for all characters. This implies that there will be selection against the extreme phenotypes in populations and hence that it will be to the advantage of the population, if reproductive wastage is to be avoided, to achieve uniformity of phenotype. At the same time the population must maintain genetic variability if it is to be flexible enough to meet long-term evolutionary challenges.

Mather believes that successful species have resolved this conflict between the demands of immediate fitness and long-range flexibility through the building of systems of linked genes. The key to the genetic model visualized by Mather to meet the challenge of these opposing demands on the genetic system lies in the distinction he makes between *free* and *potential* variability. Even in the simplest genetic situation, where the variability depends on a single gene pair, the variability will be entirely free and exposed to selection only when all individuals are homozygous and no heterozygotes occur. If any heterozygotes exist, part of the variability will not be free because, although a heterozygote can resemble one or the other of the homozygotes or can lie between them, it cannot resemble both of them at once. The genetic variability that is freely expressed as a phenotypic difference between homozygotes is therefore hidden in the heterozygote as potential variability, although it must always be partly freed by segregation in the next generation.

Where the variability depends on two or more gene differences affecting the same character, a second type of potential variability will occur. If two genes have reinforcing effects, the genotypes *AABB* and *aabb* will produce extreme phenotypes against which natural selection will act when the environment is stable. In the homozygotes *AAbb* and *aaBB*, the two genes will have counterbalancing effects, and any difference between the two phenotypes will not represent the difference in the genotypes in the sense that the genes in question can be reassociated to give phenotypic differences of greater magnitude. Potential variability is therefore present in the difference between *AAbb* and *aaBB* but it is a homozygotic potential variability. The release of this type of "bound" variability will be slow because it must first be converted into heterozygotic potential variability by crossing before it can be converted into free variability by segregation.

This additional means of storing variability may play an important role whenever a character is governed by many genes.

Genetic storage systems of this type will clearly be affected by linkage. Suppose we are concerned with a character that has been adjusted by natural selection so that the population mean represents maximum fitness, and extreme phenotypes in either direction are rejected by natural selection. If this character is governed by four genes it will be to the advantage of the population if these four genes are linked according to the third of the following arrangements:

$$(1) \quad \frac{ABcd}{abCD} \qquad (2) \quad \frac{AbcD}{aBCd} \qquad \text{or} \qquad (3) \quad \frac{AbCd}{aBcD}.$$

Arrangement 1 requires only a single strategically placed crossover for total release of the potential variability, whereas arrangement 2 requires two crossovers and arrangement 3 requires three crossovers. In terms of total flexibility the three arrangements are equal. However, in terms of immediate fitness, arrangement 3, or rather its progeny, will have an advantage over the other arrangements by virtue of producing fewer extreme and unfit phenotypes.

This genetic model proposes that many genes scattered along the chromosomes constitute the genotype for metrical characters, and it is to the advantage of the population if genes with plus and minus effects are linked to each other. When many genes are involved, and the linkages are tight ones, much of the variability in the population will be in bound form. As long as some mechanism exists to keep these organized blocks in the gene pool, crossing over can occur to transform some of this potential variability into free variability on which directional selection can act. Since, however, genes governing many different characters will be intermingled on the chromosomes, selection for one character will also produce responses in other characters. It was presumably for this reason that selection for chemical composition affected many other characters in corn. Attempts to shift populations too rapidly for specific characters will be counter-balanced by natural selection directed toward maintaining a harmonious relationship among all fitness-determining characters. This behavior is postulated to be a product of the entire evolutionary history of the species.

According to this genetic model, plateaus such as the ones observed in selection for chaetae number, are the result of a balance between the effects of directional selection and the opposing force of natural selection for overall fitness. Sudden bursts of response after a long period of stability are supposedly brought on by the occurrence of crossovers which produce the free variability necessary for developing new balanced combinations of genes and hence new stable equilibrium points.

Alternative Hypotheses of Population Structure

The classical concept of population genetics holds that the major process of evolutionary significance consists of substitution at each locus of more favorable for less favorable alleles. According to this concept, Mendelian populations should therefore be made up of individuals homozygous at most loci, with heterozygous loci in the minority. The store of variability in populations should trace to the following causes: (1) alleles that produce phenotypes that are adaptively neutral, slightly advantageous, or slightly disadvantageous depending on fluctuations in the environment; (2) the occurrence at individually low frequencies of deleterious mutants, ranging from subvital to lethal in their effects (most of these mutants in the common gene pool will be recessives shielded from selection by being situated in heterozygotes); (3) balanced polymorphism due to multiple alleles maintained in the population by fluctuations in environment so that first one and then another in the series will be favored as a result of temporary changes in environment; and (4) rare desirable mutants that have not had time to displace their alleles.

This concept of population structure adopts the allele as the basic unit for studies of populations and defines a population as a pool of individual genes. The simple models based on gene frequencies by which we introduced the study of Mendelian populations in the previous chapter followed naturally from this concept. Many investigators have had serious reservations, however, as to whether some of the observed properties of populations are consonant with the idea that populations are no more than aggregations of individuals whose characteristics can be specified by gene frequencies. These investigators hold that the properties of Mendelian populations transcend those of their component members and that populations, although, of course, composed of individuals, have a genetic cohesion that is a property of the population, not of individuals. The gene pool of a population is thus visualized as an organized system with complex integrative properties.

In the development of these integrated systems, the formation of complexes of genes, held together by linkage, are supposedly important factors in the balance between the components of the population genotype. The fate of a gene or a gene complex under selection depends chiefly on its effect in combination with other genes or gene complexes present in the common gene pool. In such a system selection should not be limited to simple fixation of genes but should tend to affect the entire population by repatterning of the gene pool. Such systems are postulated to have been built up over long periods of time and to be integrated and balanced in such a way as to produce maximum fitness for the population; superiority of individuals is thus subordinated to the welfare of the population unless individual superiority also contributes to general fitness. Once established, these systems are self-regulating, and they are expected to display great resistance to modification of individual characters by selection of any type that destroys their overall balance. This conservatism of Mendelian populations has been called *genetic homeostasis* by Lerner (1953, 1954).

Describing these two views of population structure does not imply that they are mutually exclusive. Indeed it is becoming increasingly apparent that a great variety of population structures exist in different organisms, and it is probable that every intergrade of conditions between these two genetic systems occurs. It is, of course, possible that additional and perhaps entirely different models will have to be constructed as more and more is learned about the reproductive biology of populations.

REFERENCES

Dobzhansky, Th. 1951. Mendelian populations and their evolution. In *Genetics in the 20th Century*, pp. 373–389. The Macmillan Co., New York.
Dobzhansky, Th. 1955. A review of some fundamental concepts and problems of population genetics. *Cold Spring Harbor Symposia on Quant. Biology* 20: 1–15.
Fisher, R. A. 1930. *The genetical theory of natural selection.* Clarendon Press, Oxford.
Lerner, I. M. 1953. The genotype in Mendelian populations. *Proc. Ninth Intern. Congr. Gen.*, 124–138.
Lerner, I. M. 1954. *Genetic homeostasis.* Oliver and Boyd, London.
Mather, K. 1943. Polygenic inheritance and natural selection. *Biol. Rev.* 18: 32–64.
Mather, K., and B. I. Harrison. 1949. The manifold effects of selection. *Heredity* 3: 1–52, 131–162.
Robinson, H. F., R. E. Comstock, and P. H. Harvey. 1955. Genetic variances in open-pollinated varieties of corn. *Genetics* 40: 45–60.
Woodworth, C. M., E. R. Long, and R. W. Jugenheimer. 1952. Fifty generations of selection for protein and oil in corn. *Agron. Jour.* 44: 60–65.
Wright, S. 1951. The genetical structure of populations. *Ann. Eug.* 15: 323–354.

17
Systems of Mating
and Their Genetic Consequences

Basically there are only two things breeders can do to improve genetically variable populations. First, they can decide which individuals will be allowed to produce the next generation and, within limits, which of these chosen individuals will be allowed to have many and which few offspring. This is selection. As we saw in the two previous chapters, selection has considerable power to change gene frequencies and the frequency of different combinations of genes, and hence the genetic value of populations.

Second, breeders can decide how the individuals they have selected will be mated to each other. There are five basic types of mating systems which, following Wright, we shall designate: (1) random mating, (2) genetic assortative mating, (3) phenotypic assortative mating, (4) genetic disassortative mating, and (5) phenotypic disassortative mating. In random mating, individuals are mated together at random or by chance. The other four systems all represent deviations from random mating based either on mating like to like or on the mating of unlikes. The criteria of likeness or unlikeness are either relationship (ancestry) or appearance (phenotypic resemblance or distinction).

In genetic assortative mating, individuals are mated which are more closely related by ancestry than if matings were at random. In the broad sense, this type of mating is inbreeding, although that term is usually reserved for matings between closely related individuals. In the third system, phenotypic assortative mating, plants are mated which resemble each other more closely phenotypically than the rest of the population. The fourth system, genetic disassortative mating, calls for mating of individuals which are less closely related than they would be under random mating. This system is commonly called

197

outbreeding. Finally, if mates are selected on the basis of contrasting phenotypic character, we have phenotypic disassortative mating.

Since selection and the mating of the selected individuals according to one of these five mating systems are the basic tools by which breeders change the inheritance of their populations, it is important to know what kinds of genetic change each system is capable of producing. One of the primary postulates of population genetics is that the distribution of genes among the individuals of a population is affected by the mating system in force. In general, the zygotic proportions $q^2AA : 2q(1 - q)Aa : (1 - q)^2aa$ expected under random mating no longer pertain under other systems of mating. In the discussion to follow, we shall emphasize how the effect of different mating systems on zygotic proportions is related to genetic variability in populations, degree of homozygosity or heterozygosity, and genetic relation or correlation between close relatives (probability that close relatives will inherit the same genes).

The most important investigations of the consequences of different mating systems on the genetic composition of populations are those of Sewall Wright (1921). Lush (1943), Lerner (1950), and Li (1955) have emphasized the applications of Wright's classical works. Theory and the derivation of formulas not given here can be found in these sources.

Random Mating

Strictly speaking, the term random mating should be applied only to breeding systems that meet these two requirements: (1) that each member of the population have an equal chance to produce offspring, and (2) that any female gamete be equally likely to be fertilized by any male gamete. It is doubtful whether this theoretical form of random mating is ever met in plant breeding, because some form of selection is almost always practiced, thereby invalidating the first requirement. For example, in the selection experiments considered in Chapter 16, the next generation was produced from seed resulting from randomly fertilized female gametes of *selected* plants. Since only the second condition above was fulfilled, this system of mating is appropriately called *random mating with selection*. We can note in passing that in practice random mating with selection is closely akin to phenotypic-assortative mating. Under random mating with selection, the selected individuals are usually chosen because they exhibit extreme phenotypes for some character such as color, protein content, or high yield. Thus, in both cases, phenotypically similar individuals are mated, but mates are determined by chance in random mating with selection and according to some definite scheme under phenotypic-assortative mating.

We can also note in passing that there is doubt whether even the second requirement above is ever strictly fulfilled in plant populations. Differences in time of pollen shed and ovule maturity, location in the nursery with respect to prevailing wind direction, various types of genetic incompatibility, and the like, make it improbable that fertilization is ever completely at random. The general importance of such deviations from random mating and their effects on the genetic composition of populations are not known at present.

If, however, we are willing to assume that both of the conditions necessary to the theoretical form of random mating are fulfilled, the following predictions can be made about the genetic composition of populations from generation to generation. First, gene frequency will remain constant. Second, the amount of genetic variability in the population will not change. Finally, the genetic relationship between individuals will also remain constant, as will the degree of homozygosity and heterozygosity.

Random Mating under Selection

When selection is combined with random mating, the effect on gene frequencies can be very different from the expected effect we have just noted for the theoretical form of random mating. Thus, as we saw in Chapters 15 and 16, the combined effect of random mating and selection is usually a shift in gene frequencies which is accompanied by a shift of the mean of the population in the direction of the selection. If random mating with selection changes gene frequencies, does it also change the other attributes of population structure?

First, it is obvious that directional selection will tend to reduce the variability of a group of individuals selected as parents. Lush gives the following examples of the magnitude of this reduction. If the breeder eliminates 10 per cent of the extreme individuals from one tail of a normally distributed population and retains the remaining 90 per cent as parents, the standard deviation of the retained group will be reduced by 16 per cent. Eliminating 20 per cent reduces the variability of the remiander by 24 per cent, and eliminating 50 per cent reduces the variability by 40 per cent. There is no doubt that selection has a striking effect in increasing the uniformity of a group of individuals selected to be parents.

The key question, however, is the effect of the selection on the uniformity of the next generation. When the selected and now more uniform group of parents are mated together at random, the variability of their offspring will be altered in two ways contrasted to an unselected group. First, effective selection will have altered gene frequencies. If q is small, that is, the gene is rare—the effect of increasing its frequency is also to increase the variance of the population (Chapter 15). If q is large the effect is to decrease the population variance. However, as we have seen, these effects are always small for quantitative characters.

The second effect of directional selection is to produce an excess of intermediate gametes by eliminating gametes produced by individuals at one extreme of the curve. This narrowing effect of selection is also slight. For example, if half of a previously unselected and random mating population is saved as parents, and furthermore if heritability is 100 per cent and all gene effects are additive, the standard deviation of the next generation will be 17 per cent less than that of the original population. But if the heritability is 50 per cent the reduction will be only 4 per cent. And if the heritability is 30 per cent the reduction is only 2 per cent. It is clear that directional selection under random mating can be expected to have little effect on the variability of populations as regards most production characters.

As we saw in Chapter 15 random mating with selection has considerable powers to change gene frequencies, particularly for simply inherited and highly heritable characters. This system of mating tends to increase the frequency of the alleles for which selection is practiced and consequently to increase the frequency of homozygotes. With more and more complexly inherited characters, and with lower heritability, however, the ability of this system of mating to increase homozygosity declines rapidly until, for most production characters, it can be shown to have very low powers of fixation. Similarly, random mating with selection has almost no effect on the genetic correlation between close relatives.

In summary, it is seen that the main effect of random mating with selection is to change the mean of the population and that it has little effect on population variance, homozygosity, or the genetic correlation between relatives.

Random Mating in Small Populations

In our discussions of the factors disturbing the Hardy-Weinberg relationship, we have assumed that we were dealing with large populations. When population size is limited, which is likely to be the case in plant breeding, two additional factors must be taken into account. These factors, chance and inbreeding, can be very important in plant breeding, and we shall consequently consider them in some detail.

The role of chance can be illustrated by considering two hypothetical populations, one made up of 10 and the other of 1000 individuals in each generation. Suppose that in each population a gene is represented by two alleles, A and a, whose frequencies are equal so that $q = 1 - q = .5$. The smaller population arises in each generation from a sample of 20 gametes and the larger one from 2000 gametes. The expected numbers of A and a gametes in a sample of 2000 gametes is

$$1000 \pm \sqrt{\frac{q(1-q)}{N}} = 1000 \pm \sqrt{\frac{1000 \times 1000}{2000}} = 1000 \pm 22.4.$$

Similarly, in the small population the expected numbers are

$$10 \pm \sqrt{\frac{10 \times 10}{20}} = 10 \pm 2.24.$$

In the large population the standard deviation is only 1.1 per cent but in the small population it is 11 per cent of the gamete number. Whereas the proportions of A and a might be expected to remain constant from generation to generation in the larger population, they might drift widely in the small population. This random fluctuation of q arising due to accidents of sampling in small populations is called *genetic drift* because the value of q drifts about without approaching any particular value. Gene frequency in a small population ultimately becomes zero or one for all genes after a sufficient number of generations. In other words, the population ultimately becomes homozygous for all loci (excepting the effect of mutation) even though random mating has been the practice. The effect of drift will clearly be accentuated by intense selection. Drift is ordinarily beyond the control of the breeder except as selection pressure and population size can be manipulated to increase or decrease the rate of chance fixation.

The other factor affecting the genetic composition of small random-mating populations is the inbreeding that results from reduction in the number of effective parents. In any small population, it is inevitable that more consanguineous matings will occur than in infinitely large random-mating populations. The decrease in heterozygosity per generation as a result of this inbreeding will be $\frac{1}{2}N$, where N is the number of monoecious individuals whose gametes unite completely at random (including self-fertilization). For example, in a population of 10 individuals in which 100 loci are heterozygous,

$$\frac{1}{2N} = \frac{1}{20} = 5 \text{ per cent}$$

or 5 loci are expected to become fixed in the first generation of random mating. In each subsequent generation, 5 per cent of the remaining heterozygous loci are expected to reach fixation.

The situation is only slightly different in monoecious plants where self-fertilization is prevented (e.g., by self-incompatibility) or in dioecious plants where the population is made up of equal numbers of males and females. Here the rate of decrease in heterozygosity is approximately $1/(2N + 1)$, where N is again the breeding size of the population. The factor $1/(2N + 1)$ is not much different from $1/2N$ except when N is very small.

When there are different numbers of male and female parents, the effect is equivalent to reducing the effective breeding size of the population. This reduction can be estimated by

$$\tilde{N} = \frac{4N_f N_M}{N_f + N_M}$$

where N_f and N_M are the number of female and male parents respectively, and $\tilde{N}$ is the effective population size. The rate of decrease in heterozygosity is then given by $1/2N$. Suppose, for example, the next generation is reproduced from 50 seed parents and from 450 pollen parents. Heterozygosis in this population will be reduced at the same rate as if there were only

$$\tilde{N} = \frac{4(50)(450)}{50 + 450} = 180$$

breeding individuals, instead of 500, in the group. Effective population size obviously depends more on the sex that is fewer in number than on the other sex. With a limited number of seed parents but a large number of pollen parents, as might often be the case in selection experiments in plants, $\tilde{N}$ will be only slightly larger than $4N_f$. Thus the rate of decay of heterozygosity is approximately $1/8N_f$. Conversely, it is approximately $1/8N_M$ if there are many female but few male parents.

In summary, we see that variations in gene frequencies due to sampling accidents can be appreciable in small populations. The rate of decay of heterozygosity is $1/2N$ per generation, where N stands for effective population size. The effective population size can be smaller than the actual breeding size, which in turn may be smaller than the total number of individuals in the population. Aside from random fluctuations associated with small population size, random mating has poor powers of fixation of genes, with or without selection. The utility of random mating in breeding is therefore greatest for special purposes such as preserving desirable alleles which might be lost by chance under mating systems which increase homozygosity. (See, for example, Chapter 23.)

Genetic Assortative Mating

This mating system is usually referred to as *inbreeding*. In various chapters of Sections Two and Three we discussed in some detail the effects of self-fertilization, the most extreme type of genetic-assortative mating possible in higher plants. Some other aspects of inbreeding will be considered in the next two chapters and in Chapter 23. At this point only the more general features of genetic-assortative mating will be considered.

The primary effect of genetic assortative mating is to increase the probability that offspring will inherit the same genes from both their parents. This tends to lower the percentage of heterozygosity in the population, leading to fixation of alleles and hence also to fixation of phenotype to the extent that it is under genetic control. Genetic assortative mating is the most powerful of all mating systems in this regard, because the determination of genetic relationship does not depend on the ability to recognize which individuals have the same genes but can be determined from simple pedigree records. In other words, genetic assortative mating is not limited by mistaking the effects of environment for genes or by the difficulties caused by dominance or epistasis. Inbreeding is therefore especially effective in fixing genes governing characters with low heritability, which are the characters that are usually most important to breeders and also most difficult to handle in practical breeding.

Intense inbreeding leads to an ever-increasing number of noninterbreeding groups, some of which must be discarded each generation if population size is to remain within reasonable bounds. Intense inbreeding without selection is therefore virtually impossible. Furthermore, this necessary selection cannot be independent of genetic relationship, as it is under phenotypic assortative mating.

The effect of genetic assortative mating on the total variability also depends on the selection practiced. Inbreeding without directional selection increases the total genetic variance. The increase is entirely among lines (or families) because, as homozygosity increases due to inbreeding, genetic variance within families decreases toward zero (for the more intense forms of inbreeding). At the same time, the among-line variance increases as differences among families increase owing to random fixation of different alleles in different families. It follows that if directional selection is practiced, the resulting families will be similar phenotypically and total genetic variance will be decreased. In the extreme, selection of a single homozygous line will reduce total genetic variance to zero.

Genetic assortative mating also affects the genetic correlation between close relatives in a population. Genetic correlation among relatives is important in breeding because it is closely related to *prepotency,* which can be defined as the ability of an individual to impress characteristics on its offspring so that they resemble that parent *and each other* more closely than usual. Obviously if close relatives are unusually like each other this aids the breeder by allowing him to predict more accurately what the results of various matings will be. Differences in prepotency depend on homozygosity, dominance, epistasis, and linkage; among these factors homozygosity is the most important and the only one under the control of the breeder. Homozygotes are highly prepotent because they produce only one type of gamete. Multiple heterozygotes, which produce many types of gametes, are generally not highly prepotent. Clearly, a homozygote dominant at all loci would excel in prepotency because its close relatives would be unusually alike.

The effect of inbreeding, with or without selection, is to increase the genetic correlation between close relatives. As inbreeding progresses, this correlation increases until, with complete inbreeding, the genetic correlation between any two relatives within a single family, regardless of their degree of relationship, is perfect. The important practical consequence of this high genetic correlation is the high prepotency it produces which is obviously advantageous in practical breeding.

Quantitative Measures of Inbreeding

Since the main effect of genetic assortative mating is to increase homozygosity, the most useful quantitative measure of inbreeding is one which shows the decrease in heterozygosity expected under various types of genetic assortment. The *Coefficient of Inbreeding* (F), devised by Wright, is based primarily on the number and closeness of ancestral connections between male and female parents. The formula for the Coefficient of Inbreeding is

$$F_X = \sum_A (\tfrac{1}{2})^{n_1+n_2+1}(1 + F_A), \qquad (17\text{--}1)$$

where A is an ancestor common to the pedigree of both male and female parents of individual X, where n_1 is the number of generations between the male parent and A, n_2 is the number of generations between the female parent and A, and F_A is the inbreeding coefficient of ancestor A. For self-fertilization $n_1 = n_2 = 0$, and the formula reduces to

$$F = \tfrac{1}{2}(1 + F'), \qquad (17\text{--}2)$$

where the prime indicates the coefficient of the preceding generation. It is seen that F increases with each generation of self-fertilization.

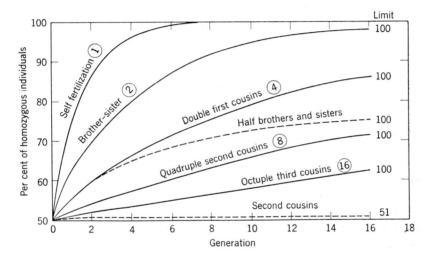

FIGURE 17-1. Percentage of homozygosity in successive generations under various systems of inbreeding. Systems in which the population breaks up into a number of lines containing a constant number of individuals are represented by solid lines. The circled figures represent the constant number of individuals per generation. The percentage of homozygosity after an indefinite number of generations is given on the right for each system. (Adapted from Wright, 1921.)

The coefficient of inbreeding has a value of zero in random-mating populations and increases toward one as the probable proportion of homozygosity goes toward zero. Figure 17–1 shows the percentage increase in homozygosity in successive generations under various systems of inbreeding. Self-fertilization (one individual in each generation in each family) leads to very rapid increase in homozygosity

(F exceeds .90 in the third generation). Under continued mating of full sibs (two individuals per generation in each family), F exceeds .90 in the eighth generation. With continued mating of double first cousins (four individuals in each generation) F does not pass .90 until the seventeenth generation. Rates of increase in F with quadruple second cousins (eight individuals per generation) and octuple third cousins (sixteen individuals per generation) are still slower. The rate of increase in homozygosity in populations of more than sixteen individuals is so slow, when maximum avoidance of inbreeding is practiced, that it is of little significance.

Since F, the coefficient of inbreeding, measures the probable degree of heterozygosity, it can be used to describe the properties of populations relative to random-mating populations. Thus zygotic proportions in an inbred population are given by

$$q^2 + F(q)(1-q)AA : 2q(1-q) - \\ 2F(q)(1-q)Aa : (1-q)^2 + F(q)(1-q)aa = 1. \quad (17\text{-}3)$$

Under random mating, $F = 0$ and 17–3 reduces to 15–1. With intense inbreeding, such that $F = 1$, the zygotic proportions obviously become $(q, 0, 1-q) = 1$ for any given gene frequency.

As a numerical example, consider a large random mating population at equilibrium with respect to the gene pair Aa. If the breeding population is restricted so that effective population size is reduced to 10, heterozygosity will be reduced in each generation by $1/2N = 1/20$. In other words F is increased from 0 to .05 in the first generation and will reach about .25 in 6 generations. For gene pair Aa ($q = 1 - q = 0.5$), zygotic proportions will no longer be $.25AA : .50Aa : .25aa$, but $.3125AA : .3750Aa : .3125aa$. If q had been .9 and $1 - q$ .1, the original genotypic frequencies would have been $.81AA : .18Aa : .01aa$, but with $F = .25$, these proportions would be altered to $.8325AA : .1350Aa : .0325aa$. With gene frequency of .5, inbreeding increased the proportion of recessive genotypes from 25 to 31.25 per cent of the population, an increase of 25 per cent. But when the frequency of the recessive allele was .1, the proportion of recessive genotypes increased from 1.0 to 3.25 per cent, an increase of 325 per cent. This relation can be expressed as

$$\frac{(q)^2 + F(q)(1-q)}{(1-q)^2},$$

which is the ratio of recessive genotypes with and without inbreeding. This ratio is small when $1 - q$ is near .5 but becomes progressively larger as $1 - q$ becomes smaller. Hence the proportion $(1 - q)$ of rare recessives in a random-mating population is increased strikingly by inbreeding whereas the proportion of common recessives is scarcely altered.

Phenotypic Assortative Mating

It can be seen intuitively that the genetic effect of mating like to like on the basis of phenotypic characteristics depends on two factors: gene number and heritability. If only one gene pair, Aa, is involved, and if there is no dominance or environmentally induced variation to obscure hereditary values, the breeder can identify AA and aa genotypes positively. Matings of only the extreme types $AA \times AA$ and $aa \times aa$ can then be made with the following effects: (1) concentration of the population at the two extremes with accompanying increase in total variability if both extremes are maintained; (2) achievement of homozygosity in a single generation; and (3) achievement of perfect genetic correlation between members of a family in a single generation. Once random mating is resumed the population will of course return to its original composition provided selection has not altered gene frequencies

If assortative mating is based on the three possible phenotypes, instead of only the two extreme types, the series of percentages for successive generations is $\frac{1}{2}, \frac{1}{4}, \frac{1}{8}, \frac{1}{16}$, and so forth.

If dominance is complete, but there is no other complication, the final result is the same but is achieved more slowly. The series is $\frac{1}{2}, \frac{1}{3}, \frac{1}{4}, \frac{1}{5}$, and so forth, the same as for self-fertilization. If, however, heritability is incomplete due to causes other than dominance, the population will come to equilibrium at a point short of fixation of the extreme types, that is, some heterozygotes will remain.

The effect of mating like to like, without errors of classification, for a character governed by two pairs of equally effective genes which lack dominance was worked out by Wright (1921). If the assortative mating is based on only the two extreme phenotypes, fixation is achieved in one generation. But if it is based on assortment among the five possible phenotypes there is a gradual decrease in the frequency of intermediate phenotypes and an increase in the two extreme phenotypes (Table 17–1). Ultimately all but the two extreme phenotypes disappear. If heritability is not complete so that some errors of classification are made, progress is slower, and the ultimate result is an equilibrium in which some intermediate phenotypes remain in the population. Just as is true of one gene, mating together individuals which are alike in having extreme phenotypes results in concentration

TABLE 17–1.   Changes in Phenotypic Proportions in a Population in Which Perfect Assortative Mating of Like to Like is Carried Out for a Character Governed by Two Equally Effective Genes, $q_A = q_B = 0.5$

(Modified from Lush, 1943)

Genotypes	aabb	Aabb aaBb	AAbb aaBB AaBb	AABb AABb	AABB
Phenotypes	1	2	3	4	5
	Per Cent	Per Cent	Per Cent	Per Cent	Per Cent
Original population	6.2	25.0	37.5	25.0	6.2
Generation 1	13.6	20.8	31.2	20.8	13.6
Generation 2	19.3	16.5	28.4	16.5	19.3
Generation 3	23.9	13.8	24.6	13.8	23.9
Generation ∞	50.0	0.0	0.0	0.0	50.0

of the population at the extremes, an increase in homozygosity, an increase in the likeness of close relatives, and enhancement of the variability of the population (provided both extremes are kept). The main difference between the one- and two-gene cases is in the rate of progress toward the final result, progress being slower with two genes (when heritability is equal).

The effect of further increases in the number of genes governing the character, heritability remaining constant, is progressively slower progress. So long as heritability is complete so that perfect assortative mating is possible, there can be an approach to fixation ($F = 1$), but it will be very slow unless the number of genes involved is small. When heritability is incomplete so that mistakes of classification are made, F cannot reach 1. The interrelations of these two factors, gene number and heritability, in affecting progress toward homozygosity are illustrated in Table 17–2.

For economically important characters, gene numbers are likely to be large and heritability low. Mating like to like is therefore likely to exert the full limit of its effects within a few generations and is unlikely to produce any real fixation of type. It does, however, tend to increase the resemblance between close relatives, and it is a powerful tool in increasing extreme diversity in a population. Whereas inbreeding tends to fix intermediate as well as extreme families, phenotypic assortative mating tends to concentrate the population at the extremes to the exclusion of intermediate types. Once random mating replaces

TABLE 17–2. Percentage of Heterozygosity in Successive Generations of Mating Like to Like for Various Numbers of Genes and Various Heritabilities

(After Wright, 1921)

| Gener-ation | Perfect Assortative Mating Heritability, 100 Per Cent | | | Imperfect Assortative Mating | | | | Perfect Disassortative Mating Heritability, 100 Per Cent | |
| | | | | Heritability, 80 Per Cent | | Heritability, 50 Per Cent | | | |
	1 Gene	4 Genes	10 Genes	1 Gene	4 Genes	1 Gene	4 Genes	1 Gene	4 Genes
0	0.500	0.500	0.500	0.500	0.500	0.500	0.500	0.500	0.500
1	0.250	0.438	0.475	0.300	0.450	0.375	0.469	0.750	0.563
2	0.125	0.406	0.462	0.220	0.430	0.344	0.461	0.625	0.531
3	0.063	0.379	0.451	0.188	0.414	0.336	0.456	0.687	0.535
4	0.031	0.354	0.439	0.175	0.402	0.334	0.453	0.625	0.533
5	0.016	0.330	0.428	0.170	0.391	0.333	0.450	0.672	0.533
10	0.001	0.233	0.376	0.167	0.357	0.333	0.445	0.667	0.533
15	0.000	0.165	0.330	0.167	0.343	0.333	0.444	0.667	0.533
∞	0	0	0	0.167	0.333	0.333	0.444	0.667	0.533

The right-hand columns are for disassortative mating. No dominance is assumed.

phenotypic assortative mating, these effects are dissipated rapidly, except as selection accompanying assortative mating may have permanently altered gene frequencies.

Disassortative Mating

Genetic disassortative mating is rarely practiced in closed populations of the type we have been considering. Its only real application in breeding is in connection with the crossing of different strains, and we will therefore postpone discussion of this system of mating until we take up varietal and species crosses in later chapters.

Phenotypic disassortative mating, on the other hand, finds some applications in closed populations. It is practiced to compensate for defects by choosing contrasting parents each of which compensates for weaknesses of the other. It may also be used when the desired type is an intermediate and existing types are too extreme in opposing directions. Phenotypic disassortative mating is probably most useful in plant breeding in maintaining diversity in populations which serve as sources of genes. For example, drift can rapidly dissipate the diversity of an open-pollinated variety of corn if a small population is allowed to mate at random. But the mating of unlike types in the population counteracts this erosion of diversity and helps to maintain the usefulness of the population as a source of genes.

The reasons for these uses of phenotypic disassortative mating become obvious when one considers the genetic implications of the mating of unlikes. First, it can be seen from Table 17–2 that this system of mating tends to maintain heterozygosity in a population, and that, when highly heritable characters are governed by few genes, it may even increase heterozygosity slightly above the level expected with random mating. Second, phenotypic disassortative mating tends to decrease the population variance, since the offspring of opposing extreme types tend to be nearer the population mean than the offspring of random matings. Finally, the mating of unlikes tends to reduce the genetic correlation between relatives. Disassortative mating is therefore the most conservative of the mating systems and the one which best holds a population together.

General Considerations

The major factors influencing the effect of different systems of mating on the genetic composition of populations are gene number and heritability. With production characters, which are likely to be governed by many genes and are also likely to be of low heritability, the effects of the various mating systems can be summed up as follows:

Random mating changes neither gene frequency, variability of the population, nor the genetic correlation between close relatives. Random mating with selection tends to change gene frequencies, and hence the mean of the population, but it has little effect on homozygosity, the population variance, or the genetic correlation between relatives.

Close inbreeding breaks populations up into numerous noninterbreeding groups and automatically brings about fixation of type (homozygosity) and an increase in prepotency. Under inbreeding without selection, genetic variance is increased. But inbreeding with selection favoring only one type of homozygote reduces genetic variance to zero as the type becomes fixed. Close inbreeding is the only effective method of bringing to light hereditary differences in characters which are determined largely by factors other than heredity.

Phenotypic assortative mating tends to concentrate the population toward the two extremes but is inefficient in fixation of type or in changing prepotency. It produces most of its effect in a few generations after it is begun, and its effects are likely to disappear almost immediately upon resumption of random mating.

Disassortative mating is the method which best holds populations together.

Each mating system has its advantages and disadvantages for particular purposes. If the goal is the production of homozygous lines, inbreeding is the obvious choice, even though it will almost certainly lead to less than optimum performance. If the goal is the development of an extreme phenotype, assortative mating with selection is appropriate. Random mating has its place in plant breeding, for example, in progeny testing. So also does disassortative mating, for example, in maintaining source populations in which maximum stability is the goal.

REFERENCES

Lerner, I. M. 1950. *Population genetics and animal breeding.* Cambridge University Press.
Li, C. C. 1955. *Population genetics.* The University of Chicago Press.
Lush, J. L. 1943. *Animal breeding plans.* Iowa State College Press.
Cruden, D. 1949. The computation of inbreeding coefficients. *Jour. Hered.* 40: 248–251.

Wright, S. 1921. Systems of mating. I. The biometric relations between parent and offspring. *Genetics* 6: 111–123.

Wright, S. 1921. Systems of mating. II. The effects of inbreeding on the genetic composition of a population. *Genetics* 6: 124–143.

Wright, S. 1921. Systems of mating. III. Assortative mating based on somatic resemblance. *Genetics* 6: 144–161.

Wright, S. 1921. Systems of mating. IV. The effects of selection. *Genetics* 6: 162–166.

Wright, S. 1921. Systems of mating. V. General considerations. *Genetics* 6: 167–178.

Wright, S. 1923. Mendelian analysis of pure breeds of livestock. I. The measurement of inbreeding and relationship. *Jour. Hered.* 14: 339–348.

18

General Features
of Inbreeding Depression
and Heterosis

There has been a tendency since earliest recorded history to associate inbreeding with unfavorable biological effects. This tendency probably arose from more or less casual observations that inbreeding is often accompanied by smaller size, lessened vigor, reduced fecundity, an increased number of defectives, and a general weakening of the stock and, on the contrary, that crossing with an unrelated stock results in a restoration of vigor. One of the results of this attitude toward inbreeding has been the pronouncement of forbidden degrees which prohibit marriages between relatives of specified closeness of kinship.

Inbreeding has not, however, been uniformly regarded as something to be avoided. There is, for example, ample evidence that consanguineous matings were favored by the ancient Egyptians, Hebrews, Greeks, and Nordic tribes, sometimes for the purpose of preventing the dilution of "superior" bloodlines, but other times for the purpose of maintaining property within families. Another example of deliberate, widespread, and long-continued inbreeding is provided by the development of modern breeds of livestock. Starting about 1700, it became common practice to mate outstanding bulls to their daughters and granddaughters. Although great advancements were made, the inbred stock sooner or later decreased in fecundity, and it became necessary to outcross to maintain lines. The system thus became one of inbreeding to concentrate desirable qualities, coupled with crossbreeding to prevent degeneration, followed by more inbreeding. Selection for type was practiced at all times. Darwin (1868) described this work in detail, concluding, "although freecross-

213

ing is a danger on one side which everyone can see, too close inbreeding is a hidden danger on the other."

Our knowledge of the effects of outcrossing in plants dates from the experiments of the plant hybridists of the eighteenth and nineteenth centuries. The observations of Koelreuter in 1763 and Sprengel in 1793 are particularly significant. Koelreuter recognized that hybrids were often possessed of the most striking and unusual vigor. Sprengel studied the relation between flowers and insects in great detail, reaching the conclusion that nature usually intended that flowers should not be pollinated by their own pollen and that peculiarities of flower structure can be understood only when studied in relation to the insect world. Darwin, in his *Cross and Self Fertilization in the Vegetable Kingdom,* which appeared in 1876, reviewed the voluminous earlier literature and also recorded his own experiments with various species. Most of these studies indicated that the offspring arising from self-fertilization were less vigorous than those obtained from cross fertilization. Darwin, impressed by the numerous structural and functional adaptations by which plants with hermaphroditic flowers avoid self-fertilization, was led to the conclusion that self-fertilization was an unnatural and harmful process. The observations upon which much of the breeding of cross-pollinated crops is based had thus been made, although not understood, well before the start of the twentieth century. In summary, these observations were that: (1) inbreeding often leads to loss of vigor and other evidences of deterioration; (2) hybridization between unlike types is often accompanied by great vigor; and (3) crossbreeding must be important biologically because so many species go to elaborate lengths to encourage cross fertilization.

Before we proceed to consider the breeding of cross-pollinated species, discussion of the interrelation between inbreeding depression and hybrid vigor would be desirable. In this chapter we shall review the observational facts of inbreeding and heterosis. In Chapter 19 we shall attempt explanation of these facts in terms of the genetic mechanisms involved.

Experiments in Inbreeding

We have already noted that the deterioration of populations subjected to continued consanguineous mating had been observed long before the era of Mendelian genetics. The evidence that inbreeding in cross-pollinated species leads, on the average, to a loss of vigor and fecundity has since become overwhelming, and a full compilation of the literature on this subject would certainly represent many thousands of titles. Hence only a small sample of the data on inbreeding will be reviewed for the purpose of gaining some appreciation for the variety and magnitude of the responses that have been recorded in representative crop species.

Maize is a logical choice as a first example because this plant has been more widely inbred than any other organism, and the effects of inbreeding upon it are exceptionally well documented. Apparently the first inbreeding experiments with maize were those reported by Darwin in 1876. This inbreeding was, however, continued only a single generation, and the results were consequently not very informative. It was not until near the end of the first decade of the twentieth century that precise data on the effects of inbreeding maize became available as a result of the independent work of East (1908) and Shull (1909). The more important effects of continued self-fertilization reported by these investigators can be summarized as follows: (1) A large number of lethal and subvital types appear in the early generations of selfing. (2) The material rapidly separates into distinct lines, which become increasingly uniform for differences in various morphological and functional characteristics—for example, height, ear length, and maturity. (3) Many of the lines decrease in vigor and fecundity until they cannot be maintained even under the most favorable cultural conditions. (4) The lines that survive show a general decline in size and vigor.

This decline is illustrated in Figure 18–1, where the height and grain yield of three of East's lines are shown for 30 generations of inbreeding. Notice that fixation for height occurred after 5 generations of inbreeding but that yield continued to decline for at least 20 generations. The yielding ability of these 3 lines was finally reduced to about one third of that of the open-pollinated variety from which they were derived. Kiesselbach estimated that about 100,000 inbred lines of maize had been produced in the United States by 1950. The number is certainly substantially greater now. The best of these lines yield only about half as much as the open-pollinated varieties from which they were derived, and the remainder grade down to ones so weak they are scarcely able to survive. Even these surviving lines give a biased estimate of the effects of inbreeding on maize, because many potential lines are lost during the inbreeding process.

Despite the conspicuous deterioration of maize upon continuous selfing, this species is nevertheless more tolerant of inbreeding than alfalfa. Upon selfing open-pollinated plants of this leguminous species, a veritable rain of subvital and lethal types appear. Furthermore,

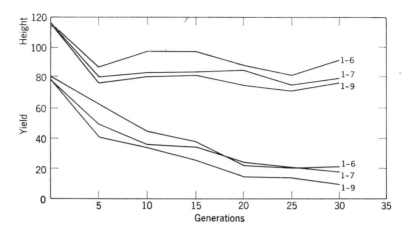

FIGURE 18-1. A comparison of three lines of maize, derived from the same variety, self-fertilized for thirty generations. Height of stalk is measured in inches and yield of grain in bushels per acre, both plotted on the same scale. There were originally four lines, but it became impossible to maintain one of them beyond twenty generations of inbreeding. (After Jones, *Genetics* 24:462. 1939.)

FIGURE 18-2. Heterosis followed by decline in height after inbreeding in maize. Representative plants of the two parental inbred lines are on the left, followed by the F_1 hybrid, then the F_2 to F_8 selfed generations toward the right. (After D. F. Jones, *Genetics* 9:405–418. 1924. Photograph courtesy Connecticut Agricultural Experiment Station.)

the rate at which alfalfa deteriorates in general vigor and in productivity upon continued selfing is appalling. The number of lines that can be maintained beyond the third selfed generation is small, and the small percentage of lines that manage to survive are greatly reduced in forage yield, as shown below:

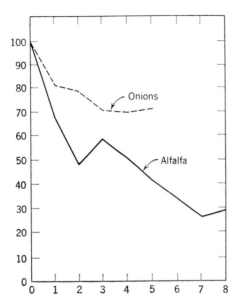

FIGURE 18-3. Yields of self-fertilized lines of alfalfa and onions in per cent of open-pollinated parental varieties. The curve for alfalfa is obviously biased because the later generations are represented by only a few exceptional lines which survived more than three generations of inbreeding. Elimination was less severe in onions, so the bias is not as great in that case. (Based on data of Tysdal et al., 1942, and Jones and Davis, 1944.)

A few species are apparently even more intolerant of inbreeding than alfalfa, for example, the hayfield tarweed investigated by Babcock (unpublished). In that species inbreeding uncovers large numbers of morbid types—albinos, dwarfs, rootless seedlings, floral abnormalities that prevent reproduction, and the like. Reduction in vigor and fecundity is so great that many lines do not survive more than two generations of sib mating, the closest form of inbreeding possible in this self-incompatible species.

Among cultivated plants, the carrot is another species that deteriorates drastically upon inbreeding.

The onion is a normally cross-pollinated species that is quite tolerant of inbreeding. In this species some varieties show no apparent reduction in vigor on continuous selfing whereas others show slight to moderate inbreeding depression. Few deleterious recessives appear upon artificial selfing in onions, and lines only rarely decline in vigor to the point where they are difficult to maintain. The average reduction in bulb weight observed over a number of varieties is shown in Figure 18–3. Taking into account the biases caused by the differences in mortality of inbred lines, we can see from comparison of Figures 18–1 and 18–3 that onions are much less depressed in vigor by inbreeding than alfalfa or maize.

Some other cross-pollinated species that are fairly tolerant of inbreeding are sunflowers, rye, timothy, smooth bromegrass, and orchardgrass. With all these species the number of recessive abnormals appearing on inbreeding appears to be somewhat less than in maize. Also in all or most of these species some extremely uniform lines as thrifty or nearly as thrifty as the open-pollinated stocks from which they are derived are found after several generations of inbreeding.

Although the majority of higher plants show inbreeding depression of greater or lesser degree, there are some species in which inbreeding can be carried on indefinitely with seeming impunity. The self-pollinating species obviously fall in this category, but so also do a number of the normally cross-fertilizing species of plants. The cucurbits, although monoecious and cross-pollinated, have already been mentioned as an example of a group of species in which certain lines appear to lose little vigor on inbreeding. Hemp, a dioecious species, is another example.

Not all families in the animal species with which careful inbreeding experiments have been carried out show a decrease in size and vigor. In the strain of rats studied intensively by King, some families developed by twenty-five generations of brother-sister matings compared favorably with crossbred stock maintained as controls. Fully vigorous lines of mice can also be maintained by continued brother-sister mating. In *Drosophila*, inbreeding usually results in a rapid loss of vigor, but some strains compare favorably with outbred populations after long-continued inbreeding. The classic study and analysis of the effects of inbreeding were done with the guinea pig; they were carried out by a succession of workers in the United States Department of Agriculture. The results were typical in showing a decline in vigor under inbreeding, but Wright's brilliant analysis made it clear that sufficient selection pressure for fitness can overcome the deleterious effects of inbreeding. Thus the effect of selection, conscious or unconscious, may be important in those experiments in which no apparent decline in fitness accompanies inbreeding.

Heterosis

Hybrid vigor, or *heterosis*, can be regarded as the converse of the deterioration that accompanies inbreeding. The beneficial effect of crossing is, however, a more widely recognized phenomenon than inbreeding depression because it is observed in nearly all F_1 hybrids between parents that are neither closely nor distantly related. Even species that seem to show little or no harmful effects from inbreeding frequently benefit from crossing. It is not surprising, therefore, that the early plant hybridists noticed and emphasized hybrid vigor more than inbreeding depression. This is understandable when one considers that the beneficial effects of crossing appear immediately in the F_1 and have their maximum expression in that generation, whereas inbreeding depression is less noticeable because its effects are delayed and offset by selection. At any rate the plant hybridists of the eighteenth and nineteenth centuries described hundreds of heterotic crosses, and at present the literature on this subject would strain a work of monumental proportions. It will again serve our purpose merely to sample the literature in order to gain some appreciation of the types and the magnitude of heterotic effects in different species.

In maize, the first studies on artificial hybridization for which yields were reported were those of Beal in the period 1877–1882. Beal made hybrids between different open-pollinated varieties, and although he did not give detailed data, he stated that the yields of the hybrids were larger than those of the parents by as much as 40 per cent. Subsequently many other studies of intervarietal hybrids were made. By 1922, when Richey summarized the results, he reported that in 244 comparisons of hybrids with their open-pollinated parents, 82 per cent of hybrids yielded more than the average of the parents, and 18 per cent yielded less. The recent data of Robinson et al., based on replicated experiments conducted at 3 locations in 2 seasons, give a more accurate picture of the precise degree of heterosis in varietal hybrids of maize than the usually unreplicated experiments of the earlier workers. In this study (Table 18–1) 12 out of 15 crosses exceeded their high parent in yield, and the mean of all hybrids relative to their high parent was 111.5 per cent.

The summary of the earlier work by Richey emphasized one of the important aspects of intervarietal crosses in maize. When more or less haphazard crosses were made between varieties, the chances were

TABLE 18–1.   Yield of Open-Pollinated Maize Varieties in Bushels (First Column) and in Per Cent of the Mean of All Open-Pollinated Varieties (Diagonal)

(After Robinson et al., 1956)

Varieties	Yield (Bu./Acre)	Per Cent					
		Jarvis	Indian Chief	Weekley	Simpson	Biggs	Latham
Jarvis	43.4	110.7	129.1	119.3	116.6	106.6	109.9
Indian Chief	38.0	(121.2)	96.9	127.5	124.1	132.2	146.2
Weekley	44.9	(117.3)	(117.8)	114.5	115.4	104.6	110.0
Simpson	40.4	(112.6)	(120.5)	(109.6)	103.1	109.2	125.0
Biggs	38.0	(100.0)	(132.2)	(96.6)	(106.1)	96.9	123.0
Latham	30.6	(93.7)	(131.9)	(92.5)	(109.8)	(111.0)	78.1
Mean	39.2						

Yield of hybrids in per cent of the average yield of the two parents (above diagonal) and in per cent of their higher-yielding parent (below diagonal).

Standard error of the percentage difference between each cross and its midparent, 7.3 per cent.

Mean of crosses relative to midparent, 119.9 per cent.

Mean of crosses relative to high parent, 111.5 per cent.

found to be about equal of obtaining a cross that was better or not so good as the better parent. But when the parents differed considerably in type, the yields of the hybrids were with few exceptions substantially greater than those of the better parent. In general the crosses that produced the largest actual yields were those between parents that were themselves high yielding and represented contrasting combinations such as flint × dent, or flour × dent. The data of Hayes and Olsen (Table 18–2) give an idea of the difference in performance of varietal hybrids. The yields are expressed in percentage of Minnesota 13, a dent variety that was the staminate parent of all the crosses. Note that the greatest increases in yield came in the dent × flour and dent × flint crosses. Thus, crosses between parents of presumably different origins gave greater heterosis than crosses between parents that were presumably more closely related.

When crosses are made between inbred lines in maize, the degree to which vigor and productivity are restored has also been found to be a function of the origin of the lines. Shull (1909) found that crosses between sibs within a self-fertilized line show little improvement over self-fertilization in the same family, and that the difference between selfing and sib crossing becomes progressively less as the number of generations of inbreeding increases. When, however, inbreds

TABLE 18–2. Comparison of Yield of Parental Varieties and Their F_1 Hybrids with Minnesota 13, a Dent Variety

(After Hayes and Olsen, 1919)

Pistillate Parent	Yield as Percentage of Minnesota 13 (Staminate Parent)	
	Pistillate Parent	Hybrid
Flour		
Blue Soft	96.7	132.5
Flints		
Smutnose	110.8	127.6
King Phillip	100.0	119.9
Longfellow (NR)	100.5	119.6
Longfellow (Bwls)	104.9	114.1
Mercer	91.8	97.3
Dents		
Northwestern	105.9	115.7
Chowen	99.0	114.9
Rustler	112.1	112.4
Minnesota 23	96.7	110.9
Silver King	100.9	106.7
Murdock	80.3	101.6

All data are expressed in percentage of a standard variety.

derived originally from different open-pollinated plants are crossed at random (including selfing), the average response is a return to the vigor and productivity of the original stock before the inbreeding was commenced. Some hybrid combinations regularly show greater improvement than the average, and others regularly show less improvement than the average. But there is no overlapping of the hybrids as a group and the inbred parents as a group, the hybrids in all cases being superior to the inbreds.

It has subsequently been shown that hybrids between inbreds of diverse parentage generally give greater hybrid vigor than hybrids between inbreds derived from the same or similar open-pollinated varieties. This subject is directly related to the development of hybrid varieties, and detailed consideration will therefore be deferred to the chapter on that subject.

Responses to hybridization in other species are not as well known as in maize, but the information that is available suggests that the range of responses is more or less comparable. In onions, for example, Jones and Davis (1944) found that two intervarietal hybrids produced much larger yields than the better of the parental varieties as measured in replicated trials at five locations in California. Many intervarietal hybrids in onions, however, produce no greater yields than the parental varieties. Very few precise data are available for the onion regarding the yielding ability of inbred lines compared to single crosses between inbred lines. This lack of data apparently stems from the realization of breeders that inbreds yield poorly. Consequently they often do not include inbreds in yield trials. Among the fifty or more hybrid varieties of onions now in commercial production, many are not superior to nonhybrid varieties in yielding ability. Often the success of hybrid varieties in onions derives from better keeping ability, more even maturity, and improvement in other specific attributes important in onion production, rather than from superior yield.

Data on hybrid vigor in alfalfa are meager. Tysdal et al. (1942) compared the forage yields of 28 different single-cross hybrids between inbred lines with the parental inbred lines and with standard varieties. Taking the yield of the standard varieties as 100, the F_1 hybrids varied from 60 to 139. The mean of all hybrids was 96, slightly less than the standard varieties used as checks, whereas the parental inbreds yielded only half as much as the average of the checks.

The results of Odland and Noll with cabbage provide another example of the measure of heterosis found in cultivated species. In replicated trials, they obtained the following yields in pounds per plot: All Seasons, 241; Ballhead, 118; F_1 hybrid, 283. In the same test seven standard varieties produced an average of 205 pounds per plot whereas average production from seven F_1 hybrids was 268 pounds per plot. Two inbred lines derived from Penn State Ballhead had head weights of 3.38 and 3.82 pounds whereas their F_1 hybrid produced heads weighing 4.46 pounds on the average.

Various data are available concerning the degree of heterosis in varietal crosses in self-pollinated species such as the small grains, beans, and tomatoes. It will suffice to say here that these data indicate generally smaller responses on the average than those observed in cross-pollinated crops, even though some F_1 hybrids display considerable excess of vigor over the superior parent.

Although the discussion of hybrid vigor up to this point has been centered on increased size and productiveness, it should be emphasized that heterosis can be manifested in many other ways. In certain F_1 hybrids, in beans, for example, the number of nodes, leaves, and pods is greater than in the parents, but the gross size of the plant is not greater. In other hybrids the growth rate is increased but is not accompanied by larger size at maturity. Another manifestation of heterosis is earlier maturity in the F_1 than in either parent, sometimes accompanied by an actual decrease in total plant weight. Other heterotic effects include instances where hybrid organisms have been reported to have greater resistance to diseases and insects, increased tolerance to rigors of climate, and various other manifestations of better fitness.

In this connection, Dobzhansky (1952) has distinguished between luxuriance and true heterosis. Luxuriance—the excessive size and vigor found in some species hybrids and in certain crosses between strains of self-pollinating species—is not necessarily an expression of superior adaptation, but is regarded as an accidental condition with quite different genetic antecedents and implications from true heterosis. A similar distinction between heterosis and luxuriance should also be made in agricultural practice because the largest and most vigorous types are not necessarily the most valuable commercially.

REFERENCES

Beal, W. J. 1876–1882. *Report Michigan State Board Agric.*
Dobzhansky, Th. 1952. Nature and origin of heterosis. In *Heterosis,* pp. 330–335. Iowa State College Press.
East, E. M. 1908. Inbreeding in corn. *Rept. Connecticut Agric. Exp. Sta. for 1907,* 419–428.
East, E. M., and D. F. Jones. 1919. *Inbreeding and outbreeding.* Lippincott Co., Philadelphia.
Hayes, H. K.. and P. J. Olsen. 1919. First generation crosses between standard Minnesota corn varieties. *Minn. Agric. Exp. Sta. Tech. Bull.* 123: 5–22.
Jones, D. F. 1939. Continued inbreeding in maize. *Genetics* 24: 462–473.
Jones, H. A., and G. N. Davis. 1944. Inbreeding and heterosis and their relation to the development of new varieties of onions. *U. S. Dept. Agric. Tech. Bull.* 874: 1–28.
Kiesselbach, T. A. 1951. A half-century of corn research. *Amer. Sci.* 39: 629–655.
Odland, M. L., and C. J. Noll. 1950. The utilization of cross-compatibility and self-incompatibility in the production of F_1 hybrid cabbage. *Proc. Amer. Soc. Hort. Sci.* 55: 391–402.
Robinson, H. F., R. E. Comstock, A. Khalil, and P. H. Harvey. 1956. Dominance versus overdominance in heterosis: evidence from crosses between open-pollinated varieties of maize. *Amer. Nat.* 90: 127–131.
Shull, G. H. 1909. A pure line method of corn breeding. *Rept. Amer. Breeders' Assoc.* 4: 296–301.
Tysdal, H. M., T. A. Kiesselbach, and H. L. Westover. 1942. Alfalfa breeding. *Nebr. Agric. Exp. Sta. Res. Bull.* 124: 1–46.
Zirkle, C. 1952. Early ideas on inbreeding and crossbreeding. In *Heterosis,* pp. 1–13. Iowa State College Press.

19

Genetic Basis
of Inbreeding Depression
and Heterosis

The Dominance Hypothesis

The most widely accepted genetic hypothesis to explain inbreeding depression and the conversely related phenomenon of hybrid vigor had its beginnings within the first decade of the rediscovery of Mendel's work. This hypothesis, proposed by Davenport in 1908, by Bruce (1910), and Keeble and Pellew (1910), starts with the assumption that cross-fertilizing species consist of a large number of genetically different individuals, many of which carry deleterious recessive genes concealed in heterozygotes. When such individuals are inbred, there is an increase in homozygosity and various morbid homozygous recessive types appear—plants that lack chlorophyll, have abnormal flowers, defective seeds, and the like—so that they cannot reproduce themselves and are eliminated. Other characters also come to light that do not lead to direct extinction—deficient root systems, dwarfness, and partial chlorophyll deficiency—but nevertheless cause severely handicapped development. These same characters are also observed in small numbers in open-pollinated populations, but not at frequencies which reduce productivity severely. The increase in their frequency on inbreeding appears to provide an explanation for part of the injurious effects of inbreeding. Unfavorable genes segregate on inbreeding and, upon fixation brought about by homozygosity, produce lines that possess different genes or gene complexes. Some lines receive more of the favorable genes than others, accounting for differences observed in the degree of inbreeding depression in different lines. Inbreeding depression is thus not a process of degeneration but a consequence of

224

Mendelian segregation. The injurious effects of inbreeding are not produced by the *process of inbreeding itself*, as believed by many early biologists (including Darwin), but are directly related to the number and kinds of Mendelian characters heterozygous in the original population.

Under this hypothesis, the intercrossing of inbred lines should lead to the formation of hybrids in which deleterious recessives contributed by one parent are again hidden, as in the original open-pollinated stock, by dominant alleles contributed by the other parent. The precise degree of response to crossing should therefore be a function of the genotypes of particular inbreds. Some genotypes should complement each other nicely to produce hybrids better than the average of the original open-pollinated variety, whereas some will not "nick" well by virtue of the particular combination of dominant and recessive genes they happened to receive during segregation. This hypothesis seemed to be consonant with the observed facts and appeared to explain the effects of inbreeding and outcrossing in their relation to one another.

Objections to this hypothesis were raised on two grounds. First, if the hypothesis is correct, it should be possible to obtain individuals homozygous for all the dominant factors. Such lines should be like the F_1 in vigor, but they should be true breeding. High-yielding homozygous lines were not found, however. Jones (1917) reconciled this apparent discrepancy with the hypothesis by pointing out that many genes probably affect growth and that each chromosome would be expected to contain several of these genes. A single linkage group would be expected to include some favorable dominants and some unfavorable recessives. Thus a series of precisely placed crossovers would be required to obtain all of the dominant alleles in one gamete.

The other objection was directed at the symmetrical distributions that were observed for heterotic characters in the F_2. It was pointed out that if heterosis were due solely to dominance of independent factors, the F_2 distribution curve should be skewed, rather than symmetrical, since the dominant and recessive phenotypes would be distributed according to the expansion of the binomial $(\frac{3}{4} + \frac{1}{4})^n$. Jones was also able to reconcile the symmetrical distributions actually observed with the hypothesis on the basis that linkage between groups of favorable and unfavorable genes would lead to theoretically symmetrical distributions. In 1921 Collins pointed out that even in the absence of linkage, skewness would be difficult to detect and moreover the chances of recovering a completely homozygous type would be remote so long as the number of genes involved were at all large.

These several ideas came to be called the *dominance* or the *dominance of linked genes* hypothesis.

The Overdominance Hypothesis

An alternative hypothesis, also proposed in the first decade following the rediscovery of Mendel's work, has competed with the dominance hypothesis through the years. This hypothesis, proposed independently by Shull and East in 1908, assumed there is a physiological stimulus to development that increases with the diversity of the uniting gametes. In Mendelian terms, this means that there are loci at which the heterozygote is superior to either homozygote and that vigor increases in proportion to the amount of heterozygosis. This idea has been called single-gene heterosis, superdominance, cumulative action of divergent alleles, and stimulation of divergent alleles. The term most commonly applied to this idea, however, is overdominance, and the hypothesis to which it applies is generally known as the *overdominance hypothesis of heterosis*.

Reduced to simple terms, this hypothesis supposes that the heterozygous combination a_1a_2 of the alleles at a single locus is superior to either of the homozygous combinations a_1a_1 or a_2a_2. The implication is that a_1 and a_2 perform different functions and that the sum of their different products is superior to the single product produced by either allele in the homozygous state. East in 1936 elaborated the idea further by proposing a series of alleles $a_1, a_2, a_3, a_4, \ldots$ of gradually increasing divergence of function. Heterozygotes were postulated to become increasingly more efficient as their component alleles diverged more and more in function. Thus $a_1a_2 < a_1a_3 < a_1a_4$, and so forth.

At the time when East and Shull first formulated this hypothesis, there was no direct evidence of any locus at which the heterozygote lay outside the range of the homozygotes. Now, however, there is at least some evidence that different alleles at the same locus can indeed do different things (references in Whaley, 1952). For example, with certain of the R alleles in maize the heterozygote is more pigmented than either homozygote, and in animals members of multiple allelic series have been found that produce different blood antigens. The evidence most directly related to the case in point, however, is that of Gustafsson who worked with spontaneous mutants in pure lines of barley. He reported at least two instances where heterozygotes for chlorophyll mutants produced more and larger seeds than homozygous normal plants. Nevertheless, clear-cut cases of single-gene heterosis are rare and this has been a deterrent to general acceptance of this hypothesis.

In most situations, the dominance and overdominance hypotheses lead to exactly the same expectations. With the dominance hypothesis, the decline in vigor is expected to be proportional to the decrease in heterozygosity regardless of the relative number of dominant and recessive genes and of the degree of dominance. The same decline in vigor with increasing homozygosity is expected with overdominance. In both hypotheses, outcrossing should lead to a recovery in vigor. The chief point of difference lies in the impossibility of obtaining homozygotes as vigorous as heterozygotes if single-gene overdominance is important in heterosis; evidence on this point, however, will be extremely difficult to obtain. Despite extensive experimentation over more than forty years, conclusive evidence favoring one or the other of these hypotheses has not been forthcoming. There are, in fact, no good reasons for believing that both systems cannot operate simultaneously in producing heterotic effects. If these two hypotheses are not mutually exclusive, neither do they together exhaust all of the possibilities. There is every reason to believe that the genes concerned with heterotic effects may be fully as complex in their interactions as the qualitative genes of classical genetics. If this is indeed true, then we must expect all sorts of complex interactions in heterosis and must entertain the possibility that both hypotheses, at least in their simplest form, are gross oversimplifications of the actual situation.

Population Structure and Heterosis

Inbreeding is regularly deleterious in some species (alfalfa, maize), sometimes deleterious in others (onions, sunflowers), and has little or no ill effects in still others (cucurbits and the self-pollinated species). Some investigators believe these differences depend on the mating system and the influence it has on the kind, the amount, and the way genetic variability is regulated in populations.

In species that reproduce by self-fertilization, homozygosity is the normal situation for most genes. Recessive deleterious mutants become homozygous soon after their origin and are eliminated promptly. Accordingly, these species become adapted to homozygosity and develop a genetic organization that Mather (1943) calls *homozygous balance*. One of the attributes of this genetic organization is that inbreeding leads to no diminution in vigor. At the same time F_1 hybrids between lines are fully vigorous and may even display heterosis. This is explained on the basis that inbreeding species have descended from outbreeders. During the outbreeding portion of their evolutionary history these species were presumably endowed with the type of genetic organization that characterizes heterozygous populations. Thus inbreeding species have not only good *homozygous balance* but also good *heterozygous balance* because inbreeding fixes some of the heterozygous balance at the same time that it leads to the rise of homozygous balance. According to this view heterosis in self-pollinated species is a relic of an outbreeding past. These ideas have been discussed by Mather (1943) and Darlington and Mather (1949). Stebbins (1950) has presented evidence to indicate that inbreeders are often descended from outbreeding ancestors.

The type of population structure characteristic of self-pollinated species may also exist in some cross-pollinated species that have comparatively small effective population sizes. In some domesticated crops, for example, a single farmer may require only a few plants to satisfy his needs. Thus effective population size may be restricted to a few interbreeding individuals. Under such circumstances, many genes could be expected to be homozygous most of the time (Chapter 17). Homozygous balance could therefore develop in such populations. This may conceivably explain why some outcrossing species such as sunflowers and cucurbits do not show much inbreeding depression.

It has been shown many times that populations that suffer inbreeding depression carry large numbers of recessive genes sheltered in heterozygotes. In *Drosophila pseudoobscura,* for example, nearly every fly tested has at least one concealed lethal, and lethals are by no means the only component of the load of unfavorable recessives carried by the species. The load of mutations in open-pollinated varieties of alfalfa and maize may be as great. There is nevertheless considerable doubt in the minds of some investigators whether heterotic responses of the size that have been observed in various species can be accounted for satisfactorily on the basis of the dominance hypothesis. First of all, there are no convincing data to indicate that the commonly observed mutants have much effect on yield. Wentz and Goodsell (1929), for example, compared the yields and the frequencies of seed, seedling, and mature-plant recessives in nineteen corn varieties and were able to detect no relationship. Experiments by Woodward (1931) also have a bearing on this point. A large number of vigorous plants of three open-pollinated varieties of maize were self-pollinated, and the resulting S_1 lines classified for mutants. Selfed ears from lines free from defectives were then composited to reconstitute these varieties and the yields of the "purged" stocks compared with the yields of the original stocks. The yields were found to be almost exactly the same.

Crow (1948) attempted to calculate the maximum heterosis possible under the dominance hypothesis and reached the conclusion that if all recessives were replaced by their dominant alleles, the vigor of an equilibrium population would be increased by about 5 per cent. This can be interpreted as the average maximum improvement in vigor, as measured in terms of selective advantage, that could occur because of hybridization. Sprague (1946) believes this expectation to be in good agreement with data on varietal hybrids in maize. Crow concluded that

> . . . it seems probable that [the dominance hypothesis] may explain a major part of the loss in vigor with close inbreeding of random mating strains and its recovery on crossing. On the other hand . . . it cannot account for more than a small increase in the vigor of hybrids whose parents are from populations which are at equilibrium. Also it cannot account for increase in vigor following the crossing of artificially inbred strains much beyond the level of the equilibrium population from which the hybrid lines were derived.

Crow also reached the conclusion from a similar computational procedure that "it would not require very many loci in which the heterozygote is superior to give a considerable selective advantage to a hybrid heterozygous for these loci . . . if as many as one percent of the gene loci were of this type, their effect on the population would be greater than all the loci at which there is a detrimental recessive. . . ."

Brieger (1950) has also applied principles of population genetics to the problem of heterosis. He reached the following conclusions: (1) ". . . there exists no effective breeding system which would tend to accumulate a sufficient number of recessive subviable or lethal mutants, as required by the dominance hypothesis"; (2) ". . . the dominance hypothesis is not sufficient to explain all the results of inbreeding"; and (3) "the hypothesis of heterotic gene interaction is capable of explaining in a satisfactory manner all known facts."

The idea of superior heterozygotes has also been advanced by Hull (1945). His argument for overdominance is a simple one, namely, that usually the hybrid between inbred lines of maize yields more than the sum of the two inbreds. This would be possible with dominant genes acting in a completely additive fashion only if inbreds with no favorable dominants had a negative yield. The validity of this argument depends upon additivity of effects between loci—that is, that interallelic interactions are of no consequence.

The Nature of Heterotic Loci

Accumulation of evidence that the dominance hypothesis, at least in its simplest form, is inadequate to account for the heterosis actually observed in various species has led to increased interest in the nature of "heterotic loci" and their relation to the selective advantage of heterozygotes, the production of balanced phenotypes, and the development of homeostatic systems (Lerner, 1954a; Dobzhansky and Wallace, 1953; Lerner, 1954b; Dobzhansky, 1956; Thoday, 1956). The earliest view concerning the nature of heterotic loci was independently formulated in 1908 by East and by Shull in rather vague physiological language. Shull's view was that "hybridity itself—the union of unlike elements, the state of being heterozygous—has, according to my view, a stimulating effect on the physiological activities of the organism." This idea, restated in modern genetic terminology by East in 1936, attaches to heterozygotes a versatility of development that is expressed in terms of alternative pathways in metabolism. These presumably arise in biochemical versatility stemming from the existence of different alleles at the same locus. Since this versatility is not open to homozygotes, it must be associated with particular loci, which can then be called heterotic loci. A necessary corollary of this view is that heterozygosity is essential for vigor.

The other hypothesis attaches no special physiological virtue to the heterozygous state. It emphasizes, rather, the genetic advantage of heterozygosity in providing flexibility by permitting segregation. Mather (1956) in particular has supported this concept stating that ". . . such physiological arguments are indeed untrustworthy guides to understanding genetical situations . . . heterozygotes are not always more stable in development. Such superior stability as they show in outbreeding species has been imposed on them by natural selection; it is not innate in the state of heterozygosis, for it is not found as a regular feature in inbreeders." Thus the first hypothesis (East, 1908; Shull, 1908; East, 1936; Robertson and Reeve, 1952; Haldane, 1954) leads to the view that outbreeding mechanisms result from the physiological advantages of heterozygotes. The second hypothesis reverses the causal sequence and makes the physiological virtues observed in many heterozygotes the consequence of selection under an outbreeding system. According to this second hypothesis, "relationally balanced" (Mather, 1943) or "co-adapted chromosomes" (Dobzhansky and Wallace, 1953) are produced as a result of long-term evolutionary development of balanced complexes of genes integrated into the population gene pool (Chapter 16). Heterozygosity is thereby assigned only its classic genetic virtue of providing for segregation—mating systems involving outcrossing and heterozygous balance follow as secondary features. Once established, the heterozygous balance will help to maintain the outbreeding system and endow the population with genetic homeostasis (Lerner, 1954a, 1954b).

Exactly what a "heterotic locus" represents in terms of physical structure is a matter of speculation. Studies of pseudoallelism (Green and Green, 1949; Lewis, 1951) suggest that the unit of specific physiological action is often not a single locus in the classical sense but a complex of tightly linked loci. Additional evidence for this type of physical structure is provided by the "complex loci" governing incompatibility and various morphological characteristics in heterostyled plants (see Darlington and Mather, 1949). Mather (1952) believes no distinction is possible between complementary action of different genes and complementary action of alleles except for the degree of linkage. This was in a way implied by Jones in 1917 when he formulated his hypothesis of linked dominant factors; with such close linkage between dominant and recessive genes that crossing over practically never occurs, the situation becomes one of heterozygosis of chromosome regions instead of single genes. The inversion heterozygotes of *Drosophila* are cytological devices that cause large numbers of loci to behave as units in inheritance. These units often appear to have the properties of "heterotic loci." The balanced lethal systems of *Oenothera* represent the ultimate in such systems because, in the extreme, the entire karyotype of a population may consist of only one "locus" with two "alleles."

Many investigators now consider the action of such heterotic complex loci to be equivalent to that of complementary genes in producing the specific effects that contribute to heterosis. This viewpoint usually assumes further that the specific effects of these complex loci depend also on the genetic background in which they act, as Mather (1943) showed for the incompatibility locus in heterostylic *Primula*.

We are thus brought back to the same picture of population structure to which we were led by the results of the selection experiments discussed in Chapter 16. In this view the population genotype is complexly integrated, and the formation of complexes of genes, held together by various genetic devices, is an important factor in the balance between the components of the population genotype. The fate of a gene or a gene complex depends chiefly on its effect *in combination* with other genes, or gene complexes, in the common gene pool. The population genotype is complexly interrelated with the reproductive biology of the species and especially with effective population size. As a result of many different environmental and genetic factors influencing the reproductive biologies of different species over long periods of time the population genotype may have acquired different organizations in different species. The differences observed in mutational loads, degree of inbreeding depression, heterosis, and response to selection depend on these variations in the amounts and kinds of gene action in different species or in different populations in the same species. It is not surprising that so many different ideas exist for the explanation of heterosis. It has no single or simple explanation.

REFERENCES

Brieger, F. G. 1950. The genetic basis of heterosis in maize. *Genetics* **35**: 420–445.

Bruce, A. B. 1910. The Mendelian theory of heredity and the augmentation of vigor. *Science* **32**: 627–628.

Collins, G. N. 1921. Dominance and the vigor of first generation hybrids. *Amer. Nat.* **55**: 116–133.

Crow, J. F. 1948. Alternative hypotheses of hybrid vigor. *Genetics* **33**: 477–487.

Crow, J. F. 1952. Dominance and overdominance. In *Heterosis,* pp. 282–297. Iowa State College Press.

Darlington, C. D., and K. Mather. 1949. *The elements of genetics.* The Macmillan Co., New York.

Davenport, C. B. 1908. Degeneration, albinism and inbreeding. *Science* **28**: 454–455.

Dobzhansky, Th., and B. Wallace. 1953. The genetics of homeostasis in *Drosophila. Proc. Nat. Acad. Sci. Wash.* **39**: 586–591.

East, E. M. 1936. Heterosis. *Genetics* **21**: 375–397.

Haldane, J. B. S. 1954. The statics of evolution. In *Evolution as a process,* pp. 109–121. Allen and Unwin, London.

Hull, F. H. 1945. Recurrent selection and specific combining ability in corn. *Jour. Amer. Soc. Agron.* **37**: 134–145.

Jones, D. F. 1917. Dominance of linked factors as a means of accounting for heterosis. *Proc. Nat. Acad. Sci. Wash.* **3**: 310–312.

Keeble, F., and C. Pellew. 1910. The mode of inheritance of stature and of time of flowering in peas (*Pisum sativum*). *Jour. Genetics* **1**: 47–56.

Lerner, I. M. 1950. *Population genetics and animal improvement.* Cambridge University Press, Cambridge.

Lerner, I. M. 1954a. The genotype in Mendelian populations. *Proc. 9th Int. Congr. Genetics* pp. 124–128.

Lerner, I. M. 1954b. *Genetic homeostasis.* Oliver and Boyd, Edinburgh.

Lewis, E. B. 1951. Pseudoallelism and gene evolution. *Cold Spring Harbor Symposia on Quant. Biology* **16**: 159–174.

Lush, J. L. 1943. *Animal breeding plans.* Iowa State College Press.

Mather, K. 1943. Polygenic inheritance and natural selection. *Biol. Rev.* 18: 32–64.

Mather, K. 1956. Response to selection. *Cold Spring Harbor Symposia on Quant. Biology* 20: 158–165.

Roberston, F. W., and E. C. R. Reeve. 1952. Heterozygosity, environmental variation and heterosis. *Nature* (London) 170: 296.

Shull, G. H. 1952. Beginnings of the heterosis concept. In *Heterosis,* pp. 14–48. Iowa State College Press.

Stebbins, G. L. 1950. *Variation and evolution in plants.* Columbia University Press, New York.

Wallace, B., and M. Vetukhiv. 1956. Adaptive organization of the gene pools and *Drosophila* populations. *Cold Spring Harbor Symposia on Quant. Biology* 20: 303–310.

Wentz, J. B., and S. F. Goodsell. 1929. Recessive defects and yield in corn. *Jour. Agric. Research* 38: 505–510.

Whaley, W. G. 1952. Physiology of gene action in hybrids. In *Heterosis,* pp. 98–113. Iowa State College Press.

Woodward, C. M. 1931. Illinois corn breeding report. *Purnell Corn Imp. Rept.,* pp. 48–49.

20

Systems of Pollination Control in Crop Plants

There is now a formidable body of evidence that heterozygosity is intimately connected with the efficient functioning of the genetic systems of Mendelian populations. This carries the implication that populations which are to maintain high fitness cannot afford to dispense with the heterozygous condition and hence must provide means to encourage or enforce cross pollination. Many systems to accomplish this purpose exist in higher plants; a brief outline of their main features has been given in Chapter 5. The systems encountered today are obviously the ones that have been successful in the past, and since the mechanisms governing mating systems occupy a special and key position in the genetic system, it is not surprising to find evidence that these systems have become intimately interwoven into the genetic organization of many species. The mating system is one of the best indicators of the ways populations are likely to respond to different types of manipulation, and it gives clues as to the limitations and possibilities of different breeding methods. Knowledge of natural mating systems is also important to breeders in the very practical sense that such information is essential to the development of efficient ways of controlling pollination according to the requirements of different breeding programs.

The mating systems of some crop species encourage but do not enforce cross pollination. Examples are systems involving monoecy, protandry, and protogyny. Some degree of self-fertility is usual with these systems with the result that artificial selfing can be used to develop inbred lines. Since self-pollination is permissible, and may be frequent, some form of artificial control of pollination may also be required when selected inbred lines or other materials are recombined into hybrid or synthetic varieties. None of these systems, which only encourage cross pollination, involves genetic differentiation among individuals. Genetic issues therefore do not arise when artificial control of pollination is practiced, and the problems that are encountered are mechanical problems of time and space as related to the characteristics of individual species or populations. The systems that enforce complete or nearly complete cross pollination are, however, genetically controlled, and this must be taken into account when pollination is manipulated artificially. Systems involving genetic differentiation among individuals fall into two main types, dioecy and incompatibility. It is with these two systems and with male sterility, which has some similar features in the artificial control of pollination, that this chapter will be mainly concerned.

Dioecy

Dioecy is not very common in plant species in general and occurs in only a few cultivated species. The most important are hops, the date palm, asparagus, spinach, and hemp. Among these dioecious species, hops and the date palm are propagated vegetatively in commercial practice. For these species breeding consists largely of the production and evaluation of large numbers of hybrids. When a suitable hybrid is found it can, of course, be multiplied indefinitely to produce commercial plantings. In hops the commercial product is the dried pistillate inflorescence, the male inflorescence having no commercial value. Seeds are superfluous, contributing nothing except additional weight to the crop. A premium is paid for seedless hops and, as a consequence, only female plants are grown in most commercial hopyards. With the date palm, on the other hand, male plants are essential because the commercial product, the fruit, fails to develop without fertilization. The male plants produce no fruit, and their only function is to produce pollen. For this reason, together with the fact that quality of fruit is influenced by metaxenia (making it necessary to pay careful attention to the pollen source), pollination is performed by hand so that the number of male plants can be kept to a minimum.

Asparagus, spinach, and hemp differ from the above dioecious crops in that they are propagated by seed, and both male and female plants produce usable product. Although the improvement of these species in the past has been by mass selection, other methods of breeding are a distinct possibility in the future.

There is good evidence that staminate asparagus plants outyield pistillate plants by about 25 per cent. Although staminate plants

A   B

FIGURE 20-1.   Sex expression in castor beans.   *A*, normal inflorescence with staminate flowers in the bottom half and pistillate flowers in the top half.   The proportion of staminate and pistillate flowers can be altered by selection to produce completely staminate and completely pistillate plants.   *B*, a completely pistillate inflorescence of the *N* type, in which a single recessive gene (*nn*) conditions the pistillate condition.   *NN* and *Nn* plants have mixed pistillate and staminate flowers in the inflorescence.   Can you devise a scheme to make use of this type of genetic control of sex expression in developing inbred lines?   How can such inbred lines be useful in producing hybrid seed on a commercial scale?

produce greater total yield, the female shoot is larger and has a more desirable appearance.   Yet, for the sake of higher yields alone, a method that would assure all-male plantings might be desirable.   The genetic studies of Rick and Hanna (1943) suggest a technique by which all-male populations might be produced conveniently.   These investigators established that sex in asparagus is monogenically inherited with maleness fully dominant.   Rudimentary organs of the opposite sex appear in both staminate and pistillate flowers, and although usually abortive, the pistils in staminate flowers may rarely function to produce seed-bearing fruits. This hermaphroditic development can be detected only by virtue of the conspicuous bright red color of the maturing berries. Since the seed in these rare hermaphroditic fruits develop by self-fertilization, the zygotic ratio is one homozygous male to two heterozygous males. Although homozygous males are phenotypically indistinguishable from heterozygous males, they can be identified by progeny tests. If homozygous males are isolated with normal females and insect pollination is allowed to proceed as usual, the resulting seeds yield staminate plants only. By selecting male and female plants as parents on the basis of their nicking ability, this procedure achieves not only the advantages of an all-male population but also the advantages of hybrid varieties.

Although spinach and hemp are usually considered to be dioecious, there is a continuous range of types regarding the proportion of pistillate to staminate flowers on each plant. Sex determination appears to be under the control of a single major gene. In addition, however,

FIGURE 20-2. Developing inbred lines in castor beans by mating heterozygous normal (*Nn*) and pistillate (*nn*) sibs. Note differences among sib-4 families in time of flowering, plant color, height, and other characteristics. Are sib-4 progenies expected to show more variability within or among families? Castor beans are strongly dichogamous. How might this influence the decision to inbreed by sib mating rather than selfing? How can inbred lines developed by sib mating in *N* type monoecism be used to produce hybrid seed on a commercial scale? Under certain environmental conditions the recessive type produces some staminate flowers. How might this be useful in increasing the efficiency of producing hybrid seed on a large scale?

there are apparently many modifying genes because it has been possible to select lines that are monoecious in various degrees. Thus although these species are nominally dioecious, selfing is not precluded, and it can be used to produce inbred lines for later recombination into hybrid or synthetic varieties. If the goal of an inbreeding program is a hybrid variety, it might be necessary to select both strictly staminate and strictly pistillate lines during the inbreeding process in order that hybrid seed can be produced conveniently at the end of the inbreeding. In such a program the breeder could well be obliged to use brother-sister matings or still less severe forms of inbreeding in developing inbred lines. As we shall see shortly, similar problems arise in connection with the self-incompatible species. The relative efficiency of various mating systems in producing homozygosity was discussed in Chapter 17.

Incompatibility

Among hermaphroditic plant species there are several genetically controlled systems for enforcing cross pollination that operate through the incompatibility of pollen and style. The incompatibility reaction appears to be a biochemical process under rather simple genetic control. The incompatibility process can operate at any stage between pollination and fertilization. In rye, the cabbage, and the radish, incompatible pollen usually fails to germinate, and if it does germinate, the pollen tube fails to penetrate the stigma. In some species of this type, the inhibition is lost if the stigmatic surface is removed, showing that the incompatibility reaction is localized in the stigma. In other species, incompatible pollen germinates, and the pollen tube grows down the style, but so slowly that it rarely reaches the ovary in time to effect fertilization. The rate of growth of incompatible pollen tubes can be influenced by the background genotype in which the major gene (or genes) governing the incompatibility reaction operates. Also, when there are several alleles at the incompatibility locus, these different alleles are not necessarily equally effective in preventing incompatible matings. In a few instances, incompatible pollen tubes have been found to grow at the same rate as compatible tubes, but nevertheless are not able to effect fertilization.

Incompatibility is very common in the plant kingdom. About a quarter-century ago, E. M. East estimated that it occurs in more than 3000 species among 20 families of flowering plants. Subsequently many additional cases were found, leading Lewis (1954) to the conclusion that East's estimate was much too low. Lewis attributes the fragmentary knowledge of incompatibility to the fact that breeding tests are required for its detection. Dioecy and nonsynchronous development of staminate and pistillate floral organs, on the other hand, are easily detected morphological characters often recorded in taxonomic treatises. Incompatibility can be as efficient as strict dioecy in enforcing cross pollination and it has the distinct advantage that every plant bears seed and thereby contributes directly to the propagation of the species.

Classification of Incompatibility Systems

Lewis, in a recent comprehensive review of incompatibility systems in plants, has classified these systems in various ways. In the discussion to follow we shall first distinguish between heteromorphic and homomorphic and then between gametophytic and sporophytic systems.

Heteromorphic systems are characterized by differences in the morphology of the flowers of different plants. The simplest and best known of the heteromorphic systems is found in many species of *Primula,* in which one type of plant, thrum, has a short style and highly placed anthers, and the other, pin, has a long style and short anthers. The only pollinations that are compatible are those between anthers and stigmas of the same height, that is, between pin and thrum and thrum and pin. In addition to these differences in floral morphology and incompatibility, pin and thrum plants also differ in some other characteristics, for example, pollen size and size of stigmatic cells. The genetic control of this complex of characters is by a single gene, S, thrum being Ss and pin ss. The pollen of thrum, despite the fact that it can be genetically either S or s, all behaves as if it were of S constitution. In other words, the incompatibility reaction of the pollen is impressed on it by the genotype of the parent plant, that is, by the previous sporophytic generation. This system is then a heteromorphic and sporophytic system. Heteromorphic systems are unimportant among crop plants.

In homomorphic systems, differences in floral morphology do not accompany the incompatibility. Also the incompatibility reaction of the pollen may be determined either by the genotype of the parent plant (sporophytic determination), as in the heteromorphic systems, or it may depend on the genetic constitution of the pollen itself (gametophytic determination). The homomorphic systems will be described in some detail because of their importance in crop plants.

The Gametophytic System

The gametophytic-incompatibility system, originally called the oppositional-factor system, was discovered in 1925 by East and Mangelsdorf in *Nicotiana sanderae*. In this system incompatibility is controlled by a single gene, S, which is usually characterized by the very large number of allelic forms in which it exists. In the gametophytic system, pollen-tube growth is usually very slow in a style that contains the same allele of S; consequently plants are virtually always heterozygous at this locus. The situation of two alleles with gametophytic control and no dominance is, of course, impossible because all plants would be incompatible and the species sterile.

The gametophytic system gives rise to three main types of pollinations: (1) fully incompatible ($S_1S_2 \times S_1S_2$), in which both alleles are common; (2) half the pollen is compatible ($S_1S_2 \times S_1S_3$), in which one allele is different; and (3) all the pollen is compatible ($S_1S_2 \times S_3S_4$), in which both alleles differ. Results from the compatible patterns are shown graphically in Figure 20–3.

In 1929 East and Yarnell found a self-compatibility allele at the S locus. Plants having the genotype S_fS_x, where S_x is any other allele in the series, are self-compatible and on selfing give self-compatible plants of the constitution S_fS_f and S_fS_x. Another allele with a special effect was found by Anderson and deWinton in 1931. This allele, S_F, is able to inhibit the growth of pollen tubes carrying the S_f allele.

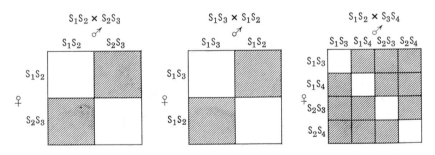

FIGURE 20-3. Incompatibility patterns obtained with the gametophytic system. Shaded areas represent compatible matings and unshaded areas incompatible matings. Note that the half-compatible cross $S_1S_2 \times S_2S_3$ and its reciprocal $S_1S_3 \times S_1S_2$ in combination reproduce both parental genotypes and one new genotype. In the fully compatible cross $S_1S_2 \times S_3S_4$ all four genotypes are new, but the parental genotypes will reappear in the next generation as a result of certain sib matings. (After Lewis, 1954.)

Lundquist in Sweden has recently presented evidence that two loci, each with multiple alleles, govern incompatibility in rye. Hayman in Australia has also found evidence for two-locus gametophytic systems in grasses. Since two-locus systems are not only highly effective in preventing selfing, but provide for wide cross compatibility, it would not be surprising if they turn out to be widely distributed in wind-pollinated plants.

The Sporophytic System

The sporophytic system is similar to the gametophytic system in that the genetic control is by a single gene with multiple alleles. It differs from the gametophytic system in that the incompatibility reaction is imparted to the pollen by the plant upon which the pollen is borne. The sporophytic system also differs from the gametophytic system in that alleles may show dominance, individual action, or competition in either pollen or styles according to the allelic combination involved. This leads to a great complexity of incompatibility relationships giving rise to many different patterns of incompatibility. According to Lewis (1954), the main features of the sporophytic system that distinguish it from the gametic system are:

"1. There are frequent reciprocal differences.
2. Incompatibility can occur with the female parent.
3. A family can consist of 3 incompatibility groups.
4. Homozygotes are a normal part of the system.
5. An incompatibility group may contain 2 genotypes."

The sporophytic system was first recognized in 1950 by Hughes and Babcock, working with *Crepis foetida,* and Gerstel, working on *Parthenium argentatum* (Guayule). The reader is referred to papers by these authors and to Lewis (1954) for a more complete discussion of the complexities of the sporophytic system of incompatibility control.

Efficiency of Incompatibility as an Outbreeding Mechanism

The reason for the existence of incompatibility is of course to prevent inbreeding or conversely to assure outcrossing. It is of interest, therefore, to compare the relative efficiency of the various systems in accomplishing this result. In considering efficiency it is particularly important to consider the amount of selfing permitted because of the power of this mating system in producing homozygosity. Mather (1942) has stressed the importance of the separation of male and female gametes by diploid stylar tissue in the angiosperms. This highly effective sieve of diploid tissue causes all systems of incompatibility to be highly effective in prohibiting selfing in the higher plants. In this regard, then, incompatibility is as efficient an outbreeding device as strict dioecy, and it is in fact more efficient than the imperfect dioecy characteristic of the nominally dioecious crop species, particularly when the regularity with which the incompatibility mechanism acts in many species is taken into account.

Regarding the cross compatibility of sib matings and general cross compatibility in a population, all that need be said is that with triallelic control only half of the matings are compatible, whereas with multiallelic control, cross compatibility rises rapidly as the number of alleles increases, exceeding 90 per cent with five or more alleles. All the incompatibility systems are therefore efficient at preventing selfing, and the multiallelic systems in particular are seen to be efficient at allowing all or nearly all plants, even in small populations, to set a good crop of seeds.

Incompatibility and Plant Breeding

Plant breeders are often interested in modifying the breeding system in crop plants in order to substitute, either permanently or temporarily, some mating system that serves their purpose better than the natural system of the species. Nowhere has this need been recognized longer than in certain of the self-incompatible orchard species, where fruitfulness depends upon including in orchards two or more crosscompatible varieties. Cross pollination is largely dependent on bees, which are ordinarily introduced into orchards in large numbers at the time of blossoming. Even then, in cold wet weather, the bees' activity may be reduced to the point where insufficient pollinations are made to produce a full crop. The introduction of self-fertility alleles into such clonally propagated crops would not only regularize the pollination process but would permit the planting of orchards to single varieties.

Lewis (1954) believes that mutations of the S gene by irradiation offers a solution to the production of self-fertile plants. As a result of irradiating pollen mother cells of *Prunus avium* and using the mature pollen on the mother tree on others of the same incompatability group, fruit set was increased from a normal 0.1 per cent to 2.0 per cent. Many of the seedlings produced by the 2 per cent of seed were fully self-fruitful. By crossing these seedlings to established varieties, many progeny were obtained which carry a self-compatibility allele S_f.

In time it may be possible to select a range of varieties that are fully self-fruitful.

There are many crops in which some inbreeding to fix type may be a desirable preliminary to the development of hybrid or synthetic varieties. Even though the inbreeding can be accomplished by sib mating, selfing is much more efficient in producing homozygosity. Various stratagems are open to the plant breeder who wishes to self-pollinate species that have incompatibility mechanisms. Self-compatibility alleles such as those described for *Nicotiana sanderae* have been described in a number of species, for example, in red and white clovers. In *Antirrhinum majus* a fertility gene exists that overrides the S system but is not allelomorphic with it. *Primula sinensis*, a distylic species, has gradually become more self-fruitful during the century or more that it has been cultivated as a greenhouse plant. This gradual reduction in the precision of the incompatibility system is believed to have resulted from modification of the polygenic background during many generations of selection of the more fertile plants.

Self-fertilization can also be achieved by taking advantage of the pseudofertility which occurs under the special conditions mentioned earlier. One of the most successful methods has been bud pollination—that is, placing mature pollen on immature stigmas. This sometimes allows the slow-growing incompatible pollen tubes to reach the ovary prior to floral abscission. Pollinating at low temperatures sometimes accomplishes the same purpose. Pseudofertility may also occur late in the season. In tobacco, for example, a plant may be self-incompatible throughout the entire growing season except for the last few days. In this period of end-of-the-season compatibility, self-fertilization occurs readily.

A number of ways in which incompatibility can be used to advantage in producing hybrid varities have been discussed by Lewis (1954, 1956). Odland and Noll (1950) have given a detailed procedure to make use of incompatibility to produce hybrid cabbage varieties from mildly inbred lines of high combining ability.

Male Sterility

In classifying plants according to why they may fail to set seed, it is desirable to distinguish between incompatibility and sterility. With incompatibility the pollen and ovules are functional, and unfruitfulness results from some physiological hindrance to fertilization, usually manifested either through failure of the pollen to germinate on the stigma or through slow pollen-tube growth down the style. Sterility, on the other hand, is characterized by nonfunctional gametes. It is caused by chromosomal aberrations, gene action, or cytoplasmic influences that cause abortion or modification of entire flowers, stamens, or pistils, or that upset the development of pollen, embryo sac, embryo, or endosperm. The particular type of sterility of concern at the moment is the type in which the male gametes are rendered nonfunctional as a result of the effects of mutant genes or cytoplasmic factors or by the combined effects of these two factors.

Male sterility, unlike incompatibility, is not a regular mechanism for controlling hybridity in natural populations. Rather, male-sterile plants appear only sporadically in populations of both self- and cross-pollinated species, presumably as a result of mutation at any one of the many loci that govern different vital steps in the formation of pollen. Although such mutants are undoubtedly deleterious in natural populations, they are interesting and useful to plant breeders, because they provide a means of emasculating plants genetically. This simpli-

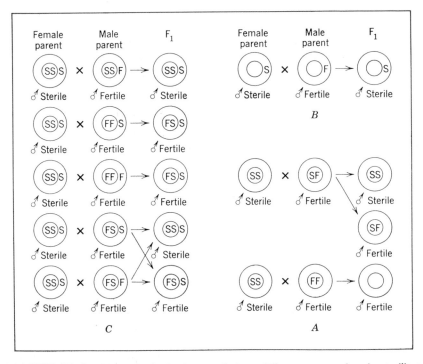

FIGURE 20-4. Method of inheritance of three different types of male sterility: A, genetic; B, cytoplasmic; and C, cytoplasmic-genetic. Letters in the inner circles represent genetic factors; letters in the outer circles show cytoplasmic factors. The S means male sterile, the F male fertile. The gene F is dominant to gene S. The cytoplasmic factors are transmitted only by the female parent. (After Sears, 1947.)

fies the making of hybrids to the point where hybrid varieties are now possible in some species in which the cost of hybrid seed was previously prohibitive.

The known instances of male sterility can be divided into three types, according to the way they are controlled genetically: *A*, genetic, *B*, cytoplasmic, and *C*, cytoplasmic-genetic. These are illustrated in Figure 20–4.

Genetic Male Sterility

Male sterility depending on a single gene has been found in many different crop species. The male-sterile condition is ordinarily recessive, and the male-sterile stock is maintained by crossing male-sterile plants with heterozygous fertile plants (Figure 20–4*A*). Half of the progeny are sterile, and half are fertile heterozygotes. In making crosses, the male-sterile line is planted in alternate rows with the intended male parent. The fertile plants in the male-sterile rows are rogued as soon as they can be recognized. If the male-sterile plants can be identified early by a closely linked gene or by some pleiotropic effect of the male-sterile gene (as is possible with one particular male-sterile gene in Lima beans), the task of producing hybrid seed is comparatively straightforward. If, however, the flowers must be examined carefully to identify the male-sterile plants, it may not always be possible to remove the fertile siblings of male steriles before they shed pollen.

In some species—for example, tomatoes, beans, and barley—pollen dispersal is poor and seed set is likely to be low on male-sterile plants. In tomatoes there are nevertheless some indications that production of hybrid seed in commercial quantities may be possible when better pollinators become available and when the areas where environmental conditions are most conducive to crossing have been identified. In those species where genetic male sterility seems incapable of providing hybrid seed in commercial amounts, it may still be useful as an aid to making the hybrids required for certain types of breeding programs and genetic studies.

Cytoplasmic Male Sterility

A second type of male sterility depends on cytoplasmic factors. Plants carrying particular types of cytoplasm are male sterile but will produce seed if pollinators are present. These F_1 seeds produce only male-sterile plants, however, since their cytoplasm is derived entirely from the female gamete. The maintenance of male-sterile lines is straightforward (Figure 20–5). The transfer of particular genotypes to the male-sterility-producing cytoplasm is also straightforward since the transfer will occur automatically if any specified genotype is used continuously as the pollinator (Figure 20–6). Cytoplasmic male

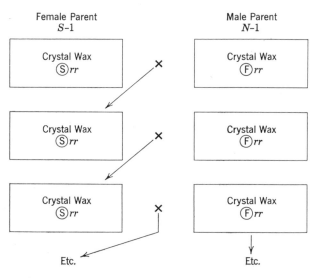

FIGURE 20-5. Method for perpetuating a male-sterile line of the variety Crystal Wax.

sterility has real advantages in certain ornamental species. This is because the offspring of all male-sterile plants are also male-sterile, regardless of the pollen parent used, and hence will remain fruitless in the absence of pollinators. These nonfruitful plants tend to bloom longer than their seeded counterparts, and the flowers remain fresh longer—obviously advantageous features in ornamentals. Cytoplasmic male sterility is also useful in producing single- or double-cross hybrids in crop species in which some vegetative part of the plant is the commercial product. It is obviously unsuitable for production of hybrid seed in crops where the fruit or seed is the commercial product, except where some satisfactory provision can be made to provide pollinators.

Cytoplasmic-Genetic Male Sterility

The third type of male sterility (type C, Figure 20–4) differs from the second type only in that the offspring of male-sterile plants are not

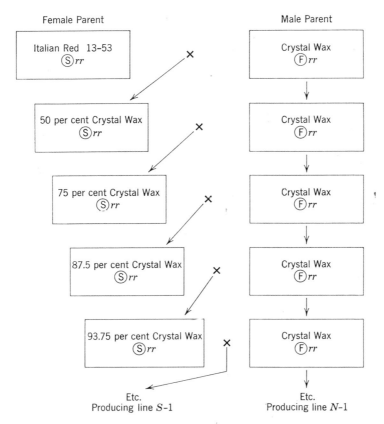

FIGURE 20-6. Method of developing male-sterile lines of Crystal Wax from male-sterile Italian Red 13-53, showing the rate at which Crystal Wax genes are incorporated into the male-sterile line by backcrossing. (After Jones and Davis, 1944.)

necessarily male sterile but can be male fertile when certain stocks are used as pollinators. These male parents that sire male-fertile F_1 progeny have been found to carry genes having the power to restore the pollen-producing ability of the male-sterility cytoplasm. Thus cases of type B male sterility are converted to the C type merely by the discovery of "restorer" genes, as they have come to be called. Study of Figure 20–4C shows the essential features of the inheritance of this type of male sterility.

The first report of the use of male sterility to produce hybrid seed was by Jones and Davis (1944), who discovered the C type of male sterility in onions. A single plant of the Italian Red variety was found that was completely male sterile. It was propagated by bulbils during the investigation of its breeding behavior. When crossed with various male-fertile parents, it gave three types of breeding behavior; some progenies were again completely male sterile, others were entirely male fertile, and still others produced both male-fertile and male-sterile plants in a 1:1 ratio. These results can be accounted for if it is assumed that there are two types of cytoplasm, fertile Ⓕ and sterile Ⓢ. All plants with fertile cytoplasm produce viable pollen, and they may have any of the genotypes ⒻRR, ⒻRr, or Ⓕrr. The gene R restores the S cytoplasm to normal. Hence when male-sterile plants (Ⓢrr) are crossed with a pollen parent of the genetic constitution Ⓢrr, a male-sterile F$_1$ results; the cross Ⓢrr × ⒻRr gives a 1:1 ratio, and the cross Ⓢrr × ⒻRR gives only male-fertile progeny.

Use of this type of male sterility in producing hybrid seed clearly requires that the female line be of the genetic constitution Ⓕrr. Fortunately male-fertile plants of the genotype Ⓕrr have been found in nearly all commercial varieties of onions, thus simplifying the development of male-sterile lines in each variety. The method of incorporating the male-sterile character of Italian Red into different varieties is illustrated graphically in Figure 20–6, using the Crystal Wax variety as an example. The rate at which Crystal Wax genes are incorporated into the male-sterile line is somewhat faster than indicated in the figure because selection is practiced for the Crystal Wax type in each backcross. As male-sterile plants cannot be selfed, the line is maintained by continued backcrossing to the normal or male-fertile line (Figure 20–5).

The next step is to make crosses between the male-sterile lines and other lines to determine which combinations produce the best hybrid. The male lines can be genetically ⒻRR, ⒻRr, or Ⓕrr but not Ⓢrr (Figure 20–7). When the male parent is ○Rr or Ⓕrr, half or all the progeny will be male sterile. The fertility of the hybrids is not important in onions because a vegetative part of the plant is harvested.

The program with onions is essentially the same as the ones now starting to be used in maize. The important difference is the necessity for a source of pollen in the commercial field of a crop such as maize, where the seed is the product harvested. In maize, two methods are being used to provide the necessary pollen: (1) blending, and (2) restorer genes.

Blending is made necessary by the fact that the appropriate combinations of cytoplasmic male sterility and restorer genes are not yet available in many of the inbred lines used to produce widely used hybrid varieties. Suppose it is desired to produce the double cross $(A \times B) \times (C \times D)$ using male sterility to avoid the expense of detasseling. It is comparatively easy to produce male-sterile A by

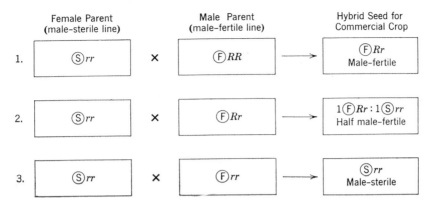

FIGURE 20-7. Method of producing hybrid seed for the commercial crop. All three methods are suitable for crops where the commercial product is a vegetative part of the plant. Only method one is suitable if the seed is the commercial product.

crossing *A* as the male to a cytoplasmic male-sterile line and back-crossing to *A*, again as the male, for several generations. The single cross *A* × *B* can then be made without detasseling to produce male sterile *A* × *B*, which in turn can be crossed with *C* × *D* without detasseling. The double cross thus produced will be male-sterile. If this seed is then blended with fertile (*A* × *B*) × (*C* × *D*), the latter provides the pollen necessary to produce a normal crop of seed on all the plants in the field. In this manner a substantial reduction in the cost of detasseling is achieved. Blending would not be necessary, of course, if the male parent *A* × *B* carried a restorer gene. However, backcrossing a restorer into an inbred line is a tedious process because the gene can be followed only by progeny tests. The use of cyto-plasmic male sterility will undoubtedly increase as cytoplasmic and restorer systems are introduced into more of the widely used inbred lines.

In the foregoing simplified discussion, some of the problems associated with production of hybrid seed by male sterility have been given inadequate attention, perhaps leading to a false impression of the ease with which the use of male sterility can be adapted to commercial practice. The first general type of problem that has arisen relates to the precision with which the male-sterility mechanism acts. It has been suggested that: (1) modifier genes may influence the precision of the male-sterility mechanism and that some of these modifiers may be lost in repeated backcrossing to male-fertile plants; (2) a small amount of cytoplasm may be carried over by the male gamete, resulting eventually in a partial breakdown of sterility; and (3) genetically male-sterile plants may produce viable pollen under certain environmental conditions. All these possibilities are apparently inconsequential in onions (Barham and Munger, 1950). In maize, however, and more particularly in sorghums, at least the third of these factors has proved to be troublesome (Briggle, 1956; Jones, 1956). Certain difficulties have also arisen in connection with the use of male sterility in some of the vegetable crops (Gabelman, 1956).

The other type of problem in adapting male sterility to commercial practice relates to the technical difficulties involved in maintaining various kinds of sterile and fertile stocks and combining them to produce large quantities of seed. An error in record keeping, for example, could lead to a large economic loss. When the numerous opportunities for blunders are considered, it can be appreciated that high competence plus careful attention to detail will be required in the practical applications of male sterility in agriculture.

REFERENCES

Anderson, E., and D. deWinton. 1931. The genetic analysis of an unusual relationship between self-sterility and self-fertility in *Nicotiana*. *Ann. Mo. Bot. Garden* 18: 97–116.

Atwood, S. S. 1945. The behavior of the self-compatibility factor and its relation to breeding methods in *Trifolium repens*. *Jour. Amer. Soc. Agron.* 37: 991–1004.

Barham, W. S., and H. M. Munger. 1950. The stability of male sterility in onions. *Proc. Amer. Soc. Hort. Sci.* 56: 401–409.

Briggle, L. W. 1956. Interaction of cytoplasm and genes in male-sterile corn crosses involving two inbred lines. *Agron. Jour.* 48: 569–573.

East, E. M., and A. J. Mangelsdorf. 1925. A new interpretation of the hereditary behavior of self-sterile plants. *Proc. Nat. Acad. Sci.* 11: 116–183.

East, E. M., and S. H. Yarnell. 1929. Studies on self-fertility. *Genetics* 14: 455–487.

Gabelman, W. H. 1956. Male sterility in vegetable breeding. *Brookhaven Symposia in Biology* No. 9: 113–122.

Gerstel, D. V. 1950. Self incompatibility studies in Guayule. II. Inheritance. *Genetics* 35: 482–486.

Hughes, M. B., and E. B. Babcock. 1950. Self-incompatibility in *Crepis foetida* (L.) Subsp. rhoedifolia (Bieb.) Schinz *et* Keller. *Genetics* 35: 570–588.

Janick, J., and E. C. Stevenson. 1955. Genetics of the monoecious character in spinach. *Genetics* 40: 429–437.

Jones, D. F. 1956. Genic and cytoplasmic control of pollen abortion in maize. *Brookhaven Symposia in Biology* No. 9: 101–112.

Jones, H. A., and G. N. Davis. 1944. Inbreeding and heterosis and their relation to the development of new varieties of onions. *U. S. Dept. Agric. Tech. Bull.* 874: 1–28.

Lewis, D. 1954. Comparative incompatibility in angiosperms and fungi. *Adv. in Gen.* 6: 235–285.

Lewis, D. 1956. Incompatibility and plant breeding. *Brookhaven Symposia in Biology* No. 9: 89–100.

Lundquist, A. 1956. Self-incompatibility in rye. I. Genetic control in the diploid. *Hereditas* 42: 293–348.

Mather, K. 1942. Heterothally as an outbreeding mechanism in fungi. *Nature* (London) 149: 54–56.

Rick, C. M., and G. C. Hanna. 1943. Determination of sex in *Asparagus officinalis. Amer. Jour. Bot.* 30: 711–714.

Odland, M. L., and C. J. Noll. 1950. The utilization of cross-compatibility and self-incompatibility in the production of F_1 hybrid cabbage. *Proc. Amer. Soc. Hort. Sci.* 55: 391–402.

Rogers, J. S., and J. R. Edwardson. 1952. The utilization of cytoplasmic male-sterile inbreds in the production of corn hybrids. *Agron. Jour.* 44: 8–13.

Sears, E. R. 1947. Genetics and farming. *U. S. Dept. Agric. Yearbook* 1943–1947: 245–266.

Wright, S. 1921. Systems of mating. II. The effects of inbreeding on the genetic composition of a population. *Genetics* 6: 124–143.

Westergaard, M. 1958. The mechanism of sex determination in dioecious flowering plants. *Adv. in Gen.* 9: 217–281.

21

Selection in Cross-Pollinated Crops

The selection procedures used in breeding cross-pollinated crops take much the same outward form as the ones used with self-pollinated crops. The results are not the same, however, owing to the different population structures of self- and cross-pollinated species. In self-pollinated crops individual-plant selections have been widely used to establish uniform, pure-line varieties. But in cross-pollinated crops, individual-plant selections are seldom if ever effective in establishing a variety, because segregation causes the progeny to deviate from the parental type and because such drastic reduction in population size usually has unfortunate effects on vigor and productivity. In cross-pollinated crops, mass selection and the closely related procedures of progeny selection and line selection are much more commonly used than single-plant selection.

Mass Selection

In mass selection, desirable individual plants are chosen, harvested, and the seed composited without progeny test to produce the following generation. Since selection is based on the maternal parent only, and there is no control over pollination, mass selection amounts to a form of random mating with selection (Chapter 17). The purpose of mass selection is to increase the proportion of superior genotypes in the population. The efficiency with which this is accomplished under a system of random mating with selection depends primarily on gene numbers and heritability, as we saw in the chapters of Section Four.

Mass selection has been effective in increasing gene frequencies for characters which are easily seen or measured. In corn, for example, it was possible by mass selection to develop varieties differing in color of grain, plant height, size of ear, placement of ear on the stalk, date of maturity, and percentage of oil and protein. Mass selection has thus been useful in developing varieties for special purposes and in changing the adaptation of varieties to fit them to new production areas.

On the other hand, mass selection has not been effective in modifying characters, such as yield, that are governed by many genes and cannot be accurately judged on the basis of the appearance of single plants. Thus, this method of breeding has proved almost powerless to affect the yield of adapted varieties, at least in short-term breeding projects.

The ineffectiveness of mass selection in increasing the yield of adapted varieties results from three main causes: (1) inability to identify superior genotypes from the phenotypic appearance of single plants; (2) uncontrolled pollination, so that selected plants are pollinated by both superior and inferior pollen parents; and (3) strict selection leading to reduced population size, which leads in turn to inbreeding depression. Recognition of these deficiencies has prompted the development of some modifications of selection procedures which tend to overcome these limitations of mass selection. *Progeny selection* and *line breeding* are procedures which can be effective in overcoming the first or third of the deficiencies above. *Recurrent selection* appears to overcome all three deficiencies; it is therefore highly interesting to plant breeders and will receive special attention in Chapter 24.

Progeny Selection and Line Breeding

The most effective way of distinguishing among single plants whose superiority is environmentally induced and those whose superiority stems from superior genotypes is by progeny testing. Growing a small progeny (perhaps 10 to 50 plants) of each individual plant selected in the previous generation generally gives a more accurate measure of the breeding value of an individual than the phenotype of the individual itself. The accuracy of progeny tests is obviously greater in replicated than in single-plot trials and still greater in replicated trials repeated over years and locations.

Various mating systems can be used in producing the progenies to be tested. The most common procedure is merely to harvest open-pollinated seed from the selected plants and use it to establish the progeny plots. This was the procedure used in ear-to-row selection in maize (Chapter 16). In other instances the progeny plots are established from seed produced by selfing the selected plants and in still other instances they are established from top crosses or other types of test crosses (the relative merits of these various types of progeny tests will be considered in detail in Chapter 23). The plants which will be composited to produce the next generation are then selected on the basis of the performance of their progeny, rather than on their own phenotypic appearance. The main point to be noted is that progeny testing, although valuable in aiding the breeder to identify superior genotypes, has no other contribution to make toward improving the efficiency of mass selection. Mass selection, whether conducted with or without progeny testing, remains a form of random mating with selection and is subject to the limitations of that system of mating in modifying the genetic composition of populations (Chapter 17).

Varieties of cross-pollinated plants are seldom developed from the offspring of a single plant. More commonly, after one or even several cycles of mass selection, the seed of a number of superior plants is composited and planted in an isolated plot where it is allowed to mate at random. The harvest from this plot then becomes the foundation seed of a new variety. When selection has been based on progeny tests and a group of progeny lines are composited, the procedure is called *line breeding*. Excessive inbreeding can be avoided in line breeding by exercising care that an adequate number of lines enter the composite and that these lines are not closely related.

Varieties Developed by Selection

Mass selection in its simplest form was undoubtedly the earliest breeding method used with cross-pollinated crops. Farmers have probably always selected from their fields the plants that struck their fancy, and it seems likely that this mass selection, carried out over very large numbers of generations, has had more overall effect on cross-pollinated crops than all other breeding methods combined. Mass selection by farmers was the principal breeding method with open-pollinated corn, and it has also been practiced with a number of other field crops, various forage crops, and in many of the vegetable species.

Many hundreds of varieties and strains were developed and maintained in these various crops by the mass-selection procedure. The power of mass selection in developing locally adapted strains can be illustrated by citing a few examples. In corn, mass selection produced such diverse types as the dent corns of the midwest, the early flints of the northern Great Plains, the tall late flints of the southwest, and the prolific types (multiple ears per stalk) of the southeast. Within each of these types occurred a profusion of open-pollinated varieties— for example, Reid Yellow Dent, Krug Yellow Dent, Hague Yellow Dent, Johnson County White and Boone County White, all adapted to the central corn belt. These, and similar open-pollinated varieties adapted to other regions, were the base populations from which hybrid varieties were later developed. A similar set of examples could be given for any of the widely grown cross-pollinated crop species.

Not only farmers but plant breeders as well have made wide use of mass selection in improving cross-pollinated crops. The simplest selection procedure available to breeders in developing new varieties is merely to harvest the seed from a locally adapted type. This procedure has been used to develop numerous varieties, particularly in the forage crops, that have been named and released for commercial production. Pennscott Red Clover is an example. It was one of a number of local Pennsylvania strains collected by H. B. Musser of Pennsylvania State College in 1937. The original seed lot was obtained from the farm of Frank Scott of Lancaster County, Pennsylvania, where it had been grown for about 20 years. Other examples where seed harvested from a field was increased and released as a named variety include: Kentucky 31 Tall Fescue; Tallarook, Mount Barker, and Dwalgenup Subterranean Clover from Australia; Akroa Orchardgrass from the Akroa peninsula of New Zealand; and Chesapeake Red Clover, released by the Maryland Agricultural Experiment Station in 1958.

Harvest *en masse* of adapted strains obviously does not permit selection of superior plants within a population, nor does it permit progressive improvement by continued selection. Many plant breeders have therefore selected for one or more generations before compositing open-pollinated seed to form a new variety. It is doubtful, for reasons discussed in the chapters of Section Four, that such selection would influence yield. But it might improve the selected population for other important characters.

Some examples of varieties developed by selection from adapted strains are Kenland Red Clover, Empire Birdsfoot Trefoil, and Buffalo Alfalfa.

Kenland Red Clover was developed from seven old farm strains from the southern part of the red clover belt. Plants from each of these strains were inoculated with southern anthracnose during each of four generations, and susceptible plants were eliminated before intercrossing among desirable plants was permitted. This variety is longer lived and higher producing when anthracnose is present and equal to other varieties when this disease is not a limiting factor. Empire Birdsfoot Trefoil is a selection from a natural stand at Preston Hollow, New York. It is a late-blooming strain that produces high-quality hay. Buffalo Alfalfa is a selection from a strain of Kansas Common Alfalfa.

Selection was first practiced in this strain in 1922. Selection for wilt resistance and good agronomic characteristics was continued until the strain was released in 1943. It is resistant to bacterial wilt and adapted throughout the central and southeastern part of the United States.

Because of the inefficiency of mass selection in improving characters of low heritability, there is currently a tendency to replace mass selection with the breeding of hybrid or synthetic varieties. Since, however, both of these procedures are more laborious than mass selection, it is unlikely that they will soon supplant mass selection except in the more important crop species. It therefore seems likely that mass selection will continue to be a major method of breeding with minor cross-pollinated species and with many forage crops. There are many species with which forage breeders work, and these species vary tremendously in habit of growth, longevity, mating system, and seed-setting ability. Only when greater effort can be expended on each of these species will it be possible to replace mass selection with more efficient methods of breeding.

Examples of Improvement by Long-Continued Mass Selection

Although there is no doubt of the importance of the role that mass selection has played over the centuries in the improvement of cross-pollinated crops, details, particularly the population sizes that prevailed and the intensity of the selection that was practiced, are rather nebulous. Despite the lack of precise records, it is nevertheless possible to gauge the impact of mass selection on a number of cross-pollinated crops in terms of the principles that we considered in the several chapters of Section Four. Since records are reasonably complete in sugar beets, corn, and alfalfa, we will attempt to reconstruct the impact of long-continued mass selection on population structure using these crops as examples.

Improvement of the Sugar Beet by Mass Selection

The sugar beet was developed as a distinct and separate crop in historical times and it provides one of the best-documented illustrations of the long-time effects of selection. Sugar beets were derived from Mangelwurzel, large, coarse types of *Beta vulgaris* that have long been grown as fodder plants in Europe. The best information available indicates that the beets first used in the manufacture of sugar early in the nineteenth century contained less than 7.5 per cent sucrose on the average. Original attempts to improve the crop emphasized selection based on size, shape, and color of root. During this period the average richness in sugar of beets delivered to factories progressed from about 8.8 per cent in 1838 to 10.1 per cent in 1868. By 1860, largely as a result of the work of Louis de Vilmorin, polariscopic analysis of the juice as a guide to selection and progeny testing came into common use, and by 1888 the average sugar content had risen to nearly 14 per cent. By 1912, a further advance to about 16 per cent had been achieved. Some present-day varieties average 18 per cent or more, but they are likely to make less gross yield of sugar than varieties which average 16 per cent sugar. Although it seems probable that at least part of this advance is due to improvements in cultural practices and in the efficiency of extraction in sugar factories, there can be little doubt that continued selection at least doubled sucrose content in about 100 years.

Another character for which long-continued selection has been practiced in sugar beets is resistance to curly top, a virus disease first described in 1900 and destined to cause enormous economic losses in the western United States before resistant varieties became available. Selection for curly-top resistance dates from 1902, and although the early work did not result in commercially acceptable strains, it demonstrated that highly resistant components could be isolated from existing varieties. Continued selection of resistant mother beets under severe disease conditions, combined with progeny tests, led by 1929 to a number of resistant lines that were massed to form the variety known as U.S. 1. Because of the emergency this variety was released before it was uniform for curly-top resistance or for certain other characters, particularly the nonbolting character. Subsequently, varieties were obtained by mass selection from U.S. 1 that represented a decided improvement in both curly-top resistance and reduced tendency to bolt.

We can infer that the original populations of Mangelwurzel contained a great store of genetic variability. Some of the potential for change was represented by rare genes which were subject to increase in frequency by selection. Potential for change also resided in polygenic combinations balanced by the past evolutionary history of the species to provide good adaptation to environment and to the uses of the species as a fodder plant. Selection for high sugar content, desirable shape of root, and other attributes favorable to its new use led to changes in gene frequencies that ultimately resulted in derived populations distinctly different from the original unselected population.   Apparently the selection pressures were mild enough and population sizes adequate to allow new balanced combinations of genes to be formed at a rate such that vigor and productivity were maintained while sugar content and other desirable characters were shifted drastically.   Possibly the recent production of types containing more than 20 per cent of sucrose represents too rapid advance in modification of sugar content to allow properly balanced gene combinations to be formed in other characters.   If so, continued selection may allow the development of types combining high yield and high sugar content.

The great diversity of types which appear in all modern varieties of sugar beets suggests that mass selection extended over more than a hundred generations has not led to gene fixation in the sugar beet. There is in fact no reason to believe that modern populations are any less heterozygous than their Mangelwurzel ancestors.   These results are exactly those we would expect from a mating system involving random mating with selection, that is, shift in the mean in the direction of selection without appreciable change in heterozygosity of the selected population.

Mass Selection in Corn

The changes that have occurred in corn in the last two centuries or more illustrate a number of effects of long-continued mass selection on populations, including the effect of selection on morphological appearance, adaptation, and yield as well as the influence of intervarietal hybridization and restriction of population size.

When the first settlers reached North America they found the Indians growing all the major endosperm types—flint, flour, gourd seed, pop, and dent—and there was a correlation in any locality between the types grown and the way they were used.   This in itself indicates that selection toward definite objectives predated Columbus. In fact it is difficult to imagine that mass selection has not been practiced in corn since earliest times, because the harvesting of each ear is an individual operation and any variation in plant or ear characteristics could hardly escape detection.   Selection based on maternal characteristics alone, without information about pollen parents or progeny performance, can hardly have been efficient.   But, for reasons we have already examined, mass selection has considerable powers in modifying plant type, maturity, kernel characteristics, and other easily recognizable characters, and it is not surprising that distinctive morphological types were found in different localities.   Mass selection was probably also effective in improving adaptation to particular environmental conditions, but it seems likely that improvement of the yielding ability of the adapted types was very slow.

Many of the superior varieties of open-pollinated corn in the United States were developed by farmers during the latter part of the nineteenth century. However, the tendency for farmers to select for specific types, and hence to increase the divergence among varieties, was well established before the end of the seventeenth century. Moreover, hybridization between varieties is known to have played a part in providing the variability from which new varieties were selected. Anderson and Brown (1952) state "archeological and historical evidence shows that the common dent corns of the United States Corn Belt originated mainly from the purposeful mixing of the northern flints and southern dents."

The history of Reid Yellow Dent can be cited as a typical example of the place of both hybridization and mass selection in the development of an open-pollinated variety. Gordon Hopkins corn, seed of which was brought from southern Ohio to central Illinois by Robert Reid, failed to mature well in 1846. The immature seed from this crop produced a poor stand in 1847, and the missing hills were replanted with an early-maturing variety—probably a flint type—generally grown in the area. The resulting crop, which presumably included many intervarietal hybrids, was the foundation stock from which the variety Reid Yellow Dent was developed by many generations of selection by James Reid, a son of Robert Reid.

Many other varieties were developed by similar mass selection from existing varieties, mixtures of varieties, and varietal hybrids. Normally enough seed was selected to plant a commercial field. Numbers might therefore be presumed to have been adequate to avoid inbreeding depression. Also, selection pressures applied were probably small enough to allow for the replacement of balanced combinations of genes by different balanced combinations at new equilibrium levels without loss of fitness.

Toward the end of the nineteenth century, fairs and corn shows began to play an important role in the popularity of particular strains within open-pollinated varieties. Originally a prize was given for corn samples with presumed superior yielding ability, but emphasis in judging rapidly shifted to emphasize highly uniform samples of ears which conformed to "show" standards. It was a common practice for the winning sample, usually ten ears, to be used by farmer-breeders as the basis for the development of a new substrain. Drastic reductions in population size produced by this procedure must, upon repetition, have led to some degree of fixation of type. At any rate, hundreds of local strains were developed during this period. Later data from state corn-yield tests showed that these substrains differed sharply in yielding ability. Experiments to measure the relation between judging points used in the corn shows and yielding ability indicated that many of the judging points were neutral or even detrimental to yield. The results of these experiments suggested that a more appropriate basis for mass selection than the fine points of ear conformation was selection for proper maturity, freedom from disease, large well-filled ears, and vigorous plant type.

Recognition of the Vilmorin principle of progeny testing stimulated C. G. Hopkins of the Illinois Agricultural Experiment Station to devise the ear-to-row method of corn breeding in 1896. As we have already seen, this modification of mass selection achieved wide popularity. By this method the selected population in each generation was usually derived from approximately forty ears chosen from about ten ear-rows. Regardless of the efficacy of the selection for yield, it seems likely that ear-to-row breeding had some effect on gene frequencies through repeated restrictions of population size alone. Although it is difficult to estimate the relative magnitude of effect of these different influences on populations of corn, their joint influence over many years was judged by Richey as "of the utmost importance in improving corn and adapting it to the varying conditions under which it is grown."

Natural Selection in Alfalfa

The changes that have occurred in alfalfa since it was introduced to the United States not only furnish additional examples of the effect of mass selection similar to the ones already cited but also illustrate the role of natural selection in modifying the genetic composition of populations of cross-pollinated crops. Although alfalfa was introduced to the United States in Colonial times and has been grown as a minor crop in the eastern parts of the country continuously, the first real success with alfalfa came from seed imported from Chile and planted in California about 1850. From there the crop gradually spread eastward to Utah and then to Kansas and Nebraska. By 1900 it had crossed the Missouri River and had become important in Iowa and Missouri, finally extending to the eastern seaboard. Alfalfa thus received its main impetus from the success it achieved in the west, and, according to Tysdal and Westover (1937), the so-called Common Alfalfa now being grown extensively originated from the introductions of Chilean alfalfa. These introductions from Chile were nonhardy types, suited to the mild climate of California but unable to withstand the rigors of a continental climate.

As Common Alfalfa made its way eastward, it is evident that it was remarkably changed by natural selection in different regions. As indicated by Westover (1934), strains evolved that had the ability to become relatively dormant in the fall and to resist cold, as do the Northern Commons. Other strains, such as the Common Alfalfa found in the south, grow until late in the fall. These changes apparently took place by natural selection, unaided by artificial selection. The major part of the genetic variability upon which natural selection acted apparently existed within the Chilean types, but hybridization with local strains may also have provided some of the variability.

At about the same time alfalfa was introduced to California, Wendelin Grimm brought his family and a few pounds of alfalfa seed from southern Germany to Minnesota. There is little reason to believe that the original crop of alfalfa sown by Grimm was uniformly hardy, because winter conditions in its source area in Germany are much less severe than those of Minnesota. Yet planting the seed from surviving plants generation after generation ultimately led to an exceptionally winter-hardy strain. This strain, grown in obscurity for nearly fifty years, greatly increased the acreage of alfalfa in the coldest parts of the United States when its value was publicized.

The point is that when a population is moved to a new environment, natural selection discriminates against unfit genotypes, such as non-hardy types in alfalfa, and prevents them from reproducing or reduces their reproductive rates relative to more fit genotypes. The pace and extent of the changes caused by this differential reproduction depend on the same factors we examined earlier for artificial selection: gene frequencies and the ways genes are distributed among individuals in balanced combinations, the number of genes governing characters related to survival, the heritability of these characters, and the mating system that prevails in the new environment. There is one important difference, however: natural selection operates only on characters that are related to the survival of the species.

The power of natural selection in modifying populations is well exemplified by the remarkable changes it was capable of producing in the hardiness of alfalfa in comparatively few generations. Natural selection no doubt also played an important role in the development of the locally adapted varieties of corn and sugar beets discussed above. This natural selective force is so powerful in its results that it no doubt exerts a consequential, even though frequently intangible, influence on all selection programs.

REFERENCES

Anderson, Edgar, and W. L. Brown. 1952. Origin of corn belt maize and its genetic significance. In *Heterosis,* pp. 124–148. Iowa State College Press.

Coons, G. H. 1936. Improvement of the sugar beet. *U. S. Dept. Agric. Yearbook,* 625–656.

Jenkins, M. T. 1936. Corn improvement. *U. S. Dept. Agric. Yearbook,* 455–522.

Richey, F. D. 1950. Corn breeding. *Adv. in Gen.* 3: 159–192.

Sprague, G. F. 1955. Corn breeding. In *Corn and corn improvement,* pp. 221–292. Academic Press, New York.

Tysdal, H. M., and H. L. Westover. 1937. Alfalfa improvement. *U. S. Dept. Agric. Yearbook,* 1122–1153.

22
Hybrid Varieties

The term *hybrid variety* will be used here to designate F_1 populations that are used for commercial plantings. Such F_1 populations can be obtained by crossing clones, open-pollinated varieties, inbred lines, or other populations that are genetically dissimilar. When hybrid varieties are feasible, they make better use of heterosis than any breeding procedure yet developed; hence the methods by which they are produced will be considered in some detail. By far the greatest development of hybrid varieties has been in maize. In that crop more than half a century of intensive research on hybrid varieties, combined with a quarter century of massive use of hybrids in farming practice, has led to information and experience that probably exceeds that of all other species combined. For these reasons this chapter will be concerned largely with hybrid varieties based on crosses between inbred lines, since this is the system upon which the spectacular success of hybrid maize has been based.

The morphological structure of the staminate and pistillate inflorescences in maize, and the way in which they are separated from each other in space, make this species uniquely suited to controlled selfing and crossing and hence to the production of inbred lines and hybrid seed. The first attempts to make practical use of the fortunate mating system of maize came late in the nineteenth century. W. J. Beal in Michigan, stimulated by Darwin's work on inbreeding and outcrossing, undertook studies which led him to suggest the use of F_1 varietal crosses for the commercial crop. The hybrid seed was to be produced by detasseling alternate rows. Sanborn in Maine, Morrow and Gardner in Illinois, McClure, also in Illinois, and several others (review in Zirkle, 1952) confirmed Beal's results, and like Beal they also recommended the use of varietal hybrids.

Some of these early reports on varietal hybrids gave detailed instructions for the production of hybrid seed by detasseling and recommended that the cross be made every year. Despite the fact that these studies established the superiority of certain varietal hybrids over their parents and suggested ways in which the hybrids could be produced, varietal hybrids were at no time planted on more than a small acreage. The practical utilization of heterosis in maize on a large scale was in fact delayed until the fourth decade of the twentieth century. Only then was it possible to overcome the practical difficulties that arose and to transmute the theoretical studies of many investigators into a large-scale commercial operation.

Many workers participated in the studies that led ultimately to large-scale use of hybrid maize. It was George Harrison Shull (1909), however, who first suggested a pure-line method of corn breeding based on inbred lines obtained by continued self-fertilization and the use of F_1 hybrids between these inbred lines for the production of the commercial crop. His proposal was to use single crosses for the commercial planting, these single crosses to be made between pairs of inbred lines selected for their superior performance in combination with each other. Shull recognized that inbreds had an advantage over open-pollinated varieties in that they were homozygous and could be counted on not only to reproduce themselves with great precision, but also to produce hybrids of exactly the same genotype year after year.

There were several reasons why this scheme did not lead to the successful use of hybrid varieties. First, inbred lines were not then available which, upon hybridization, were capable of producing hybrids sufficiently better than the best open-pollinated varieties to make them attractive to farmers. Second, hybrid seed was expensive. This was because the female parent was a low-yielding inbred line, and in addition one third to one half of the land had to be given over to the pollen parent, further reducing the yield of hybrid seed per acre. Third, the F_1 hybrid seeds were small and often misshapen, which led to difficulties in handling them in planting equipment and also to poor germination.

The double cross, suggested by D. F. Jones in 1918, made hybrid maize economically feasible. A double cross is the F_1 between two single crosses. Thus if A, B, C, and D represent inbred lines, one of the possible single crosses can be represented by $A \times B$, and one of the possible double crosses by $(A \times B) (C \times D)$. In a double-cross hybrid the seed used for commercial planting is produced on a single cross seed parent that yields two to three times as much as any inbred line. Pollen is produced in abundance by the other single cross, so that less land has to be given over to it than if the pollinator were an inbred line. In addition, the seed is normal in size and shape and produces vigorous seedlings. Although the suggestion of the double cross stimulated the initiation of programs to breed hybrid varieties in many states within a few years of its announcement, and although the first commercial production of a double-cross variety took place in 1921, there was nevertheless a considerable time lapse before hybrid maize became a significant factor in agriculture. As late as 1933 less than 1 per cent of the acreage of maize was sown to hybrids. But by 1940 more than half the acreage in the United States was planted to hybrid varieties, and by 1944 hybrid varieties made up more than 80 per cent of the acreage. In the corn belt, the acceptance of hybrids was so complete that open-pollinated varieties virtually disappeared. Corresponding to this increase in acreage planted to hybrids was a substantial increase in yields per acre. In the corn-belt states this increase has been estimated to be in excess of 20 per cent. Considering that maize is the most important crop grown in the United States, it is readily seen that the development of double-cross maize hybrids was one of the most important advances of all time in the history of agriculture.

Operations in Producing Hybrid Maize

In briefest outline, the operations that have led to the great practical success of hybrid maize are the following: (1) selecting desirable plants in open-pollinated populations; (2) selfing these plants through several generations to produce homozygous inbred lines; and (3) crossing chosen lines.

First-generation crosses between pairs of inbred lines, $A \times B$, are being used commercially in producing sweet corn for canning or for use in home gardens, where quality and uniformity are paramount in importance and outweigh the high cost of the seed. Such single crosses are for similar reasons economically feasible in the production of popcorn.

The first-generation cross between a pair of lines can be further crossed with an inbred line to produce a three-way cross, $(A \times B)C$. The single cross is used as the female parent and, to be successful, the inbred line used as the male parent must excel in pollen production.

However, the great bulk of hybrid-maize production is based on double crosses, hybrids between two single crosses, involving four inbred lines. In the production of hybrid seed, it is usual to maintain the inbred lines—particularly the one to be used as the male parent in a single cross—by hand pollination. The inbred line to be used as the seed parent is, however, frequently increased for one generation in an isolated plot and the seed obtained from this plot used to produce the single cross. Off-type plants in the female rows of the crossing plot can usually be detected by their ear characteristics. It has been found that careful selection is necessary to keep inbred lines from accumulating changes due to outcrosses and mutations, changes that may lower their value in producing double crosses.

A single cross is usually made by alternate plantings of the two inbreds in an isolated plot. The inbred chosen to be the seed parent is detasseled. Usually two rows of the seed parent are planted to each row of the pollinator inbred. Double-cross seed is produced in a similar way, but the superior pollen-producing ability of the single-cross pollinator permits planting rates of six rows of the seed parent to two rows of the pollen parent, or even four rows to one.

The advantages of cytoplasmic male sterility in producing hybrid seed were noted in Chapter 20.

It is generally agreed that the three operations mentioned at the beginning of this section have led to grain yields in excess of 20 per cent over the yields to be expected from open-pollinated varieties. In interpreting this gain in productivity, it is important to recognize that these three operations merely reproduce the same population genotype that characterizes the materials from which the inbreds were derived if: (1) the individuals chosen to be inbred are selected at random; (2) no selection occurs during the inbreeding process; and (3) inbred lines are crossed at random including selfs. In other words, if an open-pollinated variety is inbred to fixation *without selection* of any kind, and if the lines obtained are then crossed *at random,* the cycle merely reproduces the open-pollinated variety. Therefore the improvement that has actually been achieved must be ascribed to selection during one or more of these three operations.

Developing Inbred Lines

Self-fertilization has been used almost to the complete exclusion of other forms of inbreeding in the development of inbred lines of maize. The plants to be selfed are selected for their vigor, standing ability, freedom from disease, and other desirable characters. Since many attributes of importance cannot be ascertained at the time of pollination, selection is also practiced at harvest, at which time ears from undesirable plants are discarded.

Seed from the remaining ears is then used to plant ear-to-row progenies the following season. Ordinarily, progenies of twenty to thirty plants are planted in single-plant hills spaced about a foot apart. Selection is practiced both between and within progenies, and only the best plants in the best rows are selected for further inbreeding. With a continuation of the inbreeding there is a marked decrease in vigor and an increase in the uniformity of the plants within any progeny row. Many strains are lost because of gross deficiencies and others are discarded because they lack appeal, but a few outstanding lines are continued. After five or six generations of self-pollination, every plant is practically like every other plant within any line, but differences among lines are large. At this point selfing is often discontinued, and the lines are thereafter continued by sib pollination. A modification of this method suggested by Jones and Singleton in 1934 recommends the growing of only a single three- or four-plant hill from each inbred ear, thus making possible the growing of very large numbers of inbred lines in limited space. This method obviously minimizes the opportunity for selection within progenies in order to increase the opportunity for selection among progenies. This procedure has certain theoretical advantages, as we shall see.

The Value of Visual Selection during Inbreeding

Although severe selection is ordinarily practiced for various characters during the development of inbred lines, there is considerable divergence of opinion among maize breeders about the efficacy of this selection. Kiesselbach, one of the earliest workers to study this problem, reported in 1922 that there is a general relationship between the yielding ability of inbred lines and the hybrid progeny they produce. Subsequently several other investigators have presented evidence that significant phenotypic correlations exist between various attributes of inbreds and hybrids (review in Sprague, 1955). In general, however, these correlations were too small to have much predictive value. Two more recent experiments are somewhat contradictory. Sprague and Miller (1952) found evidence to indicate that visual selection for yield during inbreeding was ineffective but that selection against stalk breaking during the inbreeding process was of at least some value. Osler, Wellhausen, and Palacios (1958), on the other hand, concluded from their work that visual selection during successive generations of inbreeding resulted in significant improvement not only in ear appearance and plant appearance in a large percentage of 134 hybrid combinations, but also in yielding ability.

Regardless of the value of selection in improving hybrid performance, it is now generally accepted that selection during inbreeding does serve a useful purpose in the development of inbred lines. There is little question that selection is effective in improving the characteristics of the inbred lines themselves, and in this way it performs the important function of eliminating lines that are difficult to propagate and hence have little value in commercial practice even though they may produce desirable hybrids.

We are now in a position to examine the contribution of selection during each of the three operations by which hybrid maize has been produced. The first of these operations—that is, selecting the plants which are to be inbred—will be recognized as equivalent to a single generation of mass selection. In Chapters 16 and 21 we saw that open-pollinated varieties of maize have been developed by mass selection by many different farmer-breeders, each selecting his best plants. Carried out over a large number of generations, this was an effective method of improving highly heritable characters and perhaps even yield. Nevertheless, there was ample evidence that mass selection was incapable of producing rapid changes in yielding ability and that it certainly was unable to modify yield very much in a single generation. The dramatic increase in yield resulting from use of hybrid varieties cannot, therefore, be ascribed to the single act of selection by which the individual plants used to establish inbred lines are singled out from open-pollinated populations. The success of hybrid maize is thus not to be attributed to the fact that the foundation individuals are carefully selected and not chosen at random.

There are also difficulties in accepting the view that the success of hybrid corn results from selection during the inbreeding process. Inbreeding is highly efficient at bringing recessives, which when homozygous depress yield, into the open where the breeders can select against them. But even if the efficiency of this selection is complete, it is doubtful whether such purification of the stock has much effect on yields (see Chapter 19). This conclusion seems to be confirmed by the results of the random crossing of inbred lines. In the early days of investigations with hybrid maize, it was customary to produce and test as many single crosses as possible among the inbred lines that were available. Although all hybrids between inbred lines were higher yielding than their inbred parents, many single crosses yielded less than the original open-pollinated variety, and only a few were conspicuously superior to the original open-pollinated population. The mean yield of a group of F_1 hybrids from random combinations of inbred lines derived from a single open-pollinated variety was found to be about the same as for the open-pollinated variety from which the inbred lines were derived. Thus selection at stages 1 and 2 with no selection at stage 3 (page 265) in the production of hybrids results in little if any increase in productivity. We are therefore forced to the conclusion that selection practiced during the inbreeding process is capable of explaining at best only a small portion of the gain that has actually been achieved. By elimination, then, differences in productivity between open-pollinated varieties and hybrids of the order of 20 per cent remain to be explained by selection during stage 3 of the cycle.

In the development of methods for selecting inbreds that combine well to produce outstanding hybrids, two facts stand out. First, progress toward the development of high-producing double crosses was slow until very large numbers of inbred lines were available for testing in combination with other inbreds. Second, many early breeders attempted to develop double crosses by combining inbreds more or less at random. Although an occasional high-yielding double cross may be obtained by this method, outstanding double crosses are more likely to be obtained by combining inbred lines of diverse origin in particular ways. When methods were developed for predicting double-cross yields from the results of single crosses, selection among inbreds could be practiced efficiently and progress accelerated. The conclusion to be reached is obvious. In the improvement of cross-pollinated plants by the direct inbreeding of open-pollinated varieties, many inbred lines are essential. Creation of only a few lines leads to impoverishment of genic content compared to the original open-pollinated stock. There need be no such impoverishment, however, if a multiplicity of lines is developed. In combining these many lines the breeder is not confined to eliminating inferior recessives, as when selecting during the inbreeding process itself, but can take advantage of any of the several or perhaps many genetic forces that can produce heterosis. Both theory and the practical experiences of breeders thus indicate that conditions for selection at the third stage of the operations by which hybrid varieties are produced are more favorable for progress than in the other two stages.

The Evaluation of Inbred Lines

A hybrid between any two unrelated inbred lines of maize is certain to show at least some increase in vigor over its inbred parents. However, very few of the thousands of inbreds which have been tested show heterosis in a measure that makes them valuable economically. Kiesselbach estimated in 1951 that of the more than 100,000 inbreds which had been tested at that time, only about 60 were good enough to have been used in the commercial production of hybrid maize.

Although several hundred hybrids are being grown on a commercial scale at the present time, the number of inbreds being used to produce them is fairly small, and a few popular inbreds enter into the pedigrees of the hybrids that make up the great bulk of the acreage. Among the most popular inbreds are those designated Wf9, 38–11, Hy, R4, L317, L289, Kys, and M–14. Anderson and Brown (1952), in a review of sources of germplasm for hybrid maize, has pointed out that the best inbreds trace back to a few open-pollinated varieties, including such unimportant varieties as Lancaster Surecropper. The most widely grown of all corn hybrids remains U.S. 13, a double cross of (Wf9 × 38–11) (Hy × L317). First produced commercially in 1935, this hybrid includes inbreds developed at three different experiment stations. This example serves to emphasize the difficulty of developing and improving outstanding inbred lines.

The value of any inbred ultimately rests on its ability to produce superior hybrids in combination with other inbreds. In the early days of hybrid maize, breeding tests for combining ability (productivity in crosses) were conducted in a direct manner—that is, by crossing individual inbreds with as many other inbreds as possible. Considering that $n(n-1)/2$ different single crosses can be made from n inbred lines (ignoring reciprocal crosses), it is apparent why this system of testing broke down when a substantial number of lines became available for testing. For example, with 20 inbreds to be tested, this direct method requires measurement of the yielding ability of 190 F_1 hybrids, preferably repeated in more than one season and location. If 100 inbred lines are to be evaluated, 4950 F_1 hybrids must be tested in yield trials. These facts brought home to maize breeders in a forceful way the realization that the development of inbred lines is a minor problem as compared with evaluation of lines, and this realization prompted vigorous research for adequate testing procedures.

The suggestion that finally came to be adopted as standard practice was made by Davis in 1927. He advocated the use of inbred × variety top crosses to test the general combining ability of inbreds. The most comprehensive data relating to the value of this method were reported by Jenkins and Brunson in 1932. Their procedure was to compare the ranking of inbreds as determined by performance in inbred-variety crosses with *average* performance of the same inbreds in a number of single crosses. Inbred lines that produced low yields in top crosses were found to produce low-yielding single crosses. Conversely, the highest-yielding single crosses resulted from crosses between inbred lines that performed well in top crosses. The correlations between inbred-variety and average single-cross production, that is, between *general combining ability* and *average combining ability*, ranged from .53 to .90. On the basis of these studies they concluded it was safe to discard the lower-yielding half of the lines under test without serious risk of losing valuable material. The remaining half of the lines would then be tested in single-cross combinations. Subsequent investigations have substantiated these conclusions, and it is now generally accepted that top crosses are a satisfactory way of evaluating inbred lines for general combining ability, especially when trials are conducted in several seasons and at several locations. Sprague (1939) has described in detail the procedures to be followed in making top crosses. If the variety is used as the seed parent it is recommended that no less than ten plants be used to sample the gametes of the variety. If the cross is made by growing the inbreds in alternate rows with the variety and detasseling the inbreds, the varietal genotype should be represented adequately by the sample of pollen which falls on the silks.

Since the adoption of the inbred-variety cross as a test of combining ability, a number of testers other than open-pollinated varieties have been suggested as top-cross parents. The most desirable tester is obviously the one that provides maximum information about the performance to be expected when the lines under test are used in other combinations or are grown in other environments. The tester should also be one that is simple to use. No single tester fulfills all these requirements for all circumstances, since the value of a tester is determined to a considerable extent by the use to be made of a particular group of lines. For example, when the object is to find a replacement for an existing inbred line in a particular double cross, the most appropriate tester is the opposite single cross of the double cross. On the other hand, if the interest is in attaining a high level of general performance before attempting an evaluation in specific combinations, a tester having a broad genetic base is clearly more desirable. The literature on choice of testers for particular purposes has been reviewed by Green (1948), Keller (1949), and Sprague (1955).

Predicting Performance in Hybrid Combinations

The top cross proved to be a great boon in the testing of inbred lines for combining ability because, with its use, it was possible to identify the more promising inbred lines from a group of size n with only n crosses instead of $n(n-1)/2$ crosses. After the more promising inbreds have been selected on the basis of good *general combining ability*, as measured by top crosses with some suitable tester of broad genetic base, it is necessary to identify the particular single, three-way, or double crosses that will produce the highest yields. This aspect of combining ability was called *specific combining ability* by Sprague and Tatum in 1942. The number of combinations of inbreds taken 4 at a time, to produce double crosses, increases rapidly with increase in the number of inbreds. Thus 20 inbreds can be combined to produce $n(n-1)/2 = 190$ single crosses, and $3n!/[(4!)(n-4)!] = 14,535$ double crosses, excluding reciprocals. The task of the breeder was seen to be an almost impossible one unless some way could be found to predict the performance of double crosses from the performance of single crosses.

Several different methods of making the required predictions were investigated, and it was found that the most accurate estimate of the yield of a double cross could be made from the mean yield of the four *nonparental* single crosses. Thus the average performance of single crosses $A \times C$, $A \times D$, $B \times C$, and $B \times D$ is used to predict the performance of the double cross $(A \times B)(C \times D)$.

The data of Anderson (1938) give an idea of the accuracy of the predictions that can be made by this method. In Anderson's experiment the yields of the 10 single crosses that can be formed from 5 inbred lines were determined in replicated yield trials (Table 22–1).

TABLE 22–1.   Yields in Bushels per Acre of the Ten Possible Single Crosses among Five Inbred Lines of Maize

(After Anderson, 1938)

Cross	Yield	Cross	Yield
(23 × 24)	41.7	(24 × 27)	72.1
(23 × 26)	62.6	(24 × 28)	69.3
(23 × 27)	70.8	(26 × 27)	64.2
(23 × 28)	64.4	(26 × 28)	60.4
(24 × 26)	65.6	(27 × 28)	59.6

Difference required for significance, 6.84.

These single-cross data were used to predict the yields of the 15 double crosses that can be made up from the 10 single crosses. As an example, the predicted yield of the double cross (23 × 24) (26 × 27) was calculated as the average of the 4 single crosses, (23 × 26), (23 × 27), (24 × 26), and (24 × 27), which is 67.8 bushels per acre. The predicted yields were then compared with the actual yields of the double crosses as determined in another replicated trial grown in an area adjacent to the area in which the yields of the single crosses had been determined. These comparisons are given in Table 22–2. The correspondence between the actual and predicted yields is remarkably close. The use of single-cross performance for the prediction of double-cross performance has become a standard breeding procedure. In the larger breeding programs, punched-card equipment has been

TABLE 22–2. Actual and Predicted Yields of Fifteen Double Crosses Formed from the Ten Single Crosses Listed in Table 22–1

(After Anderson, 1938)

Cross	Actual Yield	Predicted Yield
Lines combined: 23, 24, 26, 27		
(23 × 24)(26 × 27)	68.8	67.8
(23 × 26)(24 × 27)	62.4	60.6
(23 × 27)(24 × 26)	62.0	60.2
Lines combined: 23, 24, 26, 28		
(23 × 24)(26 × 28)	65.0	65.5
(23 × 26)(24 × 28)	59.8	58.0
(23 × 28)(24 × 26)	56.0	58.5
Lines combined: 23, 24, 27, 28		
(23 × 24)(27 × 28)	71.1	69.2
(23 × 27)(24 × 28)	58.1	59.4
(23 × 28)(24 × 27)	58.0	60.4
Lines combined: 23, 26, 27, 28		
(23 × 26)(27 × 28)	68.2	65.0
(23 × 27)(26 × 28)	65.0	62.7
(23 × 28)(26 × 27)	65.7	63.4
Lines combined: 24, 26, 27, 28		
(24 × 26)(27 × 28)	70.2	66.5
(24 × 27)(26 × 28)	62.0	64.7
(24 × 28)(26 × 27)	62.7	64.4

Difference required for significance: actual, 5.26; predicted, 3.41.

utilized to lessen the computational labor involved when many predictions are to be made (Combs and Zuber, 1949).

In the prediction of hybrid performance, it might be anticipated that tests conducted in a single season at one location will not provide the information required for accurate evaluation of relative performance in other seasons and at other locations. This expectation has been borne out in practical experience with the result that most maize breeders decide upon elimination or recommendation only after repeated trials. Sprague and Federer (1951) have investigated this problem and have calculated the optimum number of years, locations, and replicates for various circumstances. In general, tests must be conducted in more seasons and at more locations when specific combining ability is of primary interest then when general combining ability is the main concern.

In Chapter 18 it was noted that more than 80 per cent of varietal hybrids in maize had higher yields than the average of their parents and, furthermore, that the greatest evidences of hybrid vigor occurred when the parent varieties were distinctly different types. This leads to an expectation that crosses between inbreds derived from different varieties would tend to be more productive than crosses between inbreds derived from the same or similar open-pollinated sources. The correctness of this assumption has been borne out by the experiences of numerous breeders. The role of genetic diversity in the performance of hybrids has been studied in detail by many investigators (references in Sprague, 1955), all of whom have stressed its importance in the development of superior hybrids.

The order of pairing in double crosses is another aspect of genetic diversity with an important bearing on the productivity of hybrids. This was first studied intensively by Eckhardt and Bryan (1940). They found that if lines A and B had been derived from one source and lines Y and Z from another, the highest-yielding double cross was likely to be obtained from pairings of the type $(A \times B)(Y \times Z)$.

Time of Testing Inbred Lines

Early testing of inbreds for general combining ability, first proposed by Jenkins in 1935, has become a matter of great interest to corn breeders. In the standard method of corn breeding, tests for combining ability are deferred until the third, fourth, or fifth generation of selfing. In the early-testing scheme, the original selections (S_0) or first generation selfed (S_1) plants are outcrossed to a tester stock, and the resulting progeny evaluated for yield and general performance. Lines with poor combining ability are then eliminated, and only the promising lines further inbred. The reasons for the interest of breeders in this scheme are obvious.

Early testing is based on the assumption that: (1) S_0 or S_1 plants differ in combining ability; and (2) these differences can be detected by a test cross despite the problem in sampling posed by the heterozygosity of S_0 or S_1 plants. Jenkins (1935) presented data that inbred lines acquire their individuality as parents of top crosses very early in the inbreeding process and tend to remain fairly stable for combining ability thereafter. Subsequently Sprague (1939, 1946),

Lonnquist (1950), Wellhausen (1952), and Wellhausen and Wortman (1954) presented data on the test-cross performance of S_0 and later-generation lines indicating that early testing aids in the detection of lines that will produce inbreds of good combining ability. In 1946 Sprague expressed the following opinion:

. . . it appears that the early testing procedure is of value where yield is an important consideration or where other important factors can be evaluated easily and efficiently by a suitable tester. When the gene frequency, conditioning some desirable characteristic, is low and where such characteristics can be evaluated visually in the inbred lines, early testing may be of limited value in the preliminary stages of a breeding program. As an example, selection within an open-pollinated variety, by means of the early testing procedure, may be inefficient where a low frequency of genes conditioning resistance to lodging may exist. The value of the method can be expected to increase as the program progresses and more desirable sources of parental material become available.

Other workers, notably Singleton and Nelson (1945), Richey (1945, 1947), and Payne and Hayes (1949), have expressed doubt about the early-testing procedure, holding that visual selection is effective in improving combining ability in early inbred generations, and if discarding is done on the basis of top-cross tests in the first or second inbred generation, many ultimately worth-while lines might be thrown away. Despite the conflicting opinions, early testing has come to be widely used to aid in the early elimination of the plants or lines unlikely to produce superior inbreds upon further inbreeding.

Advanced Generations from Hybrid Varieties

Advanced-generation seed from hybrid varieties has been used in two quite different ways, first, for parent stock in producing hybrid seed and, second, for the planting of the commercial crop. Normally the F_1's of two single crosses are used to produce double crosses, but circumstances have arisen when inadequate supplies of F_1 single-crossed seed made it necessary for F_2 seed to be used in place of F_1 seed. Double crosses produced by the hybridization of F_2 or F_3 generations derived from single crosses must, in the absence of selection that causes changes in gene frequencies, perform the same as double crosses produced from F_1 single crosses. Kiesselbach (1930) found this to be the case. Hayes, Johnson, and Doxtator (1931), however, presented evidence that changes in gene frequencies occurred when they propagated single crosses into the F_2 and F_3 generations. They found, in fact, some evidence for believing that double crosses produced from F_2 and F_3 generations might perhaps be slightly superior to the double crosses obtained from crosses between F_1 single crosses. Such evidence as there is therefore suggests that use of advanced generations of single crosses in place of F_1 single crosses in producing double-crossed seed is an acceptable procedure.

The other use of advanced-generation seed arose from the understandable desire of farmers, accustomed to producing their own planting seed, to use part of the harvest from double crosses to sow the next season's crop. There is substantial evidence that this practice is not a desirable one. Experiments by Richey, Stringfield, and Sprague, reported in 1934, indicated that about 15 per cent reduction in yield can be expected from planting second-generation double-crossed seed. Neal (1935) found 26 per cent reduction from planting the F_2 of double crosses and 36 and 48 per cent reduction from planting second generation three-way and single-crossed seed, respectively. Populations produced by planting advanced-generation seed are in reality *synthetic varieties* based on fewer component lines than can usually be recommended.

Improving Inbred Lines

In the first stages of the development of hybrid maize, inbreds of necessity had to be isolated directly from a heterozygous source. More recently the emphasis has shifted from the isolation of new inbreds to the improvement of existing inbreds. The improvement of established inbreds has usually had one or more of the following objectives: (1) to increase the productivity of the inbreds themselves to facilitate the production of hybrid seed; (2) to fix up inbreds so they will produce hybrids improved in disease and insect resistance, standing ability, or other specific characters; (3) to enhance the combining ability of specific inbreds so as to increase the yielding ability of their hybrids.

The first method of improving inbreds to be tried was "pedigree" or "second-cycle" selection. This method consists of crossing two inbred lines that complement each other in desirable attributes and selecting for desired recombinations in the segregating generations. Wu (1939), Hayes and Johnson (1939), and Johnson and Hayes (1940) have reported the value of this procedure in obtaining inbreds that were superior in their own properties and were also superior in producing hybrids improved in resistance to lodging and smut, as well as in yield.

The backcross method has much the same application to the improvement of inbred lines as it has to the improvement of self-pollinated crops. It has been widely used to improve inbreds in standing ability, resistance to diseases and insects such as smut, *Helminthosporium* leaf blight, and corn borer, and in many other specific characters. In recent years backcrossing has taken on even greater importance in modifying inbred lines because it is the most efficient method of placing the genotypes of inbreds in male-sterile cytoplasm and also for transferring restorer genes into the genotypes of the matching inbred lines.

Convergent improvement, proposed by Richey in 1927, differs from ordinary backcrossing only in that each single cross is backcrossed independently to both of its inbred parents. This is done with the view of improving both inbreds by careful selection for vigor and other favorable characteristics following each backcross. If the single-cross $A \times B$ is backcrossed to A it will be improved to the extent that favorable genes from B can be maintained by selection. Backcrossing $A \times B$ to B can be expected to lead to the similar improvement of inbred B. The first experimental data relating to the value of this method were reported by Richey and Sprague in 1931. They found evidence that three or four backcrosses followed by two or three generations of selfing produced inbreds that were somewhat superior to the original inbreds. Murphy (1942) presented additional evidence indicating that the yields of F_1 crosses in themselves can be increased by convergent improvement. The improvement was not marked, however, and convergent improvement has not been widely adopted as a method of corn breeding.

In 1944 Stadler proposed a modification of early testing, which he called *gamete selection*, as a method of obtaining improved inbred lines. This method involves crossing a good inbred line with a random sample of pollen from an open-pollinated variety. Individual F_1 plants obtained from this variety-inbred cross are then self-pollinated and also outcrossed to an appropriate tester stock. The F_1 plants whose test-cross performance indicates they have received a superior gamete from the open-pollinated variety are then continued from the selfed seed. Stadler emphasized that if superior zygotes occur in an open-pollinated variety with a frequency of q^2, superior gametes occur in the much higher frequency q. He suggested that gametic sampling therefore offers a more efficient means of sampling open-pollinated varieties than standard inbreeding procedures. Richey (1947), Hayes, Rinke, and Tsiang (1946), and others have pointed out the similarity of this method to early testing. Data indicating that the method may have some usefulness for further selection of material from open-pollinated varieties have been presented by

Pinnell, Rinke, and Hayes (1952) and Lonnquist and McGill (1954).

Hybrid Varieties in Crops Other than Corn

Hybrid varieties in maize owe their phenomenal success to the combination of exceptionally favorable floral morphology, permitting easy control of selfing and crossing, and striking heterotic response in certain F_1 combinations. Many other economic species exhibit as much heterosis as maize, and in a few species the structure of the flowers is favorable to easy inbreeding as well as controlled production of large amounts of F_1 hybrid seed. But no other species combines these features as favorably as maize.

In the cucurbits, the floral structure is such that large amounts of F_1 seed can be produced by interplanting two types to be hybridized and removing the staminate flowers of the line chosen to serve as the seed parent of the hybrid. This operation is not as economical as in maize because it must be repeated at about two-day intervals throughout the entire flowering period. It is nevertheless possible to produce hybrid seed in commercial quantities. The difficulty is that the cucurbits do not show conspicuous heterosis, and, despite the optimistic view of Curtis in 1939, F_1 hybrids sufficiently superior to open-pollinated varieties to justify the extra expense have not been developed. The possibility that lines with superior combining ability will ultimately be found cannot be discounted, however, and these species must be regarded as ones where hybrid varieties, if not likely, are at least possible.

In tomatoes, a small hybrid-seed industry based on hand-produced F_1 hybrids has developed since about 1940. Even though a single pollination may produce two hundred or more seeds, the manual transfer of pollen is economical only because certain home gardeners and producers of greenhouse tomatoes are willing to pay high prices for hybrid seed. Genetic emasculation of tomatoes by male sterility has of course been thought of as a way of reducing the labor requirement in the hybridization process, and attempts are being made to find hybrid combinations and environmental conditions that will allow an economical seed set without manual transfer of pollen. Only when this has been achieved can hybrid tomatoes be expected to assume an important place in commercial tomato production. There are some other self-pollinated species similar to tomatoes in their reproductive capacity—for example, peppers—that may be amenable to the same techniques as tomatoes.

Grain sorghum differs from many other predominantly self-pollinated species in possessing good ability to disperse its pollen. The flowers of this species are small and difficult to emasculate. Hence, even though hybrid combinations between varieties have long been known to display striking heterosis, production of hybrid seed was regarded as out of the question until the possibilities of cytoplasmic male sterility were realized. Hybrids based on single crosses between cytoplasmic male-sterile lines and pollinator lines carrying restorer genes have recently come into limited commercial production in this species. Much of the effort of practical sorghum breeders is now devoted to searching for better single-cross combinations and setting up the inbreds to be used in hybrid varieties with appropriate cytoplasm and restorer genes so that hybrids can be produced by natural crossing.

Many other species have large numbers of small, perfect flowers per inflorescence, making the manual production of hybrid seed in commercial quantities completely out of the question. Onions, carrots, and both sugar and garden beets fall in this category. Since these cross-pollinated species have good pollen dispersal mechanisms, the only major barrier to the production of cross-pollinated seed in quantity is emasculation. As we saw in Chapter 20, cytoplasmic male sterility accomplishes this emasculation effectively and has been a key force in making hybrid onions a major factor in commercial onion production. More than fifty varieties of hybrid onions have been released to growers in the United States, and these varieties occupy a major share of the acreage. The method of producing hybrid carrots and beets is basically the same as the onion scheme discussed in detail earlier, and prospects are good that a substantial portion of the acreage of these crops will soon be planted to hybrid varieties.

REFERENCES

Anderson, D. C. 1938. The relation between single and double cross yields of corn. *Jour. Amer. Soc. Agron.* 30: 209–211.
Anderson, Edgar, and W. L. Brown. 1952. Origin of corn belt maize and its genetic significance. In *Heterosis,* pp. 124–148. Iowa State College Press.
Combs, J. B., and M. S. Zuber. 1949. Further use of punched card equipment in predicting the performance of double-crossed corn hybrids. *Agron. Jour.* 41: 485–486.
Curtis, L. C. 1939. Heterosis in summer squash (*Cucurbita pepo*) and the possibilities of producing F₁ hybrid seed for commercial planting. *Proc. Amer. Soc. Hort. Sci.* 37: 827–828.

Eckhardt, R. C., and A. A. Bryan. 1940. Effect of method of combining the four inbred lines of a double cross of maize upon the yield and variability of the resulting hybrid. *Jour. Amer. Soc. Agron.* 32: 347–353.

Eckhardt, R. C., and A. A. Bryan. 1940. Effect of the method of combining two early and two late inbred lines of corn upon the yield and variability of the resulting double crosses. *Jour. Amer. Soc. Agron.* 32: 645–656.

Green, J. M. 1948. Inheritance of combining ability in maize hybrids. *Jour. Amer. Soc. Agron.* 40: 58–63.

Hayes, H. K., and I. J. Johnson. 1939. The breeding of improved selfed lines of corn. *Jour. Amer. Soc. Agron.* 31: 710–724.

Hayes, H. K., I. J. Johnson, and C. W. Doxtator. 1931. *Minn. Corn Breeding Report. Purnell Corn Imp. Rept.* p. 10.

Hayes, H. K., E. H. Rinke, and Y. S. Tsiang. 1946. Experimental studies of convergent improvement and backcrossing in corn. *Minn. Agric. Exp. Sta. Tech. Bull.* 172: 3–40.

Jenkins, M. T. 1935. The effect of inbreeding and of selection within inbred lines of maize upon the hybrids made after successive generations of selfing. *Iowa State Coll. Jour. Sci.* 3: 429–450.

Jenkins, M. T., and A. M. Brunson. 1932. A method of testing inbred lines of maize in crossbred combinations. *Jour. Amer. Soc. Agron.* 24: 523–530.

Johnson, I. J., and H. K. Hayes. 1940. The value in hybrid combinations of inbred lines of corn selected from single crosses by the pedigree method of breeding. *Jour. Amer. Soc. Agron.* 32: 479–485.

Jones, D. F. 1918. The effects of inbreeding and crossbreeding on development. *Conn. Agric. Exp. Sta. Bull.* 207: 1–100.

Jones, D. F. 1958. Heterosis and homeostasis in evolution and in applied genetics. *Amer. Nat.* 92: 321–328.

Jones, D. F., and W. R. Singleton. 1934. Crossed sweet corn. *Conn. Agric. Exp. Sta. Bull.* 361: 489–536.

Keller, K. R. 1949. A comparison involving the number and relationship between testers in evaluating inbred lines of maize. *Agron. Jour.* 41: 323–331.

Kiesselbach, T. A. 1930. The use of advanced generation hybrids as parents of double cross seed corn. *Jour. Amer. Soc. Agron.* 22: 614–626.

Kiesselbach, T. A. 1951. A half-century of corn research. *Amer. Sci.* 39: 629–655.

Lonnquist, J. H. 1950. The effect of selection for combining ability within segregating lines of corn. *Agron. Jour.* 503–508.

Lonnquist, J. H., and D. P. McGill. 1954. Gametic sampling from selected zygotes in corn breeding. *Agron. Jour.* 46: 147–150.

Murphy, R. P. 1942. Convergent improvement with four inbred lines of corn. *Jour. Amer. Soc. Agron.* 34: 138–150.

Neal, N. P. 1935. The decrease in yielding capacity in advanced generations of hybrid corn. *Jour. Amer. Soc. Agron.* 27: 666–670.

Osler, R. D., E. J. Wellhausen, and G. Palacios. 1958. Effect of visual selection during inbreeding upon combining ability in corn. *Agron. Jour.* 50: 45–48.

Payne, K. T., and H. K. Hayes. 1949. A comparison of combining ability in F_2 and F_3 lines of corn. *Agron. Jour.* 41: 383–388.

Pinnell, E. L., E. H. Rinke, and H. K. Hayes. 1952. Gamete selection for specific combining ability. In *Heterosis*, pp. 378–388. Iowa State College Press.

Richey, F. D. 1927. Corn breeding. *U. S. Dept. Agric. Bull.* 1489.

Richey, F. D. 1927. The convergent improvement of selfed lines of corn. *Amer. Nat.* 61: 430–449.

Richey, F. D. 1945. Isolating better foundation inbreds for use in corn hybrids. *Genetics* 30: 455–471.

Richey, F. D. 1947. Corn breeding, gamete selection, the Oenothera method and isolated miscellany. *Jour. Amer. Soc. Agron.* 39: 403–412.

Richey, F. D. 1950. Corn breeding. In *Adv. in Gen.* 3: 160–189.

Richey, F. D., G. H. Stringfield, and G. F. Sprague. 1934. The loss in yield that may be expected from planting second generation double cross corn seed. *Jour. Amer. Soc. Agron.* 26: 196–199.

Richey, F. D., and G. F. Sprague. 1931. Experiments on hybrid vigor and convergent improvement in corn. *U. S. Dept. Agric. Tech. Bull.* 267.

Shull, G. H. 1909. A pure line method of corn breeding. *Amer. Breed. Assoc. Rept.* 5: 51–59.

Singleton, W. R., and O. E. Nelson. 1945. The improvement of naturally cross pollinated plants by selection in self-fertilized lines. IV. Combining ability of successive generations of inbred sweet corn. *Conn. Agric. Exp. Sta. Bull.* 490: 458–598.

Sprague, G. F. 1939. An estimation of the number of top-crossed plants required for adequate representation of a corn variety. *Jour. Amer. Soc. Agron.* 31: 11–16.

Sprague, G. F. 1946. Early testing of inbred lines of corn. *Jour. Amer. Soc Agron.* 38: 108–117.

Sprague, G. F. 1946. The experimental basis for hybrid maize. *Biol. Rev.* 21: 101–120.

Sprague, G. F. 1955. Corn breeding. In *Corn and corn improvement*, pp. 283. Academic Press, New York.

Sprague, G. F., and W. T. Federer. 1951. A comparison of variance components in corn yield trials. II. Error, year × variety, location × variety, and variety components. *Agron. Jour.* 43: 535–541.

Sprague, G. F., and P. A. Miller. 1952. The influence of visual selection during inbreeding on combining ability in corn. *Agron. Jour.* 44: 258–262.

Sprague, G. F., and L. A. Tatum. 1942. General vs. specific combining ability in single crosses of corn. *Jour. Amer. Soc. Agron.* 34: 923–932.

Stadler, L. J. 1944. Gamete selection in corn breeding. *Jour. Amer. Soc. Agron.* 36: 988–989.

Wellhausen, E. J. 1952. Heterosis in a new population. In *Heterosis,* pp. 418–450. Iowa State College Press.

Wellhausen, E. J., and L. S. Wortman. 1954. Combining ability of S_1 and derived S_3 lines of corn. *Agron. Jour.* 46: 86–89.

Wu, S. K. 1939. The relationship between the origin of selfed lines of corn and their value in hybrid combinations. *Jour. Amer. Soc. Agron.* 31: 131–140.

Zirkle, C. 1952. Early ideas on inbreeding and crossbreeding. In *Heterosis,* pp. 1–13. Iowa State College Press.

23

Recurrent Selection

The standard system for producing hybrid varieties of cross-pollinated crops has involved the selection of desirable plants from heterozygous sources, inbreeding progenies of these plants to homozygosity, and utilizing the best inbreds in F_1 hybrids of one kind or another. Initial successes with this system in maize were obtained with inbred lines that were isolated directly from open-pollinated varieties. As the hybrid-corn program developed further there was an increasing tendency to bypass open-pollinated varieties as sources of inbreds because of the low frequency with which direct isolation produces outstanding inbreds. Emphasis shifted more and more to selection in segregating generations from crosses between superior inbred lines. These "second-cycle" inbred lines as a group were considerably improved *as lines* over the "first-cycle" lines, but their hybrids were not conspicuously superior in yield to the hybrids of the original lines. Doubts arose in the minds of many breeders whether the standard system of inbreeding, despite its unquestioned success, was the most efficient one possible, and, as a result, the genetic postulates upon which it was based came under careful scrutiny.

In analyzing the standard system, it appeared that the proportion of outstanding inbred lines to be expected depended on two factors: (1) the proportion of superior genotypes in the heterozygous source material from which the inbreds were isolated, and (2) the effectiveness of selection in increasing the frequency of desirable genes or gene combinations during the inbreeding process.

The proportion of superior genotypes occurring in a heterozygous population is a function of the frequency of desirable genes in the gene pool. The higher the frequency of desirable gene combinations, the greater the expectation of finding plants with high performance. The fact that second- and third-cycle inbreds had not contributed appreciably higher yields in crosses than the better lines extracted originally from open-pollinated varieties suggested the possibility that the frequency of favorable genes in the populations being sampled was too low to permit the extraction of exceptionally good combining lines with the sample sizes commonly used. If this was true, selecting and selfing more foundation plants should have improved the situation. This might not, however, be the most efficient solution, because the chance of obtaining deviates exceeding the mean by more than two or three standard deviations does not increase commensurate with increases in sample size. A more reasonable solution to the problem appeared to be some method of cumulative selection that would permit a gradual increase in the level of desirability of the populations to be sampled.*

Implicit in this idea was the assumption that a ceiling is imposed on the ultimate value of a derived line by the genotype of the foundation plant from which it was derived. The most favorable combination of genes that can be isolated from a foundation plant under the standard selfing procedure depends on the effectiveness of selection during the selfing process. If selection is ineffective, random fixation of genes will occur during selfing. On the other hand, if effective selection could be practiced during selfing it might be possible to approach the ceiling performance set by the particular sample of genes carried by the foundation plant. The evidence available suggested that selection during the selfing series was not particularly effective in increasing the frequency of desirable genes, except for genes governing the most easily observable characters. The reasons for this could be deduced from Mendelian expectations. Half of the heterozygous genes become homozygous in each generation of selfing. Hence the progress toward fixation is rapid in a selfing series—so rapid, in fact, that there is some doubt whether the most intense selection can have little more than trivial influence on the ultimate genotype of a line developed by continuous selfing.† When the small populations usually grown in each progeny generation, the hindrances to recombination imposed by linkage, and the limited effectiveness of visual selection were also taken into account, it seemed most unlikely that the potential ceiling

---

* As an example, consider a population in which 5 genes of equal effect govern the character under improvement. If q, the frequency of the more favorable of the 2 alleles at any locus, is .5 for each of the 5 genes, approximately one individual per 1000 is expected to be homozygous for the more favorable allele at all 5 loci. If, however, q could be increased by selection to .95 at each locus, approximately 600 out of 1000 individuals would be expected to be homozygous for all the favorable alleles.

† The following analogy is by Lush (1943): ". . . the difference between intense selection and no selection under self-fertilization has little more effect on the

set by the genotype of the foundation plant could ever be achieved in any inbred line developed by selfing.

Since selfing did not seem to allow adequate opportunity for selection, the suggestion was made that some less intense form of inbreeding might aid in the selection of the theoretically highest combining genotype from a given foundation plant. The reasoning was as follows. In a selfing series, a single plant is selected in each generation to propagate the line. Thus, in every generation, a new limit is imposed on the possibility for further improvement of the line because there is no possibility of regaining favorable genes lost in selecting a particular plant. If, on the other hand, milder inbreeding were to be practiced, the propagation of a given line would be based on intercrosses of more than one individual. The ceiling in this case would be the most desirable combination of genes possible from all genes present, not in just one individual, but in a group of selected individuals. Hence the chance of random loss of favorable genes would be reduced. Also, the intercrossing within the line would allow greater chance for recombination between favorable genes.

When evidence obtained from experiments in both plant and animal breeding was evaluated, it turned out that the hybrid performance of inbred lines obtained under mild inbreeding with selection was not much different from the results with selfing. There seemed to be two explanations for this result. First, the frequency of favorable genes in the original population was perhaps too low to allow the formation of enough foundation plants with superior genotypes. Second, perhaps even mild inbreeding was incapable of preventing random fixation. Recurring selection *within* mildly inbred lines therefore did not appear to be the answer to the problem. In view of this difficulty, it appeared that some breeding scheme was needed which would maintain genetic variability in the breeding population and which, at the same time, would allow the frequency of desirable genes and gene combinations to be increased by providing for recombination among lines derived from different foundation plants.

In the decade starting in 1940, these ideas were developed into a breeding system that acquired the general title of *recurrent selection.*

---

outcome than the fate of a man dropped into the Niagara River just a few yards above the falls would be affected by whether he is a good swimmer or a poor swimmer! The difference in swimming ability might be of tremendous importance if he were in comparatively still water (as is roughly analogous to the situation in which the inbreeding is mild) but would rarely make a detectable difference in the results in the presence of the much more powerful force of the swiftly moving water."

This system, as used in maize, has the following operational features: (1) plants from a heterozygous source are self-pollinated and at the same time are evaluated for some desirable character or characters; (2) plants with inferior performance for the character or characters under improvement are discarded; (3) the superior plants are propagated from the selfed seed; (4) all possible intercrosses among these superior progenies are made by hand, or, if this is impractical, the intercrosses are made by open pollination among the selected progenies; (5) the resulting intercross population serves as source material for additional cycles of selection and intercrossing.

Breeding schemes similar to this one were first suggested in 1919 by Hayes and Garber and independently by East and Jones in 1920. Critical data were published in neither case, however, and these suggestions did not lead to use of the method. The first detailed description of this type of breeding scheme was published by Jenkins in 1940 as a result of his experiments with early testing for general combining ability in maize. It was not until 1945, when Hull suggested that selection after each of several cycles of intercrossing might be useful in improving specific combining ability, that the method acquired the name recurrent selection. According to Hull (1952): "Recurrent selection was meant to include reselection generation after generation, with interbreeding of selects to provide for genetic recombination. Thus selection among isolates, inbred lines, or clones is not recurrent until selects are interbred and a new cycle of selection is initiated."

The advantage of this system is that the ceiling performance is set, not by the genotype of a single foundation plant, but by the most favorable combination of genes contained in a group of foundation plants. The chance of obtaining satisfactory individuals should therefore be increased compared to selection within selfed or mildly inbred lines because greater opportunity for recombination is present. Also, since the rate of inbreeding can, with care, be kept at a low level, it should be possible to maintain high genetic variability and hence provide for effective selection over a longer period.

In discussing recurrent selection, it is convenient to recognize four different types distinguished by the way in which plants with desirable attributes are identified. These types are: (1) simple recurrent selection, (2) recurrent selection for general combining ability, (3) recurrent selection for specific combining ability, and (4) reciprocal recurrent selection. In simple recurrent selection, plants are divided into a group to be discarded and a group to be propagated further, on the basis of phenotypic scores taken on individual plants or their selfed progeny (Figure 23–1). Since test crosses are not made, the effective use of simple recurrent selection is restricted to characters with sufficiently high heritability that an accurate phenotypic evaluation of the character can be made visually or by simple tests. It

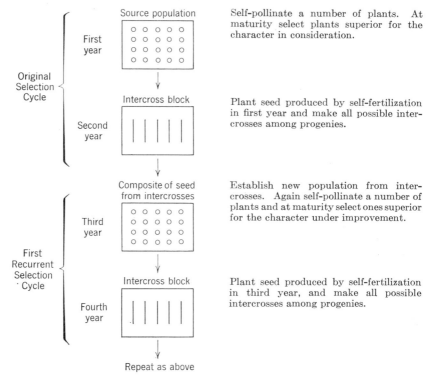

FIGURE 23-1. Diagrammatic representation of simple recurrent selection. How should the above diagram be modified to represent simple recurrent selection for a character which is expressed before completion of the sexual reproductive cycle? How would the procedure-diagrammed above be modified if the species were one in which hand pollinations are difficult? If the species can be propagated clonally?

cannot, for example, be used with much effectiveness in breeding for improved combining ability for yield.

The three remaining types of recurrent selection differ from simple recurrent selection in employing test crosses to measure combining ability. If the tester used has a broad genetic base, variations in performance in a group of test crosses will be due primarily to differences in general combining ability, and the scheme will be one of recurrent selection for general combining ability. In the third type of recurrent selection the tester, an inbred line, has a narrow genetic base. Variation in test-cross performance is therefore to be ascribed to differences in specific combining ability. The fourth type, reciprocal recurrent selection, employs two heterozygous source populations, each of which is the tester for the other (see Figure 23–6). This method, with certain limitations, provides for selection for both general and specific combining ability.

The initial heterozygous source population in which recurrent selection is practiced to concentrate genes for desirable attributes can be set up in various ways. It can be an open-pollinated variety, a synthetic variety, progeny of intercrosses among selected inbred lines, a double cross, or a single cross, depending on the breeder's judgment as to the best source of the genes he requires. Populations resulting from recurrent selection can be used in various ways to produce improved varieties. They can be inbred to produce homozygous lines to be used in producing hybrid varieties; they can be crossed with inbred lines, single crosses, or with other populations obtained by recurrent selection to form hybrid varieties; or they can be used as sources of foundation stocks for synthetic varieties.

Simple Recurrent Selection

The first convincing experimental evidence on the effectiveness of simple recurrent selection in changing gene frequencies was presented in 1950 by Sprague and Brimhall, who studied oil content in the corn kernel. Data were presented on the effectiveness of two cycles of recurrent selection, contrasted with selection within inbred lines, in increasing oil percentage. Subsequently another report including data on some unrelated materials has appeared (Sprague, Miller, and Brimhall, 1952). One of the heterozygous source populations used by these investigators was a synthetic variety designated as "stiff stalk." The procedure with the stiff-stalk synthetic was as follows. Approximately 100 ears were self-pollinated and the seed from each ear analyzed for oil percentage. The 10 ears having the highest oil percentage were used as parental material for both a selfing series and a recurrent-selection series. In the recurrent series the 10 ears having the highest oil percentage were grown in ear-row progenies and all possible inter-crosses made by hand. In the first cycle, equal quantities of seed of each combination were bulked and the composite used to plant a plot within which approximately 100 ears were selfed and analyzed individually for oil percentage. The 10 ears having the highest oil percentage were grown in progeny rows and intercrossed as before. The intercrosses at this stage were not bulked, but each was grown as a separate entry so that information could be obtained on the degree of inbreeding per cycle. In each sampling of the intercrossed population, the plants selected for selfing were chosen on the basis of vigor and general desirability. At least 200 self-pollinations were made at each sampling and approximately half discarded at harvest on the basis of vigor, disease and insect resistance, ear and kernel type, and other characters.

The selfing series was derived from the same 10 foundation ears used in the recurrent-selection series. These ears were the basis of a standard selection program for high oil content. An effort was made to keep size of plot grown, number of pollinations, and intensity of selection for phenotypic characters as similar as possible for the selfing and recurrent-selection series. Differences in results therefore provide a direct measure of the comparative effectiveness of recurrent selection versus selection under selfing.

The results obtained with the recurrent-selection series are given in Figure 23–2. The mean oil percentage was shifted rapidly upwards, being 4.2 in the original population, 4.97 for the 10 foundation ears selected from the original population, 5.2 for the first cycle, and 7.0 for the second cycle. The range in oil content was similar in first- and second-cycle populations and greater than the range in oil percentage of the original population. Standard errors follow the same pattern as the ranges, being greater for the first- and second-cycle populations (.72 and .73, respectively) than for the original population (.53).

The relative efficiencies of the selfing series and recurrent-selection series could be compared in a number of ways. The results of these comparisons can be summarized by stating that recurrent selection varied, depending on the type of comparison, from 1.3 to 3 times as effective as the selfing series. All of the comparisons tended to minimize the efficiency of recurrent selection because they did not take residual genetic variability into account. After 5 generations of self-pollination, the 10 families comprising the selfing series all appeared to be quite uniform whether judged by general phenotype or variation in oil percentage. The recurrent series was much more variable in both respects, and there was no indication that the recurrent selection had caused any important reduction in genetic variability contrasted to the original population.

One of the important factors influencing genetic variability in either breeding system was the rate of inbreeding. In the selfing series, the approach to homozygosity should follow the series $\frac{1}{2}$, $\frac{3}{4}$, $\frac{7}{8}$, . . . , with fixation of genes largely independent of the selection practiced. In the recurrent series, the rate of inbreeding should vary according to the

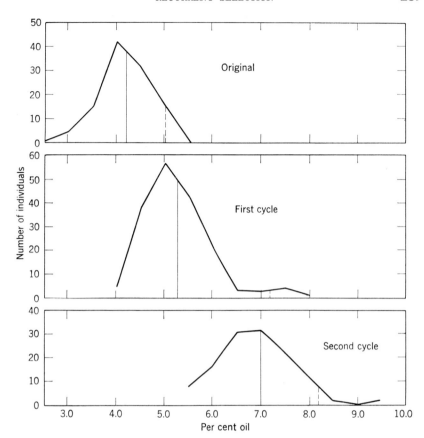

FIGURE 23-2. Frequency distributions for oil percentage in the original stiff-stalk synthetic population and after one and two cycles of recurrent selection. In each distribution the mean is indicated by a solid vertical line, and the mean of the ten ears chosen as parents for the next cycle by a broken vertical line. (After Sprague, Miller, and Brimhall, 1952.)

number of lines saved and the frequency with which each line was represented in the parentage of the plants selected from the inter-crossed population. If, in Sprague, Miller, and Brimhall's experiment, each of the 10 lines saved in cycle 1 had participated equally in the parentage of the selected sample from cycle 2, the percentage of inbreeding would have been 2.8 per cent. When this was checked in 5 different recurrent-selection experiments, it was found that the actual inbreeding coefficients varied from 4.4 to 13.6 per cent. In 3 of the 5 experiments, the inbreeding did not appear to be sufficiently high so that any considerable random fixation should occur. In 2 experiments, it was more probable that the inbreeding might lead to random fixation.

If excessive inbreeding is suspected in recurrent selection programs, it can be avoided or minimized in one of two ways. First, the intercross population can be allowed to mate at random for one generation before the population is reselected. This would almost certainly avoid excessive inbreeding but would also decrease the rate of expected gain by adding one year to each cycle. The decrease in rate of gain per year would presumably be compensated by a decreased rate of loss in genetic variability and a corresponding increase in opportunity for selection before fixation could occur.

The second procedure involves growing the intercrosses as separate progenies with maintenance of complete records of parentage. The selected sample could then be chosen to minimize relationship. This might often be done without seriously reducing the selection differential that could be maintained, but at the expense of greater labor requirement than with the first method.

Gene frequency is another factor known to have an effect on the efficiency of selection (Chapter 15). Selection is most effective when gene frequencies are near .5 and least effective when they are near zero or one. If gene frequencies are less than .5, any increasing effect that selection might have on the frequency would be expected to lead to greater variability. This was advanced as a possible explanation for the increased variability that was observed in all the first-cycle populations selected for increased oil percentage.

The method of recurrent selection illustrated in the foregoing experiment can be modified in many ways to suit special needs. Jenkins, Robert, and Findley (1954), for example, have used simple recurrent selection in concentrating genes for resistance to the leaf blight of maize caused by *Helminthosporium turcicum*. The source populations were 9 different single crosses that represented in most cases hybrids between resistant and susceptible inbred lines. The populations for selection consisted of about 250 plants that were inoculated repeatedly with a suspension of spores to induce the disease. At pollinating time the 10 most resistant plants in each population were selected for interpollination. Pollen was collected from these plants, mixed in approximately equal proportions, and placed on the silks of the same 10 plants. Seed from these hand-pollinated ears was mixed in equal proportions to produce the population of plants for the next cycle of selection.

Three cycles of recurrent selection were practiced within each of the nine progeny groups. This selection proved to be effective in improving the mean leaf-blight scores of the first-cycle population compared to the original population and also the scores of the second-cycle population compared to the first-cycle population. The third cycle of selection was less effective than the first two cycles.

These results support the conclusions of Sprague and his colleagues that recurrent selection is effective in increasing the frequency of desirable genes in populations. The difference between the two experiments lies in the apparent decrease in genetic variability caused by selection for leaf-blight resistance contrasted with the good evidence that selection for oil content was not accompanied by loss in genetic variability. It is possible that so few genes govern leaf-blight resistance that their frequency was increased to the point where little genetic variability remained after only two cycles of recurrent selection.

Recurrent Selection for General Combining Ability

In this system of recurrent selection, a number of plants which appeal to the breeder are selected from the source population. These S_0 plants are selfed and also crossed to a heterozygous tester stock to identify the S_0 individuals with good general combining ability. The selected individuals are propogated from the selfed seed, intercrossed in all combinations, and a composite of the intercrossed seed is then used to establish a population for further selection.

This type of recurrent selection developed as a direct outgrowth of studies of early testing. Early testing, first proposed by Jenkins in 1935, was based on the dual assumptions that there are marked differences in combining ability among plants in open-pollinated populations, and that a selected sample based on tests of combining ability of S_0 (or S_1) plants offers promise of yielding a larger proportion of superior lines on selfing than a sample drawn from the same population on the basis of visual selection alone. Several investigators have reported results indicating that variance for combining ability among different S_0 or S_1 plants (families) is as great or greater than the within-family variance (review in Sprague, 1946, 1952; see also Chapter 22). This suggests that selection among families in early generations of inbreeding should be more efficient than selection within families. The type of result actually obtained can be illustrated by some of the experiments of Lonnquist.

In one of Lonnquist's experiments a locally adapted strain of the Krug variety of maize was used as an open-pollinated source population. Approximately 200 plants from this population were self-pollinated, and from this sample the 36 best plants, as determined by visual inspection at harvest, were harvested. In the next season a row containing about 30 plants was sown from seed of each of these 36 selected plants. Krug was sown in alternate rows in the same block. Each S_1 progeny row was detasseled and top-crossed seed was harvested from all plants within each of the 36 progeny rows. This seed was used to measure the combining ability of the corresponding parental S_0 plant, and, on the basis of these tests, the 8 foundation plants whose top crosses exceeded the mean yield of all 36 top crosses by one or more standard errors were selected. Fifty selfed seeds from each of these 8 plants were then composited to plant a crossing block that was grown in isolation from all other maize. A similar procedure was used to establish a low-combining population based on the 7 plants whose top-cross yields were one or more standard errors below the mean yield of all 36 top-cross progenies.

Following two generations of random mating (one generation with and one generation without selection of the better-appearing plants), 152 S_0 plants from the high-combining population were selfed and simultaneously outcrossed to the single cross Wf9 $\times$ M–14 as a tester. Top crosses involving 77 S_0 plants from the low-combining population were also made. The frequency distributions for these two sets of top crosses are shown in Figure 23–3. It is obvious that the method used was successful in separating the original Krug population into two groups differing in combining ability.

Another of Lonnquist's studies of these same materials has provided information concerning the stage of inbreeding at which tests for combining ability will be most efficient in eliminating lines unlikely to be of value. This experiment was designed to measure the extent to which the combining ability of lines can be altered by selection aided by top-cross performance tests made in the S_2, S_3, and S_4 generations. Divergent selection was practiced within each selfed generation in each of the eight high lines and within each of the seven low lines. This was done by selfing several plants within each S_1 line and simultaneously crossing the same plants to a tester stock (Wf9 $\times$ M–14). On the basis of top-cross performance a high- and a low-combining S_1 plant were chosen to initiate the two directions of selection for yield within each S_1 progeny. Selection within the S_2, S_3, and S_4 generations was continued in like manner. The results obtained clearly indicate that top-cross combining ability can be modified by testing and selection (Figure 23–4). It will be noted that selection for high yield in the low group for three generations, starting in the S_1, produced S_4 lines that were the same in yielding ability as

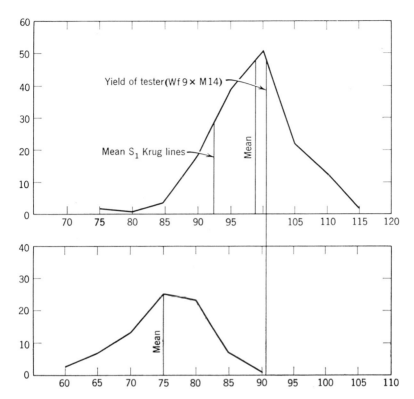

FIGURE 23-3. Frequency distributions of yields of 152 top-crossed plants from a population selected for high-combining ability (upper curve) and 77 plants from a population selected for low-combining ability. Both populations originated from S_1 lines of the Krug Yellow Dent variety. The top-crosses were grown in different years. (After Lonnquist, 1951.)

those produced by selecting for low yield in the high group for the same number of generations. This indicates that selection and testing in the low group are less efficient as a breeding procedure than selecting high-combining S_1 plants. If selection and testing were to be continued after S_1, it is most profitably confined to lines exhibiting the highest top-cross scores for yielding ability.

In still another study of these same materials, McGill and Lonnquist reported on the effects of two cycles of recurrent selection in modifying the combining ability of the populations derived from Krug. Two second-cycle populations were developed from the high-combining first-cycle population. One was based on 10 S_1 lines whose test crosses exceeded the mean of all 152 test-cross progenies by two or more

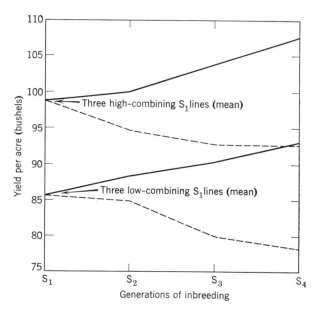

FIGURE 23-4. Comparative mean top-cross performance of three high- and three low-combining S_1 lines of Krug and their respective high- and low-combining segregates (S_2 to S_4) selected for divergent top-cross performance in each generation. (After Lonnquist, 1950.)

standard-deviation units. The second was based on 31 S_1 lines, including the 10 above, whose test-cross yields exceeded the mean by one or more standard-deviation units. A similar second-cycle population was developed from the low-combining first-cycle population, based on 11 S_1 lines whose test crosses were one or more standard-deviation units below the mean of the 77 test crosses that had been made from the first-cycle population. The second-cycle populations were handled in basically the same manner as the first-cycle populations. These three populations were then compared with each other and with the original open-pollinated Krug variety by making test crosses of plants from each population to the single cross Wf9 $\times$ M–14. The test-cross yields are shown in graphic form in Figure 23–5.

Considering the parentage of these populations, certain logical comparisons can be made among the mean test-cross yields. The difference between the mean of the test crosses from the high-combining population based on 10 S_1 plants, KH(10), and the mean of the test crosses from the original Krug was significant. Likewise the difference between KH(31) and original Krug was significant. The mean of the Krug test crosses was significantly greater than the mean of the low-

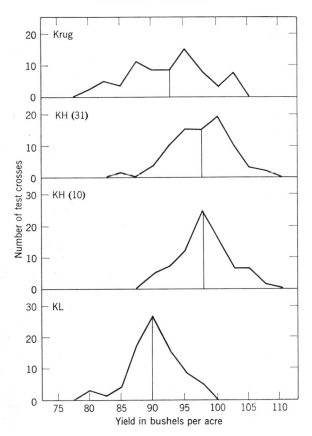

FIGURE 23-5. Frequency distributions of test crosses of Krug and three second-cycle populations derived by recurrent selection. KH(31) and KH(10) represent composites based on 31 and 10 S_1 lines, respectively, from the high-combining, first-cycle population. KL represents the composite derived from the low-combining, first-cycle population. (After McGill and Lonnquist, 1955.)

combining population. It is apparent, therefore, that two cycles of recurrent selection were effective in altering combining ability.

The results of this experiment are perhaps even more interesting for the light they throw on the effect of recurrent selection on genetic variability for combining ability. A visual comparison of the relative variability of the four populations is available in the frequency distributions of test-cross yields presented in Figure 23–5. It is obvious that the test crosses of open-pollinated Krug are more variable than those from the other populations, but it is not clear from visual comparisons whether the other populations differ from one another. Upon analysis of the component of variance due to genetic differences among S_0 plants within each population, the genetic variance for Krug was found to be significantly larger than the variances for the other populations. Differences among the selected populations were not significant, however.

Variability in combining ability is therefore smaller in the selected populations than in the open-pollinated variety from which they were derived. This reduction may have been due either to effective selection altering the frequency of yield genes, to inbreeding, or to a combination of these two factors. It is not possible, of course, to calculate changes in gene frequency, so that the effect of the first of the foregoing factors cannot be evaluated. The amount of inbreeding can be calculated, however, assuming random mating to have occurred in the intercrossing phase. It was found to be 7.8 per cent for KH(31), 10.9 per cent for KH(10), and 11.4 per cent for KL. These values were not considered to be sufficient to explain the observed reductions in variability.

McGill and Lonnquist believe the loss in variability to be due to underestimation of the amount of inbreeding that actually occurred in the intercrossing phase of the experiment. First, if there was any deviation from random mating the effect would be an increase in the inbreeding coefficient. They cite evidence that differences in date of tasseling, date of silking, length of pollen-shedding period, and plant height can affect randomness of mating. Even in stocks similar in flowering habits, random mating may not be achieved. Differences in yielding ability among S_1 lines would also increase inbreeding by giving high-yielding lines excess representation in the bulked-progeny increase. In addition, any line especially prepotent in combining ability would be expected to be represented more frequently in the sample of S_0 plants selected in the next cycle than would less potent lines. This would have the favorable effect of shifting gene frequency in the desired direction but would do so at the expense of increasing inbreeding and decreasing the chance for recombination. Differences observed among S_1 lines in germinability of seeds were another factor believed to have a significant effect in increasing the actual amount of inbreeding.

Ability to maintain genetic variability at a high level is one of the advantages assumed for recurrent selection. It is apparent, however, from the results of McGill and Lonnquist's experiment that genetic variability may be dissipated at a rate considerably in excess of expectation unless precautions are taken to hold actual amounts of inbreeding within acceptable limits. This subject will not be reopened here since remedial procedures were considered in connection with the recurrent selection experiments concerned with high-oil content of the corn kernel.

The results of the various experiments that have been performed to test the effectiveness of recurrent selection for general combining ability leave little doubt that this technique can produce substantial improvements in yield and general agricultural worth. If the increase that has been achieved in top-cross performance can be attributed to an increase in the frequency of desirable genes, recurrent selection for general combining ability would appear to have value not only as an end in itself—that is, the development of synthetic varieties for commercial production—but also as a means to an end, namely, the development of reservoirs of germplasm highly suitable for the extraction of superior inbred lines. Discussion of this first usage will be deferred to the next chapter, where it will be considered in detail. Critical evidence on the second usage is not available at present but must wait until actual extraction of inbred lines from populations improved by recurrent selection provides information on the relative frequency of superior lines obtained from such sources, compared to the proportion of superior lines obtained from other source populations.

Recurrent Selection for Specific Combining Ability

This type of recurrent selection was proposed by Hull in 1945 on the assumption that an important part of heterosis results from the nonlinear interactions of genes at different loci, from interaction between alleles at the same locus, or from both causes in combination. Hull's proposal called for tests to determine whether higher levels of specific combinability can be achieved by recurrent-selection procedures based on progeny tests with a homozygous tester line. This was to be accomplished by a breeding scheme basically the same in operational procedure as recurrent selection for general combining ability except for the difference in tester. Hull argued that evidence of significant advance in specific recombinability would indicate that (1) selection within and among inbred lines had been overemphasized in the standard breeding procedure, and (2) the inbreeding phase in the standard method of developing second-cycle inbreds could be abandoned with great saving in time and effort. The emphasis in Hull's scheme was thus upon recurrent selection to develop complementary strains by making a population steadily approach the opposite extreme in gene frequency from the line used as a tester. The practical objective of Hull's scheme was to produce lines that combine well with the tester stock. Once such highly complementary stocks had been developed, they could be used in crosses with the tester inbred to produce commercial hybrids.

Hull's plan has been criticized on the basis that it is restricted by the necessity for choosing a single homozygous tester line at the beginning of the program. This tester line must, of course, be maintained intact through several cycles of selection and serve as one parent of the commercial hybrid. Thus, if some barrier to future progress turns up in the tester partway through the program or if newer and better inbreds are developed, much effort may have to be sacrificed. Hull argues that much effort is expended in improving inbred lines by backcrossing and that lines that merit such attention may also merit the further attention necessary to build strains which are complementary to them. This problem could be overcome to some extent by substituting an outstanding single-cross tester in place of a homozygous line. This would result in some loss of efficiency in building complementary strains.

No data have been published thus far on the effectiveness of Hull's method. Sprague (1955) has stated, however, that in each of two open-pollinated varieties, using the inbred Hy as a tester, he obtained yield increases of approximately 5 bushels per acre with a single cycle of selection.

Reciprocal Recurrent Selection

Reciprocal recurrent selection was proposed in 1949 by Comstock, Robinson, and Harvey as a procedure that would be useful in selecting simultaneously for both general combining ability and for specific combining ability. The scheme outlined by these investigators (Figure 23–6) involves two heterozygous source populations, A and B, which should preferably be genetically unrelated. A number of plants from source A are self-pollinated and crossed with a sample of plants from source B. In similar fashion, a number of plants from source B are selfed and crossed with a sample of plants from source A. Selection is based on the experimental comparison of test-cross progenies in replicated-yield trials. The plants selected are then interbred from S_1 progenies derived from the selfed seed of the S_0 plants. The two resulting populations A' and B' serve as source populations to initiate the next cycle.

Sufficient plants should be selected in each generation to hold inbreeding within the two sources to an acceptable level. Otherwise the within-group variability on which selection operates may be reduced to the point where progress will be adversely affected. For

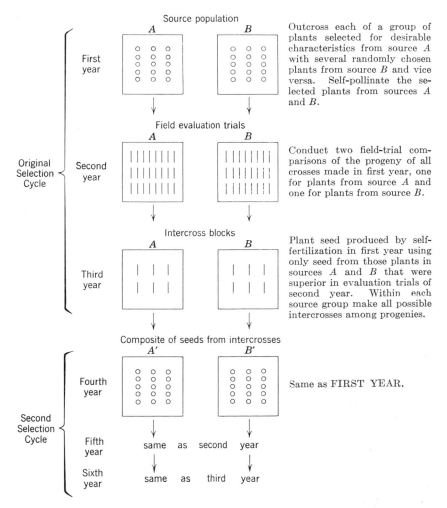

Outcross each of a group of plants selected for desirable characteristics from source A with several randomly chosen plants from source B and vice versa. Self-pollinate the selected plants from sources A and B.

Conduct two field-trial comparisons of the progeny of all crosses made in first year, one for plants from source A and one for plants from source B.

Plant seed produced by self-fertilization in first year using only seed from those plants in sources A and B that were superior in evaluation trials of second year. Within each source group make all possible intercrosses among progenies.

Same as FIRST YEAR.

FIGURE 23-6. Diagrammatic representation of reciprocal recurrent selection. How should the above diagram be modified to represent recurrent selection for general combining ability? Specific combining ability? Varieties or species in which it is difficult to make both self-pollinations and outcrosses from a single plant?

the same reason, all the plants selected in any generation should not trace back to a few matings made in a previous cycle. Methods by which this might be avoided or minimized were discussed earlier.

Populations developed by reciprocal recurrent selection would be utilized by producing commercial seed from crosses between the A and B source groups. However, the amount of inbreeding done prior to making the commercial crossed seed could vary considerably. At the one extreme standard inbreeding procedures could be used to develop inbred lines to be used in double crosses of the type $(A_1 \times A_2) \times (B_1 \times B_2)$, where A_1 and A_2 are lines from source A and B_1 and B_2 are lines from source B. The other extreme would involve crossing the two reciprocally selected populations themselves, in a manner equivalent to making a varietal cross.

In proposing this method, Comstock, Robinson, and Harvey compared its efficiency on theoretical grounds with that of recurrent selection for general combining and with that of recurrent selection for specific combining ability. The methods were compared assuming three levels of dominance: partial dominance, full dominance, and overdominance. The conclusions reached from this analysis of efficiency were as follows: (1) If dominance is incomplete, recurrent selection for general combining ability and reciprocal recurrent selection are about equal to each other, and both are superior to recurrent selection for specific combining ability. (2) If dominance is complete, the three methods are all essentially equal in efficiency. (3) If overdominance is important, recurrent selection for specific combining ability and reciprocal recurrent selection are equal to each other, and both are superior to recurrent selection for general combining ability.

These comparisons were made on the assumption that nonallelic gene interactions were nonexistent, that only two alleles were possible at any locus, and that the relative frequencies of genotypes at linked loci were at equilibrium. None of these assumptions can be considered to be generally valid. The probable effects of their nonvalidity on the efficiency of reciprocal recurrent selection relative to the other two types of recurrent selection were considered to be as follows.

(1) Effect of interactions of nonallelic genes. Nonallelic gene interaction implies that the effects of some genes are modified by genes present at other loci. Hence the importance of a gene to test-cross performance depends on genes present at other loci in the tester. When such interactions are important the advantage of testing against the same material to be used as the opposite parent in making the commercial hybrid is clear. It follows that the presence of important interallelic interaction would tend to favor recurrent selection for specific combining ability, and reciprocal recurrent selection as compared with recurrent selection for general combining ability.

(2) Effect of multiple alleles. The ultimate effect of continued selection would be to increase the most favorable gene in the series and decrease the frequency of all others. Thus, if the genetic system were the one proposed by East (Chapter 19), it would again be advantageous to use the same material for the tester that would later be

used as the opposite parent of the commercial hybrid, as in reciprocal recurrent selection or recurrent selection for specific combining ability. It would thus appear that multiple allelism should not modify seriously the conclusions reached above when the two-allele case was considered.

(3) Effect of linkage disequilibrium. If a chromosome segment behaves as a single locus, the situation is analogous to a multiple allelic series. If the frequency distribution of the genes making up these "alleles" is that expected on the basis of random distribution of the individual genes of which they are composed, the effect of selection is not different because of the linkage. If, however, the associations within the "complex locus" are predominantly in repulsion phase, certain heterozygotes would very likely be superior to any homozygote, even though the individual component genes are only partially dominant. The situation is then comparable to one of overdominance, and hence less favorable for recurrent selection for general combining ability. On the other hand, assuming that such complex loci can ultimately be broken up by recombination, and further assuming partial dominance in the action of the individual component genes (but no overdominance), recurrent selection for specific combining ability would be deficient compared to reciprocal recurrent selection.

On all counts, then, reciprocal recurrent selection appears to be either equal or superior to the other types of recurrent selection in the limit of progress it might be expected to produce.

REFERENCES

Comstock, R. E., H. F. Robinson, and P. H. Harvey. 1949. A breeding procedure designed to make maximum use of both general and specific combining ability. *Agron. Jour.* **41**: 360–367.

Dickerson, G. E. 1952. Inbred lines for heterosis tests? In *Heterosis,* pp. 330–351. Iowa State College Press.

East, E. M., and D. F. Jones. 1920. Genetic studies on the protein content of maize. *Genetics* **5**: 543–610.

Hayes, H. K., and R. J. Garber. 1919. Synthetic production of high protein corn in relation to breeding. *Jour. Amer. Soc. Agron.* **11**: 309–318.

Hull, F. H. 1945. Recurrent selection for specific combining ability in corn. *Jour. Amer. Soc. Agron.* **37**: 134–145.

Hull, F. H. 1952. Recurrent selection and overdominance. In *Heterosis,* pp. 451–473. Iowa State College Press.

Jenkins, M. T. 1935. The effect of inbreeding and selection within inbred lines of maize upon the hybrids made after successive generations of selfing. *Iowa State Coll. Jour. Sci.* **9**: 429–450.

Jenkins, M. T. 1940. The segregation of genes affecting yield of grain in maize. *Jour. Amer. Soc. Agron.* 32: 55–63.

Jenkins, M. T., A. L. Robert, and W. R. Findley, Jr. 1954. Recurrent selection as a method for concentrating genes for resistance to *Helminthosporium turcicum* leaf blight in corn. *Agron. Jour.* 46: 89–94.

Lonnquist, J. H. 1949. The development and performance of synthetic varieties of corn. *Agron. Jour.* 41: 153–156.

Lonnquist, J. H. 1950. The effect of selection for combining ability within segregating lines of corn. *Agron. Jour.* 42: 503–508.

Lonnquist, J. H. 1951. Recurrent selection as a means of modifying combining ability in corn. *Agron. Jour.* 43: 311–315.

Lonnquist, J. H., and D. P. McGill. 1954. Gametic sampling from selected zygotes in corn breeding. *Agron. Jour.* 46: 147–150.

Lonnquist, J. H., and D. P. McGill. 1956. Performance of corn synthetics in advanced generations of synthesis and after two cycles of recurrent selection. *Agron. Jour.* 48: 249–253.

Lush, J. L. 1943. *Animal breeding plans.* Iowa State College Press.

McGill, D. P., and J. H. Lonnquist. 1955. Effects of two cycles of recurrent selection for combining ability in an open-pollinated variety of corn. *Agron. Jour.* 47: 319–323.

Sprague, G. F. 1946. Early testing of inbred lines of corn. *Jour. Amer. Soc. Agron.* 38: 108–117.

Sprague, G. F. 1952. Early testing and recurrent selection. In *Heterosis,* pp. 400–417. Iowa State College Press.

Sprague, G. F. 1955. Corn breeding. In *Corn and corn improvement,* pp. 221–292. Academic Press, New York.

Sprague, G. F., and B. Brimhall. 1950. Relative effectiveness of two systems of selection for oil content of the corn kernel. *Agron. Jour.* 42: 83–88.

Sprague, G. F., P. A. Miller, and B. Brimhall. 1952. Additional studies on the effectiveness of two systems of selection for oil content of the corn kernel. *Agron. Jour.* 44: 329–331.

24
Synthetic Varieties

In the breeding of cross-pollinated crops, the basis for improvement lies in the controlled utilization of the heterosis that occurs in hybrids among certain genotypes. This controlled utilization of heterosis has had its greatest development in maize, where the floral morphology permits the large amounts of seed required for commercial production of hybrid varieties to be produced economically. More recently it has been found that male sterility allows maize methods to be extended, with appropriate modifications, to a few other species, and the prospects seem good that male sterility will ultimately allow these methods to be applied to a considerable number of cross-pollinated crops. Nevertheless, there are many crops in which the annual production of first-generation seed is impractical. When this is true, synthetic varieties seem to offer a good opportunity for the controlled utilization of an appreciable amount of heterosis.

Hayes and Garber (1919) were apparently the first to suggest the possibility of the commercial utilization of synthetic varieties. The suggestion grew out of some results they obtained with maize, from which they concluded: "The production of improved varieties through the recombination of several selfed strains has one advantage over either the single- or double-cross plan in that the farmer can save his own seed from the yearly crop and that yearly crosses need not be made. . . . It is recognized that, before recombining selfed lines for the purpose of producing improved varieties, it is necessary to determine the yielding ability of all F_1 combinations. Selfed lines which combine favorably with all others that are to be used should then be used for the recombinations." The term "synthetic variety" has come to be used to designate a variety that is maintained from open-pollinated seed following its synthesis by hybridization in all combinations among a number of *selected* genotypes. The genotypes that are hybridized to produce a synthetic variety can be inbred lines, clones,

303

mass-selected populations, or various other materials. It will be noted that this definition does not distinguish sharply between synthetic varieties on the one hand and varieties produced by mass selection or by line breeding on the other.

The key point of distinction between synthetic varieties and varieties developed by mass selection or line breeding lies in the way the constituent genotypes are chosen. A synthetic variety is synthesized from genotypes which have been tested for combining ability. Only genotypes which combine well *with each other in all combinations* are put into the synthetic variety. This prior testing of hybrid performance distinguishes a synthetic from a variety developed by simple mass selection, in that the latter is made up of genotypes that are bulked without previous testing of progeny performance or performance in hybrid combination. It also distinguishes synthetic from line-bred varieties in which progenies from superior lines are composited on the basis of the performance of lines tested individually. Thus the goal of testing in the development of synthetic varieties is to identify the genotypes that will combine well when crossed *inter se*. Many different procedures can be used to determine the combining ability of different genotypes. These procedures vary from simple visual inspection for highly heritable characters to tests of yield prepotency as the primary criterion of selection for complex ones.

The possible advantages of synthetic varieties in utilizing heterosis in cross-fertilized crops, in which floral structure causes difficulties in pollination control, are obvious. Nor have these advantages been overlooked, especially in Europe, where synthetic varieties have received wide use in the improvement of forage species in particular.

In maize, on the other hand, the success of hybrid varieties tended to suppress interest in other methods of breeding, and relatively little attention has been given to the development of synthetic varieties. However, it was noted by Jenkins and Sprague in 1943 that synthetic varieties might have value as reservoirs of desirable germplasm, and they have been used to some extent for that purpose. It was also noted: that (1) synthetic varieties might be of considerable value on fringes of the corn belt where the cost of hybrid seed is high relative to the value of the expected crop; (2) the greater variability of synthetic varieties as compared with double crosses might permit more flexibility to meet the changeable growing conditions of marginal areas; and (3) synthetic varieties might have a place where commercial acreage is too small to support a hybrid-seed corn industry. These incentives to produce synthetic varieties were not compelling, however, and synthetics have at no time been much of a factor in commercial corn production. Nevertheless, it is to this species that we must turn for an evaluation of the place of synthetic varieties in crop improvement. This is because the most precise data on this system, as with most other systems of breeding cross-pollinated crops, have been developed in experiments conducted with maize.

Factors Influencing the Performance of Synthetic Varieties

In 1922, Sewall Wright pointed out that "a random-bred stock derived from n inbred families will have $1/n$ th less superiority over its inbred ancestry than the first cross or a random-bred stock from which the inbred families might have been derived without selection." This relationship can be expressed

$$\hat{F}_2 = \bar{F}_1 - \frac{(\bar{F}_1 - \bar{P})}{n}, \tag{24-1}$$

where $\hat{F}_2$ represents the estimated performance of the F_2 generation, $\bar{F}_1$ is the average performance of all possible single crosses among the parental lines involved, $\bar{P}$ is the average performance of the parental lines, and n the number of parental lines involved. As an example, in predicting the performance of an F_2 derived from 4 parental genotypes, $\bar{P}$ is the mean performance of the 4 parents, $\bar{F}_1$ is the mean performance of the 6 possible single crosses, and n is 4. When F_1 hybrids are bred *inter se*, a single cross in F_2 will lose one-half of the excess vigor of the F_1 over the parents. When 3, 4, 5, ... n lines comprise the parents of the synthetic variety, $\frac{1}{3}$, $\frac{1}{4}$, $\frac{1}{5}$, ... $1/n$ th of the excess vigor will be lost, respectively, in the F_2 generation. According to the Hardy-Weinberg rule, there should be no further decline in vigor in succeeding generations, provided mating is completely at random and there is no differential selection, because zygotic equilibrium for any gene is reached in a single generation of panmixis.*

The performance of advanced generations of synthetic varieties in theory thus depends on (1) the number of parental lines included,

---

* Departures from random mating that might occur in the synthetic variety and the changes in gene frequency that might be caused by natural selection will also influence performance in advanced generations. Both these factors are difficult to evaluate experimentally. Selfing can probably be ignored in species such as maize or alfalfa that are almost entirely outcrossed, but it might lead to significant yield depression in less completely outcrossed species. Assortative mating due to differences in flowering habits and so forth may, however, cause considerable departures from randomness even in highly outcrossed species. If unfavorable selection causes shifts in gene frequencies, this could be overcome by reconstituting the synthetic as necessary from the original sources.

(2) the mean performance of these parental lines, and (3) the mean performance of all possible combinations among the n lines.* It would be useful to know whether the theory stated above holds in maize and whether departures from it are large enough to be of importance in predicting the performance of synthetic varieties. One line of evidence comes from experiments of the type reported by Neal in 1935. In this experiment, yields were determined for the inbred parents, the F_1 and the F_2 for each of 10 two-line, 4 three-line, and 10 four-line synthetic varieties of maize. The results are summarized in Table 24–1. The average actual loss of vigor, as represented by

TABLE 24–1. Comparison of the Actual Loss of Excess Vigor of Hybrids with the Calculated Loss

(After Neal, 1935)

Type of Synthetic Variety	Average Yield, Bu. per Acre		Average Difference $F_1 - P_1$ Bu. per Acre	Average Yield in F_2, Bu. per Acre	
	F_1	P_1		Actual	Expected
10 Single hybrids	62.8	23.74	39.06	44.2	43.3
4 Three-way hybrids	64.2	23.75	40.45	49.3	50.7
10 Double hybrids	64.1	25.00	39.10	54.0	54.3

	Loss of Vigor in Yield of Grain				
	Actual			Expected	
	Bu. per Acre	Per Cent		Bu. per Acre	Per Cent
10 Single hybrids	18.6	47.6		19.5	50
4 Three-way hybrids	14.9	36.8		13.5	33.3
10 Double hybrids	10.1	25.8		9.8	25

yield, was 47.6, 36.8, and 25.8 per cent, respectively, for single, three-way, and double-cross hybrids compared with expected losses of 50, 33.3, and 25 per cent. The actual decline in yield thus agrees remarkably well with the expected decline according to the genetic theory formulated by Wright.

* Strictly, designation of panmictic generations by F_2, F_3, . . . is incorrect usage, and it might be more appropriate to designate parents as Syn 0, F_1 hybrids as Syn 1, the panmictic progeny of Syn 1 as Syn 2, and so forth. Since the experiments to be reported are more understandable in terms of F_1, F_2, etc., and since this notation was used in the original papers, it will be followed here.

Since, in theory, genetic equilibrium is reached as soon as the hybrids reproduce, the F_3 generation should make the same yield as the F_2 generation. Neal found that the yields for the F_3 generation of the single and three-way crosses in this experiment were not significantly different from their respective F_2 yields. In similar comparisons by Kiesselbach in 1933 the yields of F_2 and F_3 generations of 21 different single crosses were found to be 38.4 and 37.8 bushels per acre, respectively, again a nonsignificant difference. This same type of result was also obtained by Sprague and Jenkins in 1943. They tested the F_1, F_2, F_3, and F_4 of one 24-line and four 16-line synthetic varieties and found the F_2, F_3, and F_4 to yield 94.3, 95.4, and 95.1 per cent, respectively, of the F_1 generation.

In an experiment of another type, Kinman and Sprague (1945) reported the yields of grain for 10 inbred lines, the 45 possible single crosses, and the F_2 generations of these single crosses. Yield data for the 10 inbred lines were used to calculate theoretical yields of synthetic varieties involving 2 to 10 inbred lines. This was done by arranging the lines in descending order according to mean-combining ability based on single-cross yields. The mean-combining ability for the 10 lines varied from 87.4 down to 68.1 bushels per acre. The calculated yield for the synthetic involving the 2 highest-combining inbreds was based on the actual yield of the single-cross and the parental inbreds. The calculated yield of the 3-line synthetic was based on the mean of all possible single crosses among the three best-combining inbreds and the yield of the parental inbreds The other calculated yields were obtained, using the highest-combining inbreds until, for the 10-line inbred synthetic, even the poorest combining inbred line and its hybrid combinations were included. The results are given in Table 24–2.

With these particular materials, the expected yield of the F_2 increased steadily as more lines were added, reaching a maximum at 5 or 6 lines, after which the expected yield decreased steadily as more lines were included. The explanation is as follows: Each increase in number of lines up to 6 led to a net gain in the yield of the synthetic because the increase in n more than compensated for the decreasing prepotency of the inbreds that had to be included—that is, for decrease in F_1. At this point, however, prepotency was so low that increases in n now failed to compensate. The result was a net loss in expected yield contrasted to the 6-line synthetic. For these particular materials, the best-yielding synthetic variety was therefore the one that included only the 5 or 6 best-combining inbreds.

This experiment illustrates the point that the yield level of a

TABLE 24–2.   Calculated Acre Yields of the F_2 Generation of Synthetic Varieties with Different Numbers of Component Lines

(After Kinman and Sprague, 1945)

Number of Lines	Yield in Bushels per Acre	
	F_1 Mean	Expected F_2 Yield
2	97.6	65.3
3	93.4	76.1
4	93.8	79.3
5	91.6	80.2
6	89.2	80.2
7	86.7	79.0
8	84.4	77.9
9	82.8	77.0
10	79.9	74.7

synthetic variety can be increased by any one or a combination of the following: (1) increasing the number of lines, (2) increasing mean F_1 yields, (3) increasing the mean yield of the parents. At first glance it might appear that the first of these possibilities would be the easiest. The difficulty here lies in finding additional lines that nick well with the ones already selected in order to maintain high F_1 mean yield. The final possibility is in fact the easiest to accomplish in most instances. Jenkins showed in 1935 that inbred lines express their individuality in combining ability very early in the inbreeding process, and potentially high-combining plants can be identified by top crossing them at the time of the first selfing. These high-combining first-generation selfed lines may yield 75 per cent as much as open-pollinated varieties whereas homozygous inbred lines yield only 25 to 40 per cent as much as open-pollinated varieties. It follows that synthetic varieties made up from a combination of relatively high-yielding parents might be expected to yield more than synthetic varieties based on more highly inbred and hence less vigorous stocks. It also follows that circumventing the inbreeding entirely would be desirable where it is possible, as in species that can be reproduced clonally.

A comparison of the effect of parental yield levels on the yield level of synthetic varieties is presented in Figure 24–1. The four yield levels assumed for the parental materials were 25, 40, 75, and 100 per cent of open-pollinated source populations. The 25 and 40 per cent yield levels include most inbred lines now being used commercially in

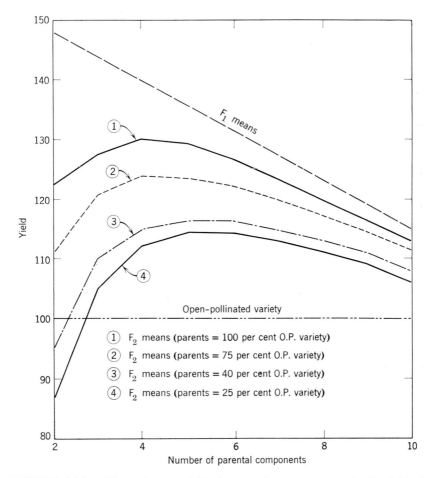

FIGURE 24-1. Effect of mean yield of parental components on the F_2 yield of synthetic varieties. (Modified from Kinman and Sprague, 1945.)

maize and would also represent the yield levels of inbred lines in many other cross-pollinated species. The 75 per cent yield level was chosen to represent superior S_1 lines. The 100 per cent yield level represents noninbred materials that might be used as parents in certain species. The F_1 mean yields are depicted as a declining series of the type usually observed when the yields of test crosses are measured in actual experiments. The calculated yields of synthetic varieties are clearly greatest when the yields of the parental stocks are high. In addition smaller numbers of parental lines are required for greatest efficiency when the parental yields are high. It should be recognized, however, that if many high-combining F_1 combinations were available, so that the highest- and lowest-yielding F_1 combinations would be more nearly alike in yield than depicted in Figure 24–1, the optimum number of lines (n) would be greater than shown in this figure.

These various considerations can be summed up as follows. First, there is close agreement between predictions made from Wright's formula and actual observations. Second, high-combining component lines can theoretically be very effective in producing a good synthetic even when the number of lines is small. Third, increasing the number of lines will improve the yield of the synthetic only as long as the increase can be made without undue sacrifice in high combining ability. Fourth, the low yield of inbred lines is a real handicap to their use in forming synthetic varieties. Use of noninbred plants or lines developed by limited inbreeding can be expected to overcome this difficulty to a considerable extent.

An account of some actual experiences in breeding synthetic varieties will illustrate these features as well as some other aspects of this approach to the controlled utilization of hybrid vigor.

Synthetic Varieties in Maize

Within a few years of the original suggestion by Hayes and Garber in 1919, that synthetic varieties might have a place in maize improvement, reports on the performance of several different synthetic varieties were published. In general these early synthetic varieties yielded about the same as open-pollinated varieties. This result is according to expectation, since, at that time, it was not possible to restrict the parentage to materials with good combining ability.

It was not until 1944 that the performance of a synthetic variety made up of inbred lines selected on the basis of good combining ability was reported. Hayes, Rinke, and Tsiang studied all of the $n(n-1)/2$ single crosses among 20 inbred lines and made up a synthetic variety from the 8 inbred lines that had given the best hybrid performance in single crosses. Equal numbers of seeds of each of the 28 single crosses among these 8 lines were mixed together, planted in an isolated plot, and the seed harvested without selection for ear or plant types. In subsequent years the variety was maintained by growing it in an isolated plot and selecting desirable plants without close selection for ear type. The predicted yield of the synthetic variety, calculated in terms of Minhybrid 403 as 100, was 102 per cent. The actual yield of this synthetic variety is given in Table 24–3. When moisture percentage was taken into account, the synthetic variety was not quite

TABLE 24–3.   Performance of a Synthetic Variety, Minhybrid 403, and an Open-Pollinated Variety at Ten Locations in the Period 1942 to 1944

(After Hayes, Rinke, and Tsiang, 1944)

Variety	Yield, Bu.	Moisture, Per Cent	Stand, Per Cent
Synthetic	69.4	27.2	94
Minhybrid 403	66.8	24.9	94
Open-pollinated	55.6	25.9	79

equal to Minhybrid 403 (a good double cross for the locality), but it was distinctly superior to a standard open-pollinated variety. In other agricultural characteristics the synthetic variety was about equal to Minhybrid 403 and superior to the open-pollinated variety. These authors emphasized "the importance of studying combining ability of the selected material used in producing a synthetic variety."

The possibility that synthetic varieties might be produced from short-term inbred lines was, as we have seen, first suggested by Jenkins in 1940. The idea was based on the earlier observation (1935) that good combining ability can be identified in S_0 by the use of top crosses, and, since line performance in top crosses is relatively stable after the S_1 generation, little would be gained by additional selfing when the goal is a synthetic variety. The steps outlined by Jenkins were as follows:

1. The isolation of one-generation selfed lines.
2. Testing these lines in top crosses for yield and other characters.
3. Intercrossing the better lines to produce a synthetic variety.
4. Repetition of the above processes at intervals after a generation or two of open pollination.

Jenkins noted that "the method has some superficial similarity to ear-to-row breeding. It differs fundamentally from the ear-to-row method, however, in that the germplasm of the tested plants is maintained by selfing the plants and this selfed seed of the superior individuals is mixed and increased in developing the new strain. In the ear-to-row method of breeding, the male parentage of the selected plants was not controlled and as a result represented the variety average."

The procedure outlined by Jenkins was followed by Lonnquist and McGill (1956) in developing several synthetic varieties. The performance of three of these synthetic varieties is shown in Table 24–4.

TABLE 24–4. Performance of Three First-Cycle Synthetic
Varieties Expressed in Per Cent of the Parental Population

(After Lonnquist and McGill, 1956)

| | Grain Yield, % | Moisture at Harvest, % | Lodging | | No. of Tests |
| | | | Root, % | Stalk, % | |
Population					
Parental open-pollinated variety	100	100	100	100	
Krug Synthetic F_2	122	91	45	120	3
Reid Synthetic F_2	109	96	32	86	2
Dawes Synthetic F_2	109	109		47	5
Means	113	99	38	84	

In each instance the synthetic variety was superior in yield to the
parental open-pollinated variety from which it was derived, and, in
addition, the synthetics were superior in resistance to lodging. Data
were also presented comparing one of these synthetics (Dawes) with
28 double-cross hybrid varieties that are offered for sale in western
Nebraska (Table 24–5). Although several of the hybrid varieties were

TABLE 24–5. Performance of the Dawes First-Cycle Synthetic,
Its Open-Pollinated Parental Variety, and 28 Double-Cross
Hybrids in 2 Locations in Western Nebraska

(After Lonnquist and McGill, 1956)

Kind of Corn	Yield, Bu. per Acre	Moisture at Harvest	Stalk Lodging	Dropped Ears
Dawes open pollinated	73.8	20.3	5.2	0.4
Dawes Synthetic	80.5	22.0	3.4	0.0
Mean of 28 hybrids	76.1	20.1	2.0	0.3
Range of hybrids	54.4–86.5	14.0–26.4	0.0–5.8	0.0–2.4

higher yielding than the Dawes synthetic, none was both significantly
higher in yield and earlier in maturity. The high cost of hybrid seed
relative to the expected crop in this area and the lack of hybrid
varieties having consistent superiority from year to year have combined to cause many farmers to continue to grow open-pollinated

varieties. The prospect for synthetic varieties would therefore seem to be good, not only in this area but also wherever hybrid varieties have, for similar reasons, been unable to replace open-pollinated varieties.

Another aspect of synthetic varieties which has been investigated by Lonnquist and McGill is the yielding ability of advanced generations derived from first-cycle intercrosses. Each of four different first-cycle synthetic varieties was maintained from the F_2 (Syn 2) through the F_5 (Syn 5) generation by selecting ears from about 150 to 200 of the better-appearing plants in open-pollinated plantings of 5000 to 10,000 plants grown in separate isolation plots. The yielding ability of each of the four synthetics in each generation was determined in replicated-yield trials with the results shown in Table 24–6. These

TABLE 24–6. Performance of Advanced Generations of Four First-Cycle Synthetic Varieties Maintained with Visual Selection for Good Type

(After Lonnquist and McGill, 1956)

Generation After Synthesis	Population				Means
	Krug	A	B	Reid	
F_2*	100	100	100	100	100
F_3	103	108	120	104	108
F_4	105	104	126		111
F_5	106	110			108

* Yield is expressed in percentage of the F_2 generation.

results suggest not only that yielding ability can be maintained in advanced generations produced by mass selection, but also that visual selection of the better plants may produce some improvement in yield.

Another important aspect of synthetic varieties is the amount of progress that additional cycles of selection are capable of producing. In this connection Lonnquist and McGill have presented data on the yields of four second-cycle synthetic varieties compared with their first-cycle counterparts. Their data are given in Table 24–7. Lonnquist and McGill concluded: "The improved performance obtained after two cycles of recurrent selection offers encouragement for further improvement from a continuation of this procedure. It is apparent that the method results in rather rapid improvement of corn varieties and might well be used in areas where development of corn hybrids

TABLE 24–7.   **Performance of First- and Second-Cycle Synthetic Varieties Compared with Double-Cross Hybrid U.S. 13**

(After Lonnquist and McGill, 1956)

Population	Grain Yield, Per Cent of U.S. 13	Moisture at Harvest, Per Cent of U.S. 13
U.S. 13*	100	100
Krug, first-cycle F_2	87	96
Krug, second-cycle F_2	98	101
A, first-cycle F_2	85	113
A, second-cycle F_2	102	112
B, first-cycle F_2	72	97
B, second-cycle F_2	80	94
Reid, first-cycle F_2	86	107
Reid, second-cycle F_2	95	113
Mean: first cycle	82	103
second cycle	96	105

*U.S. 13 averaged 104.7 bushels per acre and 18.8 per cent moisture in the grain at harvest in these tests.

and their production and distribution is not now feasible or where some sort of stop-gap procedure is needed during the development of suitable corn hybrids."

Considerable use is being made of synthetic varieties in Mexico in the cooperative corn-improvement program of the government of Mexico and the Rockefeller Foundation. There is also increasing use of synthetic varieties in many other countries where there is no well-developed seed industry.

Synthetic Varieties in Forage Crops

The breeding systems used in the improvement of forage species were originally based on simple mass-selection methods. Today most breeders agree that mass selection is effective in modifying unimproved materials, but that selection must be based on a progeny test if further improvement of adapted types is to be accomplished. This was antici-pated in at least some degree by Jenkin (1931) in the methods he employed in breeding perennial ryegrass at the Welsh Plant Breeding Station. He coined the term *strain building* to describe these methods. This term was used by him, and later by Kirk (1933) and other breeders, in a very broad sense to include any system of mating by which a strain is built up from crossing carefully selected plants. The parental plants were usually selected on the basis of a progeny test, but no particular distinction was made between the use of inbred or crossbred progeny for testing purposes. Strain building thus includes a number of breeding procedures varying from simple mass selection to methods that resemble recurrent selection. The use of so broad a term as strain building has disadvantages because of the great variation in the procedures it encompasses and the differing rates and potential limits of improvement possible under the different procedures.

More recently there have been a number of observations suggesting that the principles that have been investigated most thoroughly in maize apply also to the forage species (reviews in Frandsen, 1952; Johnson, 1952; Wexelsen, 1952; Hansen and Carnahan, 1956). For example, investigations of the yielding ability of single plants of grasses and clovers have shown that certain individuals from open-pollinated populations are able to transmit higher yielding potential to their progeny than others, in the same way that S_0 plants in maize differ in combining ability. It also seems that the majority of inbred lines, as in maize, are not materially altered in combining ability from their parental types during the inbreeding process. There have similarly been indications that careful selection among the original plants based on observation and measurements in replicated clonal nurseries should precede progeny testing. Elimination of undesirable genotypes at this stage in the breeding program aids in making possible more adequate evaluation of a smaller number of selections to be progeny tested, just as many S_0 plants in maize can be eliminated on the basis of deficiencies in readily visible characteristics. This elimination is perhaps more important in forage breeding programs than in maize, because progeny tests are in general more expensive and troublesome to conduct in the forage crops. On the basis of these and other similarities to maize, it is natural to expect that the contributions of the maize breeders to the theory of breeding systems would be of value to breeders of the cross-fertilizing forage species. Important differences between maize and forage species in ability to produce inbred and crossed seed and in methods of evaluating test-cross progenies have, however, made it necessary for the breeders to modify maize methods substantially in adapting them to their species.

Tests for Combining Ability in Forage Species

Most forage species are notably difficult to manipulate in making hybrid seed, and it has consequently been necessary to rely on natural crossing to obtain the progenies required for tests of combining ability. In general, four different types of tests of combining ability can be made on progenies obtained by natural crossing. These tests, listed in order of increasing labor requirement in obtaining crossed seed, are as follows.

1. Open-pollinated progeny test. This test depends on progenies derived from seed produced on selected plants outcrossed with other plants that happen to occur in the breeding nursery.

2. Top-cross test. The seed upon which this test depends is obtained from selected individuals (or clones) that have been planted alternately with a single tester variety. The test-cross seed thus consists partly of top crosses to the variety and partly of intercrosses with the other selected clones. The proportion of top-crossed seed can be increased by increasing the number of plants of the tester variety sown in the nursery relative to the number of plants of the selected clones that are sown.

3. Polycross test. The term polycross was proposed by Tysdal, Kiesselbach, and Westover (1942) to "designate the progeny from seed of a line that was subject to outcrossing with selected lines growing in the same nursery." The same procedure had been suggested in 1940 by Frandsen in Denmark (see Frandsen, 1952). To insure that every clone has an equal chance of being pollinated by every other clone, the clonal isolates should be replicated many times in the isolated natural crossing block in which the outcrosses are made.

4. Single-cross test. In a single-cross test, every selected clone is crossed with a number of other clones, usually by growing together in isolation each pair of clones to be crossed, or by enclosing an inflorescence from one clone in a bag with an inflorescence of the other clone. If all the $n(n-1)/2$ possible single crosses among n selected clones are made, the resulting set of crosses is called a *diallel cross*.

The first three types of test all measure general combining ability. It should be noted that if a large number of clones are included in a polycross test, the average genetic composition of the male parentage is not expected to be much different from the male parentage in open-pollinated or top-cross tests. If, however, only a few highly selected clones are included in a polycross block, the performance of the test-cross progenies might be very different from open-pollinated or top-

FIGURE 24-2. Variability in space-planted clones of ryegrass. Is it possible to tell by visual inspection which of these clones has superior combining ability? Which clones will produce progeny with superior performance in mixed stands with other species under grazing? Why is evaluation of performance difficult in most forage species?

cross progenies. Single crosses measure the combining ability of particular pairs of clones. The average combining ability of any single clone can, of course, be calculated from single-cross data as the mean performance of that clone in its crosses with all other clones with which it was crossed. Average combining ability becomes more and more similar to general combining ability with increases in the number of single crosses in which the clone was involved. In practical breeding work, where hundreds of individuals might have to be tested, it is impractical to establish a diallel cross between all individuals. The use of diallel crosses therefore appears to be restricted to the final stages of a breeding program when it might be desirable to determine the combinability *inter se* of a few strains of high prepotency.

Relationship of Tests for Combining Ability

There have been a number of studies of the relationships among these different types of tests for measuring combining ability in forage species. In a study reported by Tysdal and Crandall in 1948, poly-crosses of seven clones of alfalfa were compared with single crosses and with top crosses. The single crosses were produced by interplanting pairs of clones in isolated natural crossing blocks. Polycross seed was obtained from the same clones (plus one additional clone) grown in isolated blocks, and top crosses were produced by isolating each of the eight clones in a separate isolation block in alternate rows with the variety Arizona Common. An additional comparison was made by growing all of the eight selected clones in one block with Arizona Common planted in alternate rows between the clonal materials. Seed was harvested separately from each clone and designated as Arizona polycross to differentiate it from the other top crosses.

Table 24–8 gives the forage yields of the single crosses and polycross progenies of the seven clones tested for which single-cross data

TABLE 24–8.   Forage Yields of Single-Cross Combinations Compared to Polycross Progenies from the Same Clones

(After Tysdal and Crandall, 1948)

Clone No.	Self-Fertility, Per Cent	No. of Single Crosses*	Yield Relative to Grimm†		Rank	
			Single Crosses	Poly-cross	Single Cross	Poly-cross
1019	18	4	115	125	1	1
1124	15	3	111	109	2	2
1128	13	4	108	109	3	2
1229	39	3	106	106	4	4
1120	49	3	105	100	5	6
1112	98	5	103	96	6	7
1241	17	4	101	104	7	5

* Although six single crosses are possible, some combinations produced inadequate seed for a replicated test. Hence the number of single crosses reported is less than the number possible.

† Minimum level of significance is 5.5 per cent for single crosses and 10.5 per cent for polycrosses.

were available. The clones ranked about the same in both tests. A comparison between the polycrosses and the two types of top crosses is given in Table 24–9. Apparently it did not matter whether the clones were grown individually in isolation blocks with Arizona Common, grouped together with Arizona Common, or planted in a regular polycross nursery with only the other selected clones. The various methods, single-cross tests, top-cross tests, and polycross tests, all indicated about the same ranking for combining ability. Tysdal and

TABLE 24–9. Forage Yields of Polycrosses Compared to Two Types of Top Crosses of the Same Clones

(After Tysdal and Crandall, 1948)

Clone No.	Yield Relative to Grimm*			Rank		
	Polycross	Arizona		Polycross	Arizona	
		Top Cross	Polycross		Top Cross	Polycross
1019	121	130	117	1	1	1
1128	111	122	116	2	2	2
1112	101	117	114	3	3	4
1111	99	103	116	4	5	2
1229	97	105	91	5	4	8
1120	96	101	109	6	6	5
1124	89	101	109	7	6	5
1241	76	101	99	8	6	7

* Minimum level of significance, 15 per cent.

Crandall point out that the widest differences were obtained in the polycross nursery, although, if the best 20 per cent of the clones were selected, the same clones would be chosen on the basis of any one of the tests.

In other studies, correlations between polycross and single-cross performance have generally been less consistent than the correlations observed by Tysdal and Crandall. The correlations have generally been positive, however, and large enough to indicate that either method is a suitable one for measuring combining ability.

The most important aspect of general combining ability in forage crops is its relation to the ultimate performance of lines or clones when they are synthesized, by intercrossing, into a synthetic variety. Tysdal and Crandall have published the most extensive data on this point. These investigators studied eight different synthetic varieties, each made up of intercrosses among four or five parental clones. They found that the rank of the synthetic varieties in yielding ability followed very closely the average polycross performance of the parental clones and concluded that polycross data were valuable in identifying plants that can make worth-while contributions to the yielding ability of synthetic varieties. Johnson (1956) has also studied this problem and has presented data on both a first cycle and a second cycle of improvement in developing synthetic varieties in sweet clover (*Melilotus officinalis*). Johnson found that a first-cycle synthetic was markedly improved in combining ability over the open-pollinated source population and that a second cycle of recurrent selection for general combining ability produced another marked increase in combining ability.

The evidence available from forage-crop breeding, although not extensive, indicates that tests of progenies derived from seed produced in polycross blocks, by open pollination in breeding nurseries, or in top-cross blocks all provide a useful means of screening noninbred plants or clones to be used in producing synthetic varieties. Among these methods the polycross test has been most widely accepted by forage breeders because it combines economy of effort with estimates of combining ability that in theory at least are more relevant to performance in synthetic varieties than estimates obtained from open-pollination or top-cross tests. One of the difficulties inherent in polycross tests is nonrandomness of pollination. Murphy (1952) and Hittle (1954) in particular have stressed the importance of large numbers of replications (ten or more) of each genotype, arranged in random order, to minimize differential pollen effects in producing polycross seed. Another problem, even more serious, relates to the frequent necessity for evaluating forage types in nursery trials where the methods of handling the materials are not comparable to the methods used in commercial practice. For example, it would be desirable if tests of combining ability could be conducted to measure expected performance in mixed stands under grazing conditions. But, for reasons of economy, test progenies must usually be grown in pure stands in rows.

Despite problems of this type, a number of synthetic varieties of forage crops have been developed and released for commercial production. Among these varieties are Primus and Gloria Timothy, Scandia II and Brage Orchardgrass, Viking Red Fescue, and Victoria Perennial Ryegrass, all developed in Sweden. In England, a number of synthetic varieties of grasses and legumes have been developed and released by the Welsh Plant Breeding Station. In the United States, the timothy varieties 1777 and 4059 were bred from S_1 selections by the New York Agricultural Experiment Station, Itasca Timothy by the Minnesota station, and the alfalfa varieties Ranger and Lahontan by the Nebraska and Nevada stations respectively Some of these synthetic varieties might make suitable source populations for additional cycles of improvement to produce still better synthetic varieties. Others, however, were selected for specific attributes to meet particular emergency problems and might well have been too narrowly selected to make suitable sources for additional cycles of selection. Lahontan

Alfalfa, for example, was synthesized from five clones selected primarily for good combining ability with respect to stem-nematode resistance. Although this synthetic variety might be valuable as a source for certain components to go into a new synthetic variety, its limited base suggests that it should probably not be the exclusive source of materials for further improvement.

REFERENCES

Frandsen, K. J. 1952. Theoretical aspects of crossbreeding systems for forage plant. *Proc. Sixth Intern. Grasslands Congr.* 1: 277–283.

Graumann, H. 1952. The polycross method of breeding in relation to synthetic varieties and recurrent selection of new clones. *Proc. Sixth Intern. Grasslands Congr.* 1: 314–319.

Hansen, A. A., and H. L. Carnahan. 1956. Breeding perennial forage grasses. *U. S. Dept. Agric. Tech. Bull.* 1145.

Hansen, A. A., W. M. Myers, and R. J. Garber. 1952. The general combining ability of orchardgrass selections and their I_4 progenies. *Agron. Jour.* 44: 84–87.

Hawk, V. B., and C. P. Wilsie. 1952. Plant-progeny yield relationships in bromegrass, *Bromus enermis* Leyss. *Agron. Jour.* 44: 112–118.

Hayes, H. K., and R. J. Garber. 1919. Synthetic production of high protein corn in relation to breeding. *Jour. Amer. Soc. Agron.* 11: 309–319.

Hayes, H. K., E. H. Rinke, and Y. S. Tsiang. 1944. The development of synthetic varieties of corn from inbred lines. *Jour. Amer. Soc. Agron.* 36: 998–1000.

Hittle, C. N. 1954. A study of the polycross testing technique as used in the breeding of smooth bromegrass. *Agron. Jour.* 46: 521–523.

Jenkin, T. J. 1931. The method and technique of selection, breeding and strain building in grasses. *Imp. Agric. Bur. of Plant Gen., Herb. Plants Bull.* 3: 5–34.

Jenkins, M. T. 1940. The segregation of genes affecting yield in maize. *Jour. Amer. Soc. Agron.* 32: 55–63.

Johnson, I. J. 1952. Evaluating breeding materials for combining ability. *Proc. Sixth Intern. Grasslands Congr.* 1: 327–334.

Johnson, I. J. 1956. Further progress in recurrent selection for general combining ability in sweet clover. *Agron. Jour.* 48: 242–243.

Kiesselbach, T. A. 1933. The possibilities of modern corn breeding. *Proc. World Grain Exhib. and Conf., Canada* 2: 92–112.

Kinman, M. L., and G. F. Sprague. 1945. Relation between number of parental lines and theoretical performance of synthetic varieties of corn. *Jour. Amer. Soc. Agron.* 27: 341–351.

Kirk, L. E. 1933. The progeny test and methods of breeding, appropriate to certain species of crop plants. *Amer. Nat.* 67: 515–531.

Knowles, R. P. 1950. Studies of combining ability in bromegrass and crested wheatgrass. *Sci. Agric.* 30: 275–302.

Knowles, R. P. 1955. Testing for combining ability in bromegrass. *Agron. Jour.* 47: 15–19.

Lonnquist, J. H., and D. P. McGill. 1956. Performance of corn synthetics in advanced generations of synthesis and after two cycles of recurrent selection. *Agron. Jour.* 48: 249–253.

Murphy, R. P. 1952. Comparison of different types of progeny testing in evaluating breeding behavior. *Sixth Intern. Grasslands Congr.* 1: 320–326.

Neal, N. P. 1935. The decrease in yielding capacity in advanced generations of hybrid corn. *Jour. Amer. Soc. Agron.* 27: 666–670.

Oldemeyer, D. L., and A. A. Hansen. 1955. Evaluation of combining ability in orchardgrass. *Agron. Jour.* 47: 158–162.

Sprague, G. F., and M. T. Jenkins. 1943. A comparison of synthetic varieties, multiple crosses, and double crosses in corn. *Jour. Amer. Soc. Agron.* 35: 137–147.

Tysdal, H. M., and B. H. Crandall. 1948. The polycross progeny performance as an index of the combining ability of alfalfa clones. *Jour. Amer. Soc. Agron.* 40: 293–306.

Tysdal, H. M., T. A. Kiesselbach, and H. L. Westover. 1942. Alfalfa breeding. *Nebr. Agric. Exp. Sta. Res. Bull.* 124.

Wellensiek, S. J. 1952. The theoretical basis of the polycross test. *Euphytica* 1: 15–19.

Wellhausen, E. J. 1952. Heterosis in a new population. In *Heterosis,* pp. 418–450.

Wexelsen, H. 1952. The use of inbreeding in forage crops. *Proc. Sixth Intern. Grasslands Congr.* 1: 299–305.

Wright, S. 1922. The effects of inbreeding and crossbreeding on guinea pigs. *U. S. Dept. Agric. Bull.* 1121.

25

Variability Systems of Pathogenic Fungi

Knowledge of two closely interrelated phenomena, variations in the pathogenic capabilities of parasitic organisms and differences within host species in resistance to infectious disease, are basic to the intelligent planning of breeding programs designed to develop disease-resistant varieties. Consideration of each of these subjects in some detail is therefore an essential preliminary if we are to discuss breeding for disease resistance with perception and insight.

In the elucidation of host-pathogen relationships the idea that cultivated varieties differ in their ability to avoid disease was discussed by Theophrastus in the third century B.C. The discovery of disease resistance thus far predates not only the realization that parasites can differ in their pathogenicity, but even the concept of parasitism. Theophrastus noted the occurrence of fungi and other organisms in association with diseased plants and suggested that they had arisen as products of the decaying plants. This idea of the autogenic origin of disease prevailed into the nineteenth century, and it was an important factor in retarding the advancement of knowledge of the nature and cause of infectious diseases. The beginnings of modern concepts of host-pathogen relations did not come until 1807, when Benedict Prevost discovered that the bunt disease of wheat is incited by a fungus. The erroneous idea that fungi are a result rather than an inciting cause of disease was still so strong, however, that Prevost's concept was not widely accepted until about 40 years later, when DeBary established the parasitism of the rusts and smuts.

Mendelian Nature of the Host-Pathogen Relationship

During the last part of the nineteenth century, it gradually came to be recognized that host, pathogen, and environment all play essential roles in pathogenesis, but the variability encountered in each of these factors was so great that the knowledge and methods of the time were insufficient to support much progress toward an understanding of the mechanisms that determine host-parasite relationships. The rediscovery of Mendel's laws in 1900 provided the foundation necessary for the analysis of differing reactions of varieties to diseases. The first advances based on Mendelian genetics, which followed almost immediately upon the rediscovery, were concerned with the genetics of the host. In 1905, Biffen in England published an account of his experiments with the yellow rust of wheat on hybrids derived from crosses of susceptible varieties with Rivet, a variety long known to be resistant. In the F_2, Biffen observed approximately 3 susceptible to 1 resistant individuals. Furthermore, F_3 families appeared in the ratio of $\frac{1}{4}$ true-breeding susceptible lines, $\frac{1}{2}$ segregating lines, and $\frac{1}{4}$ true-breeding resistant lines. Reaction to this disease was clearly governed by a single Mendelian gene.

At the same time significant progress was being made in the United States in breeding varieties that were resistant to diseases. Immediately before the turn of the century, Orton initiated his experiments with the wilt diseases. He found that most varieties of cowpeas were uniformly susceptible, but that the variety Iron was uniformly resistant to cowpea wilt. In cotton the situation was different in that only occasional individuals were disease free. Selection of these disease-free plants led to populations in which the level of wilt resistance was materially increased. Similar successes in selecting resistant varieties were achieved by Bolley in flax, Essary in tomatoes, Jones and Gilman in cabbage, and by a number of other contemporaries of Orton (references in Walker, 1951).

Despite the demonstration by Biffen of the Mendelian inheritance of resistance and the success of selection experiments such as those of Orton, skepticism remained as to the heritable nature of disease resistance and particularly as to the stability of resistance. The reaction of cereal varieties to rust was particularly confusing and led to widespread attacks on Biffen's conclusions regarding the Mendelian nature of resistance and the value of Mendelism in guiding breeding work. The difficulty stemmed from the observations that varieties resistant in certain localities succumbed in others and that resistance varied widely in different seasons. Working with bromegrass species, Ward (1902) had noted considerable variations in reactions to brown rust. By growing a given strain of rust on a moderately resistant brome for several vegetative generations, he found that pathogenicity could be altered to the point where the rust would attack previously resistant brome species. To explain this observation Ward proposed the *bridging-host* hypothesis, which assumed that the pathogen is plastic and that it can be influenced by the host substrate in ways that change its pathogenicity. According to this view, breeding for resistance is futile because the fungus can produce new pathogenic types in step with the breeding of new varieties. Although Biffen (1912) countered with the observation that the resistance of Rivet Wheat to yellow rust had remained constant over many years, the bridging-host hypothesis undoubtedly discouraged many from attempting to breed for disease resistance. Studies of the inheritance of disease resistance and the breeding of resistant varieties nevertheless continued to be pursued vigorously by other investigators, and soon many additional cases of the Mendelian control of resistance were recorded.

During the period following Biffen's work when knowledge of the inheritance of disease resistance was being rapidly expanded, parallel advances were being made toward an understanding of variability in fungi. Following Blakeslee's pioneer work on mating-type differences in the Phycomycete *Rhizopus*, numerous studies of sexuality in fungi showed they possess a remarkable diversity of reproductive systems. Some fungi reproduce by asexual means only, while others reproduce parthenogenetically, by the growth of an unfertilized egg; but the vast majority have a sexual cycle. The mating systems met in many species of fungi can be regarded as systems to promote outbreeding. Sexual reproduction in outbreeding forms, of course, provides for recombination, because the fusing nuclei are expected to be unlike genetically. However, even in forms in which the mating system does not insure outcrossing, the gametes must be unlike occasionally as a result of mutation. Recombination thus provides for variability and for the plasticity that is necessary for response to changes in environment, including changes in the frequency of genes governing resistance in host species.

In the light of these advances in the knowledge of sexuality in fungi, it was possible to develop methods for genetic studies of fungi (reviews in Moulton, 1940; Catcheside, 1951; Raper, 1954). It was soon shown that fungus species are comprised of a great array of variants with different genotypes similar in stability to those of higher plants and animals. Parasitic disease could then be regarded as a product of the interaction of a plastic host with a plastic pathogen, each varying under the influences of heredity and environment. The key to understanding variability in fungi, including variability in pathogenic capabilities, clearly lies in their reproductive systems. It is therefore important to know how reproductive systems relate to life cycles, how they are controlled genetically, and in what respects life cycles and mating systems can be expected to aid or impede genetic investigations of pathogenicity and the development of resistant varieties.

Life Cycles of Fungi

The fungi are commonly considered to be haploid organisms in which nuclear fusions occur occasionally to give rise to a short-lived diploid stage. This is true of many species, but, in fact, life cycles in the fungi range from completely haploid to completely diploid (if the immediate products of meiosis are excluded). Further variation is added by the occurrence of unique nuclear associations in which two or more nuclei occupy the same cell. In all life cycles, however, the cardinal events are changes in nuclear phase.

The essential features of the sexual process are as follows (Figure 25–1). First, two sexual cells, each of which contains one or more haploid nuclei, undergo fusion. This fusion is termed *plasmogamy*.

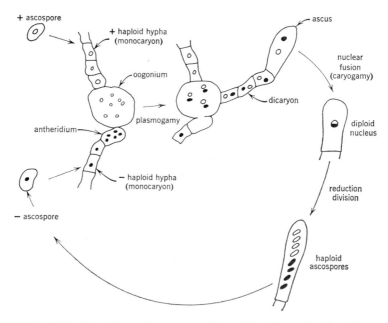

FIGURE 25-1. Semischematic representation of the life cycle of a fungus with restricted dicaryon. The fungus depicted is a heterothallic ascomycete with differentiated sex organs.

The nuclei of the fusing cells become associated in one or more pairs, these pairs being termed *dicaryons*. The dicaryotic phase of the life cycle may persist for a short period only, or it may persist more or less indefinitely as a result of repeated mitotic divisions. These mitotic divisions are termed *conjugate divisions*. Fusion of the pairs of associated nuclei, a process called *caryogamy*, ultimately establishes diploid nuclei. The sexual phase in many fungi is thus complicated by the failure of the male and female nuclei to fuse immediately after union of the protoplasts from the male and female thalli.

Raper (1954) has distinguished seven basic types of life cycles in the fungi, based on differences in the duration of the haploid, dicaryotic, and diploid phases in the life cycle.

1. Asexual cycle. In this life cycle, alternation in nuclear phase is apparently lacking, and reproduction is by asexual spores or by other specialized organs. The Fungi Imperfecti belong to this group, as do many species whose morphological features make them clearly assignable to other groups. Fungi that are exclusively asexual are reported to make up about 15,000 to 20,000 species of the approximately 80,000 species of fungi that have been described by mycologists. The actual proportion of truly asexual species must, however, be somewhat smaller than is indicated by this estimate, owing to a failure to observe the sexual stage in some species in which it occurs rarely or only under special conditions.

2. Haploid cycle. The life cycle is haploid except for a single diploid nuclear generation, meiosis occurring immediately following fusion of the sexual cells. This type of life cycle is the dominant one of the Phycomycetes and primitive Ascomycetes.

3. Haploid cycle with restricted dicaryon. The life cycle is predominately haploid, but repeated divisions of the dicaryon occur after plasmogamy and prior to caryogamy. This type of life cycle, illustrated in Figure 25–1, is characteristic of the higher Ascomycetes.

4. Haploid-dicaryotic cycle. Both the haploid (monocaryotic) phase and the dicaryotic phase are capable of indefinite vegetative growth. The life cycle is therefore made up of two more or less equivalent stages separated by a single diploid nuclear generation. This is the predominant life cycle of the Basidiomycetes (excluding some smuts). This cycle is particularly interesting in pathogenesis because the dicaryon frequently has different physiological requirements from its haploid components. For example, the haploid phase is saprophytic in some smuts whereas the dicaryotic phase is obligately parasitic. The best-known examples of differences between nuclear phases in pathogenesis occur in the rusts, where the haploid phase is an obligate parasite on one species (e.g., barberry in *Puccinia graminis tritici*), and the dicaryon is an obligate parasite on other species (wheat and related grasses).

5. Dicaryotic cycle. The life cycle is dicaryotic except during the meiotic cycle when a single diploid nuclear generation and a single haploid nuclear generation occur. The dicaryotic stage is reinitiated by fusion of the immediate haploid products of meiosis (ascospores or basidiospores). This type of life cycle occurs commonly in the smuts.

The dicaryotic life cycle differs from the previous one only in the duration of the haploid stage, and one cycle can sometimes be converted into the other by accidents of temporal or spatial relationships. For example, if compatible haploid products of meiosis in the haploid-dicaryotic cycle happen to germinate next to each other, the dicaryotic stage can be reinstated immediately after meiosis, and the haploid-dicaryotic cycle is converted into the dicaryotic cycle. On the other hand, the dicaryotic cycle can be converted into the haploid-dicaryotic cycle, under experimental conditions at least, by artificial prolongation of the haploid phase.

6. Haploid-diploid cycle. The life cycle here is similar to life cycles occurring in the algae and higher plants in that haploid and true diploid generations alternate. The mycelia of the two generations are identical except for the different reproductive organs they bear. This cycle is rare in fungi.

7. Diploid cycle. The life cycle is completely diploid except for the immediate products of meiosis, and hence is equivalent to the life cycle of animals. This cycle occurs in some yeasts and a few Phycomycetes.

In each of these cycles, asexual reproduction by spores (or other specialized structures) occurs in most species in addition to the sexual reproductive process. With rare exceptions, the asexual reproductive structures reproduce the phase of the cycle by which they are produced.

Mating Systems in Fungi

The essential features of the sexual process in fungi are the fusion of compatible nuclei to form the diploid phase and the subsequent meiotic cycle that again produces the haploid phase. Fungi can be divided into two great groups with respect to mating system, the *homothallic* group and the *heterothallic* group. In homothallic fungi, sexual fusions occur between cells produced on any two hyphae or even between cells of the same hypha. Differentiation between compatible sexual elements is thus intramycelial. •In heterothallic fungi, the differentiation between compatible sexual elements is intermycelial, and sexual fusions occur only between cells from thalli of different mating groups.

The homothallic fungi are in a sense comparable to self-compatible inbreeders among the higher plants. This might seem to imply that such fungi are deprived of the flexibility deriving from the segregation and recombination that follow sexual fusions between genetically dissimilar gametes. However, homothallism does not prevent matings between thalli with separate histories. Hence the opportunity exists for segregation to occur to much the same extent that would be expected in the progeny derived from natural crosses between, for example, two varieties of wheat. Homothallism occurs in all major groups of fungi and, with a few exceptions, is the predominant mating system in each group. The obvious success of homothallism is in itself evidence supporting the view that this system of mating provides the variability necessary for adaptation and survival.

Heterothallism was discovered by Blakeslee in 1904 when he found that conjugation between thalli of different mating types precedes the formation of zygospores in mucor. In this bread mold the identical appearance of the two mating types made their designation male and female anomalous, and for convenience they were denoted plus (+) and minus (−). The term heterothallism was introduced to designate (1) the occurrence of two types of individuals, each self-incompatible, and (2) the necessity for conjugation between mycelia of different types to accomplish sexual reproduction. Heterothallism is obviously the antithesis of homothallism, discussed above, which denotes species in which all individuals are self-compatible and sexually self-sufficient.

Extensive studies of sexuality in fungi have revealed that mating type is not necessarily correlated with true sex (defined as the existence of recognizably differentiated male and female organs). Whitehouse (1949) distinguished two general types of heterothallism: (1) morphological heterothallism, in which the conjugating thalli differ in producing dissimilar sexual organs or gametes distinguishable as male and female; and (2) physiological heterothallism, in which the conjugating thalli differ in mating type regardless of the presence or absence of differentiated male and female sex organs or gametes. Thus (taking homothallism into account) four types of segregation occur with respect to mating type and sexual differentiation: (1) segregation for mating type, (2) segregation for sexual differentiation,

(3) segregation for both mating type and for sexual differentiation, and (4) segregation for neither mating type nor sexual differentiation. The first three types are heterothallic, and the fourth type is homothallic.

Among the heterothallic species with sexual differentiation a few display simple dioecism, that is, there are only two types of thalli, each of which is distinguishable as male or female and is self-incompatible but cross compatible. In most species, however, both male and female gametangia are borne on the same thallus, but conjugation occurs only between male and female organs of thalli that are intercompatible as to mating type (or between intercompatible nuclei, + and —, of multinucleate mycelia). Although the inheritance of mating type is often simple, the genetic basis of sexual differentiation is frequently complex. The fact that sexual differentiation and mating type are superimposed one on the other in many species shows that mating-type genes are compatibility alleles not necessarily related either to sex organs or to the evolution of sex in the fungi.

The Genetic Control of Mating Type

In many fungi, mating type is governed by a single locus with two alleles, mt^+ and mt^-. Thus, in the haploid phase, only two types of mycelia occur. The diploid phase is necessarily heterozygous at the mating-type locus since fertilization occurs only between haploid thalli of unlike constitution. This type of genetic control of mating reaction is very common. It occurs frequently in the Phycomycetes, Ascomycetes, and in the heterothallic rusts (Uredinales) and smuts (Ustilaginales). Evidence has not been found of more than two mating types, so more than two alleles apparently do not occur in these groups of fungi.

In the Hymenomycetes and Gasteromycetes, there are two types of incompatibility systems: one involving a single locus, and the other two loci. Dicaryons are produced only in matings between monocaryons that differ in their alleles at one locus in the first type and in alleles at both loci in the second type. The Eubasidiomycetes differ from all other fungi in that their incompatibility mechanism consistently features multiple allelism.

The similarity between mating type in the fungi and incompatibility in the higher plants (Chapter 20) is obvious. One important difference exists, however (Mather, 1944). In the diploid plants, the action of the S gene in the style prevents any fusion of gametes from the same diploid mother. In haploid fungi, fusions between sister haploids—that is, between haploids from the same diploid parent—can occur in half the cases when a single pair of alleles is involved. Two separate series of alleles are required to reduce such fusions to a quarter. A third series (of which no example is known) would be required to reduce the frequency to one in eight. Thus the presence of the style makes incompatibility a more efficient out-breeding mechanism in higher plants than in fungi where no sieve of diploid tissue separates male and female gametes.

The several different life cycles and the several different patterns of mating type and sexuality, particularly when considered in com-bination with some patterns which fit uneasily into the scheme of organization presented here, obviously allow a bewildering array of distinct reproductive types in the fungi.

Variations in Pathogenicity—Physiological Specialization

The first unambiguous demonstration that microorganisms can include strains that differ in pathogenicity, even though morpho-logically undifferentiated, was that of Eriksson in 1894. First he noted that black stem rust taken from wheat did not infect oats, rye, and some other grass species. He next established that collections taken from various hosts each had a characteristic host range, attacking certain species but not others. On this evidence he defined several subspecies of *Puccinia graminis* based on differing physiological prop-erties as expressed in specific pathogenic properties on different grass hosts. The next important step was taken by Barrus in 1911, when he described *physiological races* on the basis of differing pathogenicity on different varieties of the same host species. Barrus distinguished two races, denoted α and β, of the bean-anthracnose organism on the basis of their different pathogenic relationships on varieties of the common bean, *Phaseolus vulgaris*. Shortly thereafter Stakman (1914) showed that Eriksson's subspecies were not pathogenically homo-geneous, but, like the bean-anthracnose organism, included physio-logical races varying in their pathogenicity on different varieties within a single cereal species. Evidence rapidly accumulated, largely through the efforts of Stakman and his collaborators, that each of Eriksson's subspecies was comprised of many physiological races. Subsequently these findings were extended to a large number of pathogenic fungi.

Physiological races are identified by their pathogenicity on a group of varieties known as differential hosts. Originally the differential hosts were chosen with knowledge that they differed in resistance, but with no information of the genetic basis of the resistance. Information about races has become more valuable as it has become possible to test them against differential hosts of known genetic constitution (as is possible for a few races at present). The way in which races are identified can be illustrated by the use of two varieties of wheat, Martin and Turkey, known to be genetically $MMtt$ and $mmTT$, respectively in respect to resistance to bunt.

Host	Race of Bunt			
	I	II	III	IV
Martin	Resistant	Resistant	Susceptible	Susceptible
Turkey	Resistant	Susceptible	Resistant	Susceptible

If only two levels of pathological reaction are recognizable (resistant or susceptible), a set of n differential hosts whose reaction to infection is governed by a single gene pair each can differentiate 2^n races. If five levels of pathological reaction can be recognized on each differential host, n differential hosts are potentially capable of differentiating 5^n races. Ten differential hosts and 5 levels of infection have been used in differentiating races of stem rust. Thus if each differential host variety is homozygous for only a single different pair of genes governing pathological reaction (an unlikely contingency), the 10 hosts are theoretically capable of distinguishing among 9,765,625 races. Several hundred different races of black stem rust of wheat have been identified using only 10 differential hosts. It is unlikely that the total number of races can ever be determined for any particular parasitic species, but the actual number is much less important than the fact that a very large number of races seem to occur in all parasites. If the plant breeder is to deal with this problem intelligently, he will need to know the genetic basis of pathogenesis for the light it may throw on host-pathogen relations.

The Inheritance of Pathogenicity in *Venturia inaequalis*

Venturia inaequalis, the organism that causes the apple-scab disease, has many advantages for the study of problems of infectious disease. This organism is an eight-spored Ascomycete in which serial isolation of spores is possible, and it is thus endowed with one of the major advantages of the eight-spored Neurosporas for genetic investigations. It also is free of one of the severe disadvantages of the obligately parasitic rusts and smuts in that it can be cultured and hybridized at will *in vitro*. Finally, it is haploid throughout the parasitic phase

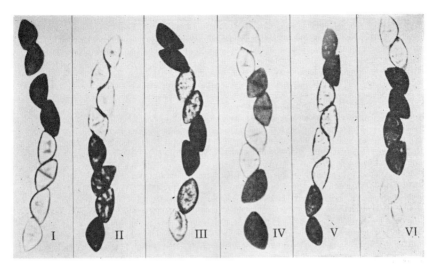

FIGURE 25-2.   Segregation of pale versus dark ascospores in *Bombardia lunata.* The eight linearly arranged ascospores provide an accurate diagrammatic representation of what has happened at meiosis.   Why?   Can you read and give a genetic interpretation of the six spore patterns above?   What advantages do the eight-spored ascomycetes have in studying the variability systems in pathogenic fungi?   (Reproduced from *The Genetics of Micro-organisms,* by D. G. Catcheside, Pitman (London), 1951, with permission of the author and his publisher.)

and in its vegetative stage *in vitro,* thus avoiding the complicating effects of heterocaryosis.   Intensive investigations of this organism have been carried out over many years by G. W. Keitt and his collaborators at the University of Wisconsin (reviews in Keitt, 1952; Keitt and Boone, 1956).

Two main types of pathogenic reaction have been encountered with wild-type lines of the pathogen.   These are "lesion" and "fleck," the former being typically pathogenic with abundant sporulation and the latter being non- or only slightly pathogenic with little or no sporulation.   With wild-type lines and a given apple variety, all the 8 ascospores in each ascus produced by a cross of lesion × lesion were found to produce mycelia that gave a lesion reaction.   Crosses of lesion × fleck, however, always gave 4 lesion ascospores and 4 fleck ascospores.   These results can be accounted for by assuming a single locus with multiple alleles controlling different pathogenic capabilities on different apple varieties, or, alternatively, they can be explained by assuming that a series of loci exist, each with two alleles.

The decision between these alternatives was made as follows.   Two monosporic wild-type lines were selected, one that incites lesions on the Haralson and Wealthy apple varieties and flecks on the Yellow

Transparent and McIntosh varieties, and one that produces the converse pathogenic reactions. When these two strains were crossed *in vitro*, the ascospores isolated in serial order, and the resulting haploid lines tested on the 4 differential host varieties, the results given in Table 25–1 were obtained. Among 35 asci analyzed, 17 first-division

TABLE 25–1. Pathogenic Reactions of Progenies Derived from a Cross of Two Strains of *Venturia inaequalis*

(After Keitt, Leben, and Shay, 1948)

Segregation Pattern	Type of Pathogenicity Reaction				Number of Asci
	Haralson	Wealthy	Yellow Transparent	McIntosh	
4 : 4	L	L	F	F	5
	F	F	L	L	
4 : 4	L	L	L	L	12
	F	F	F	F	
2 : 2 : 2 : 2	L	L	F	F	
	F	F	L	L	18
	L	L	L	L	
	F	F	F	F	
Total					35

L = Lesion.   F = Fleck.

segregations occurred, giving 4 different 4 : 4 arrangements when linear order in the ascus is taken into account. In the remaining 18 asci the arrangements were 2 : 2 : 2 : 2, indicating second-division segregation. Considering each host variety alone, the difference in pathogenicity appears to be governed by a single gene with 2 alleles. Comparisons of all 4 differential hosts indicate that pathogenicity on Haralson and Wealthy is governed by one gene, whereas another gene on a separate chromosome governs pathogenicity on Yellow Transparent and McIntosh. These results clearly rule out the hypothesis of a single locus with multiple alleles. By similar studies of other isolates and hosts, 6 separate genes governing lesion versus fleck reaction have been established and the occurrence of others postulated.

Sexuality and Pathogenesis: Synopsis

Studies such as the ones we have considered thus far in this chapter dispelled much of the confusion that attended early attempts to interpret the results of breeding for disease resistance. The following key points were clarified: (1) a great variety of life cycles and patterns of mating exist in fungi, providing various mechanisms for the storage and release of variability; (2) release of variability is through segregation and recombination following hybridization between genetically unlike types; (3) pathogenicity is a Mendelian character; (4) species of pathogenic fungi usually are made up of many physiological races that have different pathological capabilities even though they may be morphologically identical; and (5) physiological races are highly stable, but on sexual reproduction a race may produce numerous other races as a result of the segregation of genes governing virulence versus avirulence.

Although these studies were illuminating in explaining the plasticity of sexual microorganisms, they were not helpful regarding the fungi that have no sexual stage. There is ample evidence that many asexual fungi are fully as flexible in their characteristics, including pathogenicity, as the sexual organisms. We shall now examine possible explanations for this apparently anomalous situation.

Heterocaryosis

In many fungi, particularly the asexual Ascomycetes and the Fungi Imperfecti, the plant body is multinucleate, at least during the stages in the life cycle when growth is active. Furthermore, hyphal fusions in which nuclei are exchanged between different mycelia, regardless of their sex or mating type, are a regular occurrence in these fungi. Opportunity thus exists for genetically differentiated nuclei to come together in the same cell to form heterocaryons, a phenomenon known as heterocaryosis. In itself, this opportunity is not a guarantee that genetically unlike nuclei do in fact come to occupy the same cell, but it does make a *prima facie* case for regarding multinucleate fungi as potentially heterocaryotic.

The implications of heterocaryosis in providing for "segregation" without sexual fusion had been realized by mycologists for some time, but the proof that genetically unlike nuclei occur together in the same hypha was first clearly demonstrated in 1932 by Hansen and Smith in the imperfect fungus, *Botrytis cinerea*. They found that single-spore cultures produced three different kinds of mycelia. One type was characterized by abundant sporulation, and mycelia derived from these asexual spores were uniformly conidial. A second type was characterized by abundant mycelial growth but sparse sporulation. Like the "conidial" type, this "mycelial" type remained constant when reproduced asexually. The third type was intermediate in sporulation and was not constant like the first two types but continued to produce conidial, mycelial, and intermediate types. Furthermore, when conidial and mycelial strains were grown in a mixed culture, mycelia of all three types appeared even though the possibility of sexual fusion could be eliminated.

Hansen and Smith established that the intermediate type of hypha was made up of two kinds of nuclei and that it arose as a result of the frequent anastomoses of hyphae they observed in their cultures. Although these fungi are septate, the cross walls are perforated. Hence nuclei can be transmitted from one end of the mycelium to another through protoplasmic streaming. Nuclear exchange in a single cell therefore provides the opportunity for genetically unlike nuclei to exist side by side in an entire mycelium. These observations were later extended by Hansen (1938) to a large number of species of imperfect fungi, extending over thirty different genera.

Even more clear-cut evidence of the nature of the heterocaryotic relationship has come from laboratory studies of biochemical mutants (see Beadle and Coonradt, 1944). It was shown that two homocaryons, each carrying a single allele for a nutritional deficiency and hence unable to grow on a minimum medium, could be combined into a heterocaryotic individual capable of conspicuously greater growth than either of its constituent homocaryons. It is readily apparent that a heterocaryon containing different types of nuclei behaves like a genetic heterozygote and hence possesses "heterocaryotic vigor" analogous to the heterotic advantage of heterozygotes in cross-fertilizing sexual species. Furthermore, the system is clearly capable of providing for a type of "somatic segregation" based on exchange of entire nuclei during hyphal fusions.

These observations suggest that heterocaryotic mycelia may have the important property that their genetic composition can be altered by natural selection. Unfortunately, there is little evidence of the extent of heterocaryosis in wild fungi and of its importance as a genetic system. The implications of heterocaryosis as a mechanism of adaptation in fungi are potentially so important, however, that such evidence as there is justifies review.

Some of the most conclusive studies of the importance of heterocaryosis in nature are those of Jinks (1952) with *Penicillium cyclopium*. In this species, fourteen out of sixteen isolates taken from the wild turned out to be heterocaryons and only two were homocaryons. The homocaryons were found to have poorer growth rates on a minimal medium than the heterocaryons, and in this sense were similar to the laboratory mutants studied in earlier investigations. The shortcomings of the wild homocaryons were less drastic than the biochemical mutants, which were lethal on minimal medium. Apparently, therefore, heterocaryons which have a substantial advantage over their constituent homocaryons exist in the wild.

The ability of heterocaryosis to produce adaptive changes in response to altered environmental circumstances would be proved if it could be established that the ratios of the types of nuclei vary in response to changes in the conditions under which the heterocaryon is growing. Jinks was able to show that the ratio of the two kinds of nuclei occurring in one of his heterocaryons varied with the medium on which this heterocaryon was grown, and that a typical nuclear ratio occurred for each of several types of media. In general, the ' nuclear ratio was related to the comparative growth rate of the constituent homocaryons on any medium in the sense that the proportion of one type of nucleus in the heterocaryon was equal to the comparative growth rate of the related homocaryon on the same type of medium. The variation in the relative proportions of nuclei was regarded as adaptive because, irrespective of the proportions of nuclei in an initial hypha (so long as it was not homocaryotic), the balance of a colony after growth was always found to be characteristic of the substrate upon which the colony had been grown.

Similar results have been found in a few other species. These include the *Fusarium* species, *Aspergillus niger,* occurring on the gladiolus, and *Sclerotinia trifoliorum.* The work with *Sclerotinia* was particularly interesting because it established that heterocaryosis can lead to changes in pathogenicity in a plant-disease organism. An even more clear-cut demonstration that heterocaryosis can lead to new pathogenic races has been provided by Nelson, Wilcoxson, and Christensen (1955), working with the black stem-rust organism. Urediospores of two biotypes were mixed manually and dusted on a compatible variety of wheat. The first generation of urediospores from the compatible variety was transferred to resistant varieties, and one biotype, originating from a mixture of races 38 and 56, was found that was highly pathogenic on Khapli Emmer, a wheat previously resistant to all known North American races of stem rust. After several uredial generations on Khapli, the virulent heterocaryon became unstable, as indicated by gradual loss of virulence on Khapli. After twenty-five uredial generations it produced only a resistant reaction on Khapli. When single uredium isolations were made from the resistant-type pustules on Khapli, both parental biotypes were recovered and in addition two new races were obtained, each possessing some of the pathogenic properties of the two parental biotypes.

Some of the urediospores and hyphal cells of the biotype that was highly virulent on Khapli were three- or four-nucleate, but the two new races derived from it were again binucleate. The results were interpreted as indicating that the multinucleate condition is unstable in black stem rust, and that complete reversion to the dicaryophase is the ultimate fate of multinuclear associations. The results also demonstrate that genes for virulence on Khapli Emmer are present in North American races of stem rust and can be associated in different nuclei in a cell by hyphal fusions. These authors believe it highly probable that heterocaryotic biotypes are produced in nature. The extensive acreages of wheat planted to varieties susceptible to more than one race of stem rust, and the billions of urediospores produced in a single acre of wheat provide an excellent opportunity for heterocaryons to be produced and for reassociations of nuclei to occur.

For plant parasites whose hosts vary with environmental conditions and also change progressively with age, the advantages of adaptive somatic change are clear. Heterocaryotic fungi can adjust themselves to circumstances as they develop. Fungi depending solely on the sexual cycle, on the other hand, must go through the entire sexual cycle, including fertilization, segregation, and recombination, before variants adapted to the new circumstances can arise. Heterocaryosis thus provides exactly the continuous and opportunistic adaptation that a parasite which must change its requirements in response to changes in its host would be expected to need. It should also permit the parasite to adapt itself to attack plants with different types of resistance. For immediate purposes, heterocaryosis both improves on the mitotic part of the life cycle and supplements or replaces the meiotic part of the life cycle. It enables the fungus to dispense with meiosis and fertilization and may well be the reason why so many fungi have become imperfect. But like other systems that bypass the sexual cycle or render it ineffective (e.g., inbreeding, ring formation), heterocaryosis seems likely to produce a genetic rigidity that may be fatal in the long run. This is because the unit of segregation and recombination in heterocaryosis is the entire haploid nucleus, a much less flexible unit than the gene of sexual reproduction. Nevertheless, if present heterocaryons leave no descendents in the distant future, heterocaryotic species would be likely to turn up in other lines of descent if the great immediate advantages it would seem to confer are in fact real.

Before leaving the subject of heterocaryosis, we should consider its relations to the dicaryotic phase of the life cycle of fungi with extended dicaryon. The dicaryon appears to be a normal part of the process of sexual reproduction, related to the sexual cycle and to the mechanism of spore dispersal. Dicaryosis features a rigid nuclear balance that precludes the opportunistic type of somatic variation occurring in the multinucleate fungi. Dicaryons thus appear to differ in evolutionary history and present function from heterocaryosis in the sense discussed here. For these reasons Jinks (1952) has suggested that the terms homo- and heterocaryon be limited to the components of systems in multinucleate fungi and mono- and dicaryon be used to describe the condition found in the heterothallic Basidiomycetes. This distinction does not prevent the use of the term heterocaryosis to describe the temporary multinucleate associations, and reassociations of nuclei resulting from them, which occur in certain of the normally dicaryotic *Uredinales*.

The Parasexual Cycle in Fungi

Another process has recently been discovered in fungi that appears to throw light on the long-range flexibility of these organisms. It has been found that strains of certain fungi can be synthesized that carry diploid rather than haploid nuclei in their cells. Such strains arise with a frequency of approximately one in 10^7 cells during the growth and sporulation of heterocaryons. These cells give rise to *heterozygous diploid* thalli that have dominant phenotypes in respect to the loci at which they are heterozygous, and as they grow they produce patches of mycelium with the recessive phenotype at *some* of these loci. Such patches cannot have resulted from the formation of homocaryons because entire nuclei are the unit of segregation in heterocaryosis.

Investigations by Pontecorvo, Roper, and their associates have shown that two processes are required to explain these results. The first is haploidization of the diploid nuclei. It occurs with a frequency of about one cell in 10^3, apparently through some failure of regular distribution of the chromosomes during mitosis. The genotypes of the haploids show that random recombination has occurred among entire chromosomes during the process by which a diploid nucleus forms (presumably) two haploid nuclei. In other words all genes on a single chromosome behave as if they were linked. Since the frequency of haploidization is much higher than fusions to form diploid nuclei, the ultimate result of the growth of a diploid heterozygote that has arisen from fusion of unlike nuclei in a heterocaryon is a variety of haploid strains recombining the chromosomes of the diploid in all possible ways. In this process the unit of segregation is the chromosome.

The second process is mitotic crossing over similar to that described by Stern in *Drosophila*. The incidence of this mitotic crossing over appears to be about one chiasma per hundred nuclei. The gene is the unit of recombination in this second process.

It is easily seen that the combination of these two processes is qualitatively exactly equivalent to the usual sexual cycle, since together they involve diploidization (fertilization), recombination, and haploidization (reduction division). The only significant difference is the absence of a precise time sequence in the *parasexual* cycle. Crossing over and haploidization are not two parts of a single process, following in that order, as they do in meiosis. Rather, they occur independently and usually, but not necessarily, in different nuclei. The questions that remain are the importance of this parasexual cycle in nature and its efficiency as a substitute for the normal sexual cycle. These questions will not be easily answered because diploid strains are unlikely to be common in nature. In the laboratory haploids outnumber diploids as $10^{-3}:10^{-7}$, or 10,000:1, and, if the relative frequency in nature is the same, diploids will be difficult to detect and study.

Heterocaryosis in itself can give rise to strains with new characteristics in only one way, by interactions among genes carried in the two or more types of nuclei that may come into persistent association to form a heterocaryon. Heterocaryosis alone cannot give rise to new pathogenic properties, for example, by recombination. The parasexual cycle would, however, make this possible in the imperfect fungi and must therefore be regarded as a process of potential importance in the origin of new pathogenic races.

The discovery of the parasexual cycle also has implications in breeding improved strains of asexual fungi used in industry. Improvement of these organisms (e.g., *Penicillium notatum*) has heretofore depended on selection among large collections of strains taken in nature or on selection of spontaneous or artificially induced mutators within strains. The parasexual cycle permits the use of procedures akin to those used in breeding sexual species; in fact, patent rights have been applied for by the discoverers of this process.

Bacteria, Viruses, and Other Parasites

Many important plant diseases are caused by bacteria and viruses, organisms that have been amply demonstrated to be fully as resourceful as fungi in modifying their pathogenic capabilities. Limitations of space prohibit descriptions of the experiments that in recent years have given us insight into the way the variability systems operate in these organisms. However, in overall pattern the genetic systems of bacteria do not appear to differ greatly from fungi in that

1. Mutations occur at about the same rate.
2. Some bacteria possess a sexual cycle.
3. Some races are homothallic and some are heterothallic.
4. Heterocaryosis can occur.
5. Linkage can be demonstrated.

The viruses have also been shown to be complex genetically.

Other plant pathogens, for example, insects and nematodes, have also demonstrated that they can produce large numbers of physiological races. Here the sexual cycle is clearly the most important mechanism for adaptive change even though many animal pathogens, such as aphids and the nematode that incites the root-knot disease, have parthenogenetic phases in their life histories.

REFERENCES

Barrus, M. F. 1911. Variation in varieties of beans in their susceptibility to anthracnose. *Phytopath.* 1: 190–195.

Beadle, G. W., and V. L. Coonradt. 1944. Heterocaryosis in *Neurospora crassa*. *Genetics* 29: 291–308.

Biffen, R. H. 1905. Mendel's law of inheritance and wheat breeding. *Jour. Agric. Sci.* 1: 4–48.

Biffen, R. H. 1912. Studies in inheritance of disease resistance. II. *Jour. Agric. Sci.* 4: 421–429.

Catcheside, D. G. 1951. *The genetics of microorganisms.* Pitman Co., London and New York.

Hansen, H. N., and R. E. Smith. 1932. The mechanism of variation in imperfect fungi, *Botrytis cinerea. Phytopath.* 22: 953–964.

Hansen, H. N. 1938. The dual phenomenon in imperfect fungi. *Mycologia* 30: 442–445.

Jinks, J. 1952. Heterocaryosis: a system of adaptation in wild fungi. *Proc. Roy. Soc.* (London), Series B, 140: 83–99.

Johnson, T. 1947. Variation and the inheritance of certain characters in rust fungi. *Cold Spring Harbor Symposium on Quant. Biology* 11: 85–93.

Johnson, T., and M. Newton. 1946. Specialization, hybridization, and mutation in the cereal rusts. *Bot. Rev.* 12: 337–392.

Keitt, G. W. 1952. Inheritance of pathogenicity in *Venturia inaequalis* (Cke) Wint. *Amer. Nat.* 86: 373–390.

Keitt, G. W., C. C. Leben, and J. R. Shay. 1948. *Venturia inaequalis* (Cke) Wint. IV. Further studies on the inheritance of pathogenicity. *Amer. Jour. Bot.* 35: 334–345.

Keitt, G. W., and D. M. Boone. 1956. Use of induced mutations in the study of host-pathogen relationships. *Brookhaven Symposia in Biology* No. 9: 209–225.

Lederberg, J. 1948. Problems in microbial genetics. *Heredity* 2: 145–198.

Lewis, D. 1954. Comparative incompatibility in angiosperms and fungi. *Adv. in Gen.* 6: 235–285.

Lindegren, C. C. 1948. Genetics of the fungi. *Ann. Rev. of Microbiol.* 2: 47–70.

Mather, K. 1944. Genetical control of incompatibility in angiosperms and fungi. *Nature* (London) 153: 392–394.

Moulton, F. R. 1940. *The genetics of pathogenic organisms.* Science Press, Lancaster, Pa.

Nelson, R. R., Roy D. Wilcoxson, and J. J. Christensen. 1955. Heterocaryosis as a basis for variation in *Puccinia graminis* var. *tritici. Phytopath.* 45: 639–643.

Newton, M., T. Johnson, and H. T. Gussow. 1932. Specialization and hybridization of wheat stem rust, *Puccinia graminis tritici,* in Canada. *Can. Dept. Agric. Bull.* No. 160, 60 pages.

Orton, W. A. 1909. The development of farm crops resistant to disease. *U. S. Dept. Agric. Yearbook* 1908, 453–464.

Papazian, H. P. 1958. The genetics of basidiomycetes. *Adv. in Gen.* 9: 41–69.

Pontocorvo, G. 1946. Genetic systems based on heterokaryosis. *Cold Spring Harbor Symposia on Quant. Biology* 11: 193–201.

Pontocorvo, G. 1954. Mitotic recombination in the genetic system of filamentous fungi. *Proc. Ninth Intern. Congr. Gen.,* Caryologia, Vol. 6 (Suppl.), 192–200.

Raper, J. R. 1954. Life cycles, sexuality and sexual mechanisms in the fungi In *Sex in Microorganisms,* Amer. Assoc. Advancement Sci., Wash. D. C.

Stakman, E. C. 1914. A study in cereal rusts. Physiological races. *Minn. Agric. Exp. Sta. Bull.* 138.

Walker, J. C. 1951. Genetics and plant pathology. In *Genetics in the 20th Century,* pp. 527–554.

Ward, H. M. 1902. On the relations between host and parasite in bromes and their brown rust, *P. dispersa* (Erikss.) *Ann. Bot.* 16: 233–315.

Whitehouse, H. L. K. 1949. Heterothallism and sex in the fungi. *Biol. Rev.* 24: 411–447.

26

Variability in Disease Reaction of Host Species

Although it was known before the start of the Christian era that cultivated varieties differ in ability to withstand disease, no use was made of this knowledge in the breeding of resistant varieties for many centuries. During the intervening period it is probable that many varieties were discarded because of conspicuous susceptibility to diseases. It is also probable that advantage was taken of resistant plants as they occurred naturally in fields and gardens. Such selection for disease resistance no doubt helps to explain the tolerance of many land varieties to diseases, but programs designed explicitly to produce resistant varieties did not start until the nineteenth century.

It was toward the middle of the nineteenth century that the first clear-cut differences among varieties in reaction to disease were recorded, mostly in England. Thomas Andrew Knight, for example, noted differences among wheat varieties in their resistance to rust and described these differences in unambiguous language. Shortly thereafter M. I. Berkeley pointed out that whereas white-bulbed varieties of onions were seriously affected by smudge, varieties with colored bulbs tended to be free of the disease. There also was interest in the reactions of varieties to diseases in the United States at this time. One of the earliest Americans to work in the field was Chauncey Goodrich of New York State, who published a paper in 1848 dealing with the resistance of potatoes to the blight disease. The number of descriptions of differences in resistance among varieties increased sharply in the last half of the nineteenth century, and this period also saw the start of a number of breeding programs designed to develop resistant varieties in crops such as potatoes, cereals, and grapes. The stage was therefore set for the initiation of studies of disease resistance immediately upon the rediscovery of Mendel's paper in 1900.

Subsequent to the rediscovery of Mendel's paper, studies of the inheritance of disease resistance have progressed through three more or less distinct phases.

1. *Studies of resistance in varietal crosses.* The earliest studies of the inheritance of disease resistance were concerned with numbers of disease-conditioning genes in the host. As a result the number of gene pairs segregating in many individual hybrids was recorded, but little was learned of the distinctiveness or duplication of disease-resistance genes in different resistant varieties.

2. *Identification of individual disease-conditioning genes.* In this phase hybrids were tested against single physiological races of the pathogen in experiments designed to allow the identification of individual genes for resistance. Additional useful information was then obtained by determining the reaction of these individual genes to the various races of the pathogen.

3. *Genetics of the host-pathogen interaction.* This stage had its start when specific genes for virulence or avirulence were identified in the pathogen and related to specific genes for resistance or susceptibility in the host.

Progress in advancing from one stage to another has been uneven in different crops and for different diseases because of the disparity among various hosts and pathogens in suitability for precise genetic studies.

Inheritance of Resistance

Following the announcement by Biffen in 1905 that resistance to yellow rust of wheat is governed by a single recessive gene in crosses between certain susceptible varieties (Michigan Bronze and Red King) and the resistant variety Rivit, there were a large number of papers dealing with the inheritance of resistance. In a comprehensive review of the literature on this subject up to 1934, Hansen listed more than 200 titles. The papers published at that time included studies involving 18 different genera of host plants, 18 genera of fungi, 6 viruses, and 1 bacterial species. Resistance to 23 diseases in 13 different crops was described as monogenic in inheritance. Some investigators, however, reported that resistance was digenically controlled for some of the same diseases. In 88 papers dealing with 29 diseases in 15 crops, the conclusion was reached that resistance was due to action of multiple genes.

Subsequent to Hansen's 1934 review, many additional studies of the inheritance of disease resistance have been published. In the majority of studies resistance has been reported to be governed by a single gene, but instances of duplicate, complementary, and various other types of gene action have also been recorded. Although resistance governed by many genes is not uncommon in published accounts, the number of such cases reported is probably less than the number actually encountered because of reluctance to publish data that cannot be analyzed precisely. As we shall see later, complete analysis of many cases of apparently multigenic inheritance is possible under proper experimental conditions.

Classification for Reaction to Disease

Before attempting further discussion of segregation for disease resistance in progenies derived from hybrids, some of the problems encountered in classifying segregating generations for pathological reaction must be considered. When dealing with diseases such as the rusts and powdery mildew, it is possible to arrive at an opinion of the resistance of individual plants in the F_2 generation. In some instances individual plants fall into one or the other of two distinct categories, such as resistant or susceptible. Other times, however, F_2 progenies run the gamut of all possible reactions from immune to susceptible. Even with monogenic segregation it is often difficult to establish the inheritance of disease reaction with confidence from F_2 data alone. For this reason the standard practice in studies of the inheritance of disease resistance is to grow generations in addition to the F_2, usually the F_3 or backcross generations but sometimes still other generations, in order to obtain evidence to support the conclusions reached from analysis of the F_2 segregation ratios.

Diseases such as the smuts pose more difficult problems of classification. Individual plants can be scored only as diseased or healthy; and since infection of all plants is a rare occurrence, even in susceptible varieties, F_2 data in themselves have little value for analytical purposes.

Data on the percentage of infected plants in the parental types and in F_3 progeny rows are, however, useful in interpreting the inheritance of resistance. This can be illustrated with the data obtained by Briggs (1926) in a study of inheritance of bunt resistance in the hybrid between the wheat varieties Martin and White Federation. In this study 71.8 per cent of the plants of White Federation, the susceptible parent, were diseased. Martin, the resistant parent, and the F_1 of Martin × White Federation were both completely bunt free. Among F_2 plants 17.2 per cent were diseased. In the F_3 generation, 299 families were grown and classified for percentage of infection with the results shown in Figure 26–1. Seventy-three families came under the first mode of the curve, 147 under the second, and 79 under the third mode. If these families arose from homozygous resistant (*MM*), heterozygous (*Mm*), and homozygous susceptible (*mm*) F_2 plants, respectively, the fit to the expected ratio of 1 : 2 : 1 is good. The first minimum is clearly defined and for all practical purposes (as shown by later studies) accurately separates the *MM* and *Mm* genotypes. The second minimum is not as definite. The 50 per cent infection point was chosen to delimit the upper limit of the *Mm* genotype and lower limit of the *mm* genotype. Had the point of limitation been arbitrarily moved 10 per cent in either direction, the conclusions regarding the inheritance of resistance would not have been altered.

Although the resistant parent, Martin, was completely free of bunt, F_3 rows including $7\frac{1}{2}$ per cent of infected plants were classified as resistant. It was later shown that lines that are genotypically *MM* are not necessarily as resistant as Martin owing to the action of modifying genes.

When a variety that carries two dominant genes, each individually capable of producing a resistant phenotype, is crossed with a susceptible variety, the ratio expected in the F_2 generation is 15 resistant to 1 susceptible. In such crosses it may be difficult to specify the geno-

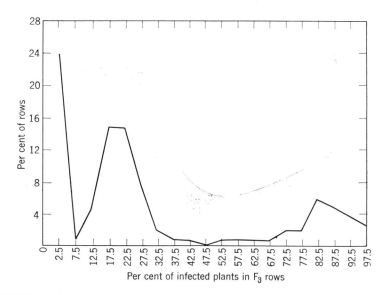

FIGURE 26-1. Distribution of F_3 rows into 5 per cent classes for bunt infection in the hybrid Martin × White Federation. (After Briggs, 1926.)

types of F_2 plants even with the aid of their F_3 progenies. An example of such a cross is given in Figure 26–2. This figure represents percentage of bunt infection in 598 F_3 families derived from Hussar, a variety completely bunt free under the test conditions, crossed with Baart, a variety in which 83.6 per cent of the plants were infected. In the F_1 and F_2 generations, 0.8 per cent and 9.7 per cent, respectively, of the plants were infected. The distribution of the F_2 plants, judged from their F_3 progenies (Figure 26–2), suggests a 15 : 1 ratio.

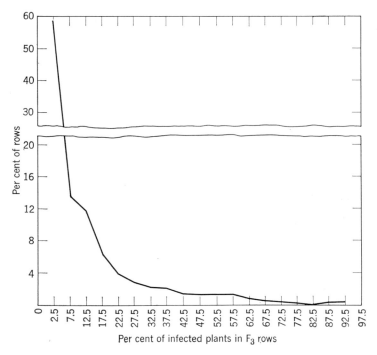

FIGURE 26-2. Distribution of F_3 rows into 5 per cent classes for bunt infection in the hybrid Hussar $\times$ Baart. (After Briggs, 1926.)

Convincing analysis of this cross requires, however, that the two genes be extracted and individually identified. This subject will be considered in more detail later.

Studies with Known Physiologic Races

In the earliest studies of the inheritance of disease resistance, variations in the pathogenic capabilities of the fungus were not taken into account. This is understandable since the existence of physiologic races within fungus species was not yet known. With the demonstration by Barrus in 1911 of physiological specialization in the bean-anthracnose organism and the researches of Stakman and his collaborators on physiological specialization in the rusts, it became clear that this factor would have to be taken into account in studies of the inheritance of resistance.

The value of working with known races in studies of inheritance was first demonstrated in 1921 by McRostie, in his studies of the inheritance of resistance to anthracnose disease in common bean varieties. McRostie found that the reactions of the White Marrow and Robust varieties of beans to the alpha and beta strains of the anthracnose organism are as follows:

Variety	Alpha Strain	Beta Strain	Mixture of Alpha and Beta
White Marrow	Resistant	Susceptible	Susceptible
Robust	Susceptible	Resistant	Susceptible

When White Marrow was crossed with Robust and the F_2 generation tested against the alpha and beta strains of the organism individually and against a mixture of the two strains, the results allowed McRostie to set up the following Mendelian explanation of reaction to the disease:

	F_2 Ratio	Genotype of		
Inoculum	(Resistant : Susceptible)	White Marrow	Robust	F_1
Alpha	3 : 1	AA	aa	Aa
Beta	3 : 1	bb	BB	Bb
Alpha and beta	9 : 7	AAbb	aaBB	AaBb

Had the existence of races not been known and the two varieties been inoculated inadvertently with a culture consisting of a mixture of the two races, the conclusion would have been reached that both varieties were susceptible to the disease. If the F_2 generation had been unknowingly inoculated with the mixture of races, it would have been concluded that hybridization between two susceptible varieties had produced a resistant type and that complementary genes had been brought together by hybridization. The separation of the fungus into two components of differential pathogenicity allowed McRostie to make a more penetrating analysis of the genetic basis of the reaction of the host to this disease than would otherwise have been possible.

Identification of Individual Genes for Resistance

Although there have now been many studies in which the inheritance of the resistance of particular varieties to specific races has been determined, all too rarely have relationships between genes in different resistant varieties been established. Sometimes inferences about these relationships can be made from the type of reaction obtained in tests with a series of physiological races. However, convincing proof whether a resistant variety carries the same or a different set of genes than other resistant varieties can be obtained only by genetic tests of appropriate hybrids. This procedure can be illustrated using some of the results of Briggs and his associates in identifying genes governing resistance to bunt in wheat and to mildew disease in barley.

The resistance of the wheat variety Martin to race T–1 of the bunt organism is governed by a single gene (Figure 26–1), and that of Hussar appears to be governed by two genes (Figure 26–2). When a hybrid was made between Martin and Hussar and the F_2 and F_3 generations tested against race T–1, all the progenies were uniformly resistant. One of the genes in Hussar is thus the same as the single gene carried by Martin.

In the next step of the analysis of Hussar, a number of resistant F_3 progenies derived from the cross Hussar $\times$ Hard Federation (a susceptible variety) were crossed with Martin and also with Hard Federation. If the original hypothesis that Hussar carries two genes for resistance is correct, some of the F_3 lines should differ from Hard Federation by a single gene and should also produce some susceptible progeny when crossed with Martin. F_3 lines with this reaction pattern presumably carry only the second gene of Hussar. Several such lines were found, among which one, designated selection 1403, was chosen as a test line for further investigations. When F_3 progeny of the hybrid of selection 1403 $\times$ Martin were studied in detail, the distribution was found to be of the same type as had been observed in the hybrid of Hussar $\times$ Baart (Figure 26–2). This can be interpreted to indicate that Hussar carries two genes for resistance, one of which is identical to the single gene of Martin (MM). The second gene in Hussar was designated (HH); hence the genotype of Hussar is $MMHH$.

Four major genes and two minor genes for resistance to race T–1 of the bunt organism have now been identified. The distribution of these genes in the wheat varieties most widely used by plant breeders as sources of bunt resistance is given in Table 26–1.

TABLE 26–1.   Distribution among Resistant Varieties of Wheat of Six Genes Governing Pathological Reaction to Race T–1 of the Bunt Organism

Resistant Variety	Number of Genes for Resistance	Dominance Relations	Genotype of Resistant Variety*
Martin	1	Dominant	MM
White Odessa	1	"	MM
Odessa	1	"	MM
Sherman	1	"	MM
Banner Berkeley	1	"	MM
Hussar	2	"	MMHH
Selection 1403	1	Intermediate	HH
Turkey (C. I. 1558)	1	"	TT
Turkey (C. I. 2578)	1	"	TT
Turkey (C. I. 3055)	1	"	TT
Oro	1	"	TT
Rio	1	"	RR
Turkey (C. I. 10015)	2	Near recessive	XXYY†
Turkey (C. I. 10016)	3	"	TTRRXX (or YY)

* The genes for bunt resistance have been named for the variety in which they were first found.   Thus M stands for the Martin gene, H for the Hussar gene, and so forth.

† The genes X and Y are weak genes for resistance to race T–1.   The former allows about 22 per cent and the latter about 45 per cent infection under conditions where the M, H, T, and R genes permit 0 to 5 per cent infection, and susceptible varieties are infected to the extent of 75 per cent or more.

Parallel methods have been used to identify individual genes governing resistance to race 3 of the powdery-mildew organism, *Erysiphe graminis hordei*.   Analysis of 13 different resistant varieties has disclosed the existence of 10 different genes for resistance.

Reaction of Resistance Genes to Additional Races

Testing against a single race is a valuable procedure in studies of the inheritance of disease resistance.   However, information obtained by this method has its limitations because pathogenic fungi are normally comprised of a large number of physiological races.   The bunt fungi (*Tilletia caries* and *T. foetida*), for example, are known from their pathogenic reactions against a set of 11 differential varieties to include at least 26 races (Rodenheiser and Holton, 1945). It would clearly be useful in the breeding of bunt-resistant varieties of wheat to know how each of the genes for resistance in the host, as identified by testing against race T–1, reacts to each of these 26 races.

In 1950 Briggs and Holton presented data showing the reaction of the 4 major genes for bunt resistance to 25 of the 26 races of bunt then known (Table 26–2). One of the conspicuous features of these data

TABLE 26–2. Reaction of Varieties of Wheat to Twenty-five Races of the Bunt Fungi, *Tilletia caries* and *T. foetida*

(Modified from Briggs and Holton, 1950)

Host Varieties	Genotype of Host Variety	Reaction Groups* of *T. caries* and *T. foetida*								
		1	2	3	4	5	6	7	8	9
Martin	MMM_2M_2	R	R	S	S	R	R		R	
White Odessa	MM	R	R	S	S	S	S		S	S
Selection 1403	HH	R	S	R	S	S	R		S	R
Turkey 1558	TT	R	R	R	R	R	R	R	S	S
Rio	RR	R	R	R	R	R	R	R	S	S

* Each reaction group includes one or more physiological races with the same pathogenicity against the M, M_2, H, T, or R genes. See text.

was that each gene controlled several races of both *T. caries* and *T. foetida*. The Hussar gene controlled 11 of the 25 races, the Martin gene 15, and the Turkey and Rio genes each controlled 23 of the 25 races. On the basis of these data it was possible to divide the 25 races into 8 pathogenicity groups as follows:

Group I. (Races T–1, T–2, T–9, T–10, L–1, L–2.) The races in this group are controlled by any one of the M, H, T, or R genes.

Group II. (Races T–3, T–11, T–16, L–3, L–8, L–10.) M, T, and R give control, but H does not.

Group III. (Races T–4, T–6, T–12, L–4.) H, T, and R give control, but M does not.

Group IV. (Races T–5, T–7, T–8, T–15, L–5, L–7.) T and R give control, but M and H do not.

Group V. (Races T–13, L–9.) This group differs from the previous group only in the reaction of Martin and White Odessa. It was postulated from this difference in reaction (Table 26–2) that Martin carries a second resistance gene not carried by White Odessa. This difference between the two varieties was not unexpected because Bressman in Oregon and Crepin, Bustarret, and Chavalier in France had observed dihybrid ratios in crosses of Martin with certain susceptible varieties. Schaller and Holton later confirmed these speculations by showing that Martin carries a second gene, M_2, which is intermediate in dominance.

Group VI. (Race T–14.) This race is controlled by M_2, H, T, and R but not by M.

Group VII. (All races except T–16 and L–8.) These races are controlled by T and R. Some are controlled by M, M_2, and H, and others are not, as shown above.

Group VIII. (Races T–16, L–8.) This group is controlled by M and M_2, but not by H, T, or R.

Two genes, the Martin gene and either the Turkey or Rio gene, control all the 25 races studied. Subsequently a number of additional races have been found, one of which (T–18) is interesting because it is not controlled by the Martin, Turkey, or Rio genes and hence constitutes a ninth group. Race T–18, is however, controlled by the H gene and also by a gene (or genes) in the variety Hohenheimer.

In developing varieties resistant to all presently known races, wheat breeders therefore need be concerned with only three major genes, because varieties that are genotypically $MMTTHH$ or $MMRRHH$ are resistant to all known races. Furthermore, experimental stocks need to be tested against only three races since either T–16 or L–8 will serve as a tester for the M gene, T–18 will serve as a tester for the H gene, and any one of the other races tests for the T and R genes. It would therefore appear that the problem of breeding for bunt resistance is simple genetically and has a simple solution as far as known races are concerned. It appears that this work has provided wheat breeders with a tool by which they can classify new races as they appear, and hence rapidly determine the steps to be taken in their control.

Genetics of Host-Pathogen Interaction

Whether a variety is resistant or susceptible to a physiological race depends on its genotype for resistance and the genotype for virulence or avirulence of the race in question. In the final analysis, therefore, pathogenic reaction involves interaction of genes conditioning resistance in the host with those conditioning pathogenicity in the parasite. To this point we have been able to take into account only one or the other aspect of the host-pathogen interaction because of the unsuitability of either host or parasite for precise genetic studies. For example, the apple-scab organism is favorable material for studies of the inheritance of pathogenicity, and much has been learned about the inheritance of parasitism in *Venturia inaequalis*. Little is known about the inheritance of resistance in the host, however, owing to the long generation time and other unfavorable features of apples for genetic investigations. In contrast, wheat and barley are hosts with favorable attributes for genetic studies of host resistance, but their principal diseases, the heteroecious rusts, the smuts, and mildew, have exacting requirements for genetic studies of pathogenicity. Thus, although the genes governing resistance are well known, the genetic basis of pathogenicity in the parasite, despite recent progress, remains obscure. There is only one study in which both the inheritance of resistance and the inheritance of pathogenicity have been established to the point where the host-pathogen interaction can be placed on a gene-for-gene basis. The establishment of this relationship is due to the classical researches of Flor on the complementary genic systems for resistance in flax, *Linum usitatissimum,* and for pathogenicity in flax rust, *Melampsora lini.*

Flax has several characteristics that adapt it to inheritance studies. It is an annual with a short life cycle. It is sufficiently prolific so that populations adequate for statistical analysis can be secured from single F_1 plants. Rust reaction on flax is usually sharply differentiated; and since the flax plant grows by elongation of the terminal bud, the reaction of single plants to several races can be determined by inoculating the successive leaves produced by the terminal bud and removing infected leaves as each inoculation develops.

Although flax rust suffers the disadvantage for genetic studies of being an obligate parasite, it has some compensating advantages. It is a euautoecious rust that produces all its spore stages on flax. The monocaryotic pycniospores borne in drops of liquid exuded from the pycnium are incapable of developing aecia without transfer to them of pycniospores of the opposing mating type. The aecia bear dicaryotic aeciospores that reinfect flax to produce dicaryotic urediospores, the repeating pathogenic phase of the organism. Thus the genotype of the clonal uredial stage can be either *AA, Aa,* or *aa* for any specific gene. Telia, the overwintering spores, develop in the older sori. The single diploid stage and reduction division occur during the germination of the teliospores, producing monocaryotic basidiospores that infect flax to produce pycnia. Since overwintering occurs only in the telial stage in colder climates, hybridization and the production of new forms in this fungus are coincident with the annual establishment of infection.

From his extensive studies of the inheritance of resistance in the host, Flor has demonstrated that rust resistance is governed by multiple alleles located at 5 different loci in the flax plant. These loci are designated as K, L, M, N, and P. Only 2 alleles are known for the K locus, but there are 11 at the L locus, 6 at the M locus, 3 at the N locus, and 4 at the P locus. The K, L, and M loci are inherited independently, but M and P are linked with a crossover percentage of 26. Additional alleles for resistance are known but have not yet been assigned to appropriate groups. Rust resistance has been found to be inherited as a dominant characteristic although in some groups the dominance was not complete.

In contrast, virulence in the flax-rust organism, with one exception, has been inherited as a recessive character, and no indication of multiple allelism has been observed at the many loci governing pathogenicity. Some of the loci for pathogenicity are linked, but all the various genes have been found in one or another race of the rust. Thus, although no race of flax rust is known that is virulent for all the genes for resistance, there appears to be no allelic or linkage relationship that prevents the occurrence of such a race.

Complementarity of Genes for Resistance and Pathogenicity

Flor has found that hybrids between races of flax rust segregate for pathogenicity in accordance with the number of genes in the differential host that condition resistance to the avirulent parent race of rust. Thus, on a host variety that is genotypically PP, the ratio of avirulent to virulent segregants in hybrids between an avirulent and a virulent strain was found to be $3:1$. If the host variety carries 2 genes for resistance, the ratio of avirulent to virulent strains is $15:1$, and on varieties with 3 and 4 genes the ratios are $63:1$ and $255:1$, respectively. This suggests that complementary genic systems in host and parasite control reaction to rust.

To show this complementary relationship of genes for resistance in the host with those for pathogenicity in the parasite, Flor has adopted the convention of indicating the specific gene for rust reaction as a subscript to the symbol A for dominant avirulence (Table 26–3). Resistance occurs when complementary genes in both host and parasite are dominant. If either or both of the pair of complementary genes are recessive, susceptibility results. Thus a variety that carries no dominant genes for resistance is susceptible to all races of the parasite

TABLE 26–3.   Designation of Complementary Genes for Rust Resistance in the Host and for Virulence and Avirulence in the Pathogen

(After Flor, 1956)

Flax Variety	Rust Reaction of Host	Genotype for Pathogenicity of Rust Fungus	Host Reaction
Winona	$nnpp$	$A_{N'}A_P$	Susceptible
Polk	$N'N'pp$	$a_{N'}a_{N'}A_PA_P$	Susceptible
Polk	$N'N'pp$	$A_{N'}A_{N'}A_PA_P$	Resistant
Koto	$nnPP$	$A_{N'}A_{N'}a_Pa_P$	Susceptible
Koto	$nnPP$	$A_{N'}A_{N'}A_PA_P$	Resistant
Redwood	$N'N'PP$	$a_{N'}a_{N'}a_Pa_P$	Susceptible
Redwood	$N'N'PP$	$A_{N'}$ or A_P	Resistant

and a variety that carries one dominant gene for rust reaction is resistant to all races carrying the dominant complementary gene for pathogenicity.

This can be made clear by considering some examples based upon complementary gene systems associated with the N and P loci for host reaction (Table 26–3). The variety Winona is genotypically $nnpp$ and is hence susceptible to all races of the parasite. The variety Polk is genotypically $NNpp$, and it is thus resistant to races of rust carrying $A_nA_na_pa_p$ but susceptible to races that are homozygous for a_na_n. The variety Koto is resistant to all races carrying the A_p gene and is susceptible to those races that are genotypically a_pa_p. Redwood, which is $NNPP$ genotypically, is resistant to all races carrying either or both of the dominant genes A_nA_p and is susceptible only to races homozygous for their recessive alleles $a_na_na_pa_p$.

The M gene is the only rust-conditioning gene in flax to which virulence has been dominant and consequently the only exception to the rule that resistance occurs only when the complementary genes in both host and parasite are dominant.

Genetic Basis of Differential Hosts

The concept of physiological races, as we have seen, is based on differences in types of infection produced on a group of selected varieties termed "host testers" or "differentials." The infection type is the visible expression of the interaction of the genotype of the host with the genotype of the pathogen in a particular set of environmental conditions. Differential varieties have usually been selected more or less empirically, and rarely has anything been known of the number, uniqueness, or duplication of their resistance-conditioning genes.

Tester varieties that possess a single gene for resistance clearly give a more valuable assessment of pathogenicity than testers that carry more than one gene. For example, a flax differential possessing two rust-conditioning genes is of no value in distinguishing between races virulent to neither or only one of these two genes. A differential with three genes for rust resistance cannot separate races attacking none, any one, or any two of its genes. To be of greatest value, each differential should possess a single resistance gene and each gene serving to condition resistance to commercial and experimental varieties should be included in the set of differential varieties.

Since most of the genes conditioning virulence in the rust are recessive, mutations in pathogenicity, especially mutations toward greater pathogenicity, are likely to escape detection. The presence of more than one resistance gene in the differentials makes the detection of mutation even more improbable. Techniques for studying the inheritance of pathogenicity are laborious and exacting, and hence severely limit the number of cultures that can be studied. Far fewer cultures are necessary to secure conclusive results of studies of pathogenicity if the differentials have only one resistance gene instead of two or more. Consequently the development of a series of differentials with single genes for resistance is almost a prerequisite to comprehensive studies of the inheritance of pathogenicity. Flor has recently been able to set up a series of eighteen differential hosts in flax, each apparently carrying a single rust-conditioning gene, to replace the original differentials which were selected more or less blindly. The development of this series of flax-rust differentials should facilitate the breeding of rust-resistant varieties, because it permits a more accurate determination of the pathogenic potentialities of native races and makes possible a more accurate inventory of resistant germplasm for disease resistance.

Genetic studies of heterocaryosis in pathogens other than flax rust, although incomplete, suggest that most genes for virulence are recessive as in flax rust. Hence pathological capabilities are masked in the heterocaryon. If this situation turns out to be general, determinations of physiological races using genetically unknown differentials lose much of their significance, particularly for pathogens that require a sexual stage for propagation from year to year, because numerous biotypes may be designated by the same race number. Other obligate parasites are too incompletely understood at present to decide whether the host-pathogen relationship is on a gene-for-gene basis as in flax. There are, however, reasons for speculating that this may be so. Since obligate parasites cannot live apart from their host, it is clear that host and parasite must have evolved together. During their parallel and interdependent evolutionary histories, opportunity would have existed for host and parasite to have developed complementary genetic systems such that, for each gene conditioning pathogenicity in the parasite, a specific gene governing resistance would have developed in the host.

Genetics of Insect Resistance

Research on the genetics of insect resistance has not been nearly so extensive as on the genetics of resistance to fungal or bacterial diseases. Nevertheless, enough is known to indicate that the inheritance of insect resistance differs in no major way from inheritance of disease resistance. In 1951, Painter summarized differences noted in the resistance of certain crop varieties to insect pests. The list includes one or more insect species for each of the major crop species. In a few cases the inheritance of resistance to physiological races has been extensively investigated, notably the inheritance of resistance to Hessian fly, an insect that attacks wheat. The approach to the problem was similar to the one used in searching for and identifying genes for resistance to the wheat-bunt and barley-mildew diseases.

The first work on the inheritance of resistance to Hessian fly dealt with the variety Dawson. Under conditions where more than 90 per cent of the plants of the susceptible varieties Big Club and Poso were infected, only about 1 per cent of the plants of Dawson were infected. The resistance of Dawson was shown by Cartwright and Wiebe (1936) to depend on two dominant genes, designated as H_1 and H_2.

Although Dawson showed a high degree of resistance in California, it proved to be susceptible to Hessian fly in Indiana (Cartwright and Noble, 1947). Flies taken from the two areas were found, however, to retain their respective infective capacities to the differential host varieties under cages in the same greenhouse. The difference in reaction of Dawson is therefore due to genetic differentiation of the insects of the two localities. Subsequently additional sources of resistance to Hessian fly were found, and three additional genes for resistance, H_3, H_4, and H_5, identified (Shands and Cartwright, 1953).

REFERENCES

Briggs, F. N. 1926. Inheritance of resistance to bunt *Tilletia tritici* (Bjerk.) Winter, in wheat. *Jour. Agric. Res.* 32: 973–990.

Briggs, F. N., and C. S. Holton. 1950. Reaction of wheat varieties with known genes for resistance to races of bunt, *Tilletia caries* and *T. foetida*. *Agron. Jour.* 42: 483–486.

Cartwright, W. B., and G. A. Wiebe. 1936. Inheritance of resistance to the Hessian fly in the wheat crosses Dawson × Poso and Dawson × Big Club. *Jour. Agric. Res.* 52: 691–695.

Cartwright, W. B., and W. B. Noble. 1947. Studies on the biological races of the Hessian fly. *Jour. Agric. Res.* 75: 147–153.

Flor, H. H. 1954. Identification of races of flax rust by lines with single rust-conditioning genes. *U. S. Dept. Agric. Tech. Bull.* 1087, 25 pages.

Flor, H. H. 1955. Host-parasite interaction in flax rust—its genetics and other implications. *Phytopath.* 45: 680–685.

Flor, H. H. 1956. The complementary genic systems in flax and flax rust. *Adv. in Gen.* 8: 29–54.

Hansen, H. P. 1934. Inheritance of resistance to plant diseases caused by fungi, bacteria, and vira. *Yearbook Royal Veterinary and Agric. Coll.*, pp. 1–74. Copenhagen, Denmark.

Kendrick, E. L., and C. S. Holton. 1958. New physiological races of *Tilletia caries* in the Pacific Northwest. *Pl. Disease Reporter* 42: 15–17.

McRostie, G. P. 1921. Inheritance of disease resistance in the common bean *Jour. Amer. Soc. Agron.* 13: 15–32.

Painter, R. H. 1951. *Insect resistance in crop plants*. The Macmillan Co., New York.

Rodenheiser, H. A., and C. S. Holton. 1945. Distribution of races of *Tilletia caries* and *Tilletia foetida* and their relative virulence on certain varieties and selections of wheat. *Phytopath.* 35: 955–969.

Schaller, C. W., and F. N. Briggs. 1955. Linkage relationships of the Martin, Hussar, Turkey and Rio genes for bunt resistance in wheat. *Agron. Jour.* 47: 181–186.

Shands, R. G., and W. B. Cartwright. 1953. A fifth gene conditioning Hessian fly response in common wheat. *Agron. Jour.* 45: 302–307.

Walker, J. C. 1951. Genetics and plant pathology. In *Genetics in the 20th Century*, pp. 527–554. The Macmillan Co., New York.

27
Breeding Disease-Resistant Varieties

The breeding of disease- and insect-resistant varieties of crop and ornamental plants has perhaps received more popular attention than any other phase of plant breeding. This stems at least partly from the fact that almost everyone is acquainted with the spectacular damage that diseases and insects can wreak on plants, and there has therefore been less difficulty in appreciating the advantages attending the growing of resistant varieties than in grasping the significance of equally important improvements in less obvious characteristics. The disproportionate publicity given breeding for disease resistance does not, however, detract from the fact that resistant varieties represent some of the greatest triumphs of modern agriculture in increasing and stabilizing supplies of food, fiber, and vegetable oils. The world supply of cereals, for example, depends in large part on the stabilizing influence of rust- and smut-resistant varieties upon production in the granaries of the world. Sugar production rests on varieties of sugar beets resistant to curly-top disease. And stable supplies of linseed oil, which is a basic ingredient of paint, come from rust- and wilt-resistant varieties of flax. These are but a few examples of diseases whose only feasible means of control has been through genetically controlled disease resistance.

Many examples could, of course, be cited where use of fungicides and other methods of disease and insect control have given effective control of diseases. Nevertheless, whenever satisfactory disease-resistant varieties are available, they have been preferred over other means of control because they add little or nothing to the cost of production. Also disease resistance is built into the plant and is there ready to provide protection, whereas unfavorable weather, mechanical failures, and the like can prevent application of fungicides.

359

Sources of Disease Resistance

Breeding programs designed to produce resistant varieties must obviously start with resistance-conferring genes. The resistance most directly useful in plant breeding is that found in varieties of the same species. In most instances, locating a satisfactory source of resistance does not pose a serious problem because, as a result of continued observations and tests made by many different plant breeders and plant pathologists, varieties or strains carrying resistance to most of the major diseases are known in the important crop species. Moreover, when new diseases or races of established diseases appear, search through the diversity of germplasm represented in the world collections of varieties of crop plants has almost always been successful in locating adequate sources of resistance. Some examples of the use of the world collections for this purpose were given in Chapter 3.

Sometimes, however, adequate resistance does not appear to exist in cultivated species, and then the breeder usually has two alternative sources to which he can turn for resistance. First, he can search for the resistance he requires in related species or genera. Discussion of the difficulties of exploiting such genes for resistance will be deferred to Chapter 33, where the general features of interspecific hybridization are considered. The alternative is to attempt to induce resistance through mutagenic agents, a subject to be considered in Chapter 35.

Methods of Breeding for Disease Resistance

Breeding for disease resistance differs in no fundamental way from breeding for other characteristics. Consequently any of the various methods of breeding appropriate for the crop in question can be used in developing disease- or insect-resistant varieties, once resistance-conferring genes have been found.

When genes for resistance occur in existing commercial varieties, selection within these varieties will almost always provide the easiest and most satisfactory method of developing resistant strains. The success in locating plants resistant to *Periconia* root rot in all the commercial varieties of sorghum is an outstanding example of the rapidity and economy with which an important disease can be controlled by this procedure (Chapter 11). Improvement of sugar beets in curly-top resistance (Chapter 21), alfalfa in mildew and leaf-spot resistance (Chapter 14), and maize in resistance to turcicum leaf blight (Chapter 23) are other examples.

When adequate resistance is not found in commercial varieties, but only in types that cannot be used commercially because of their unsuitable agricultural properties, either the backcross or pedigree methods of breeding are usually selected. With either method one of the parents is chosen for its good agronomic or horticultural characteristics, and the other parent is selected on the basis of demonstrated high level of resistance to a maximum number of races and minimum number of genes controlling resistance. If the resistant parent is a wholly unadapted type, the backcross method is the logical choice as a breeding procedure. If, on the other hand, the breeder is satisfied that the resistant parent can also contribute to improved adaptation, quality, or yield he may choose the pedigree or bulk methods of handling the segregating generations. The pedigree method has been very widely used in breeding for disease resistance, and the majority of disease-resistant varieties have been produced by this procedure. More recently the backcross method has gained in favor (Chapter 14). Many examples of pedigree breeding for disease resistance are given in Hayes, Immer, and Smith (1955), and in the United States Department of Agriculture Yearbooks for 1936, 1937, 1943–1947, and 1953, including such fascinating stories as the work of McFadden with wheat, Murphy with oats, and Hayes with several crops.

Disease Epiphytotics

Whether breeding for disease resistance or other characteristics, it is of course essential to be able to correlate genotype and phenotype. This problem is critical in breeding for disease resistance because all genotypes are indistinguishable in the absence of the parasite. It follows, therefore, that programs of breeding for disease resistance cannot proceed unless the causal organism is present to induce the symptoms that will allow genotypes conferring adequate resistance to be distinguished from susceptible genotypes.

The level of infection is often light or irregular in natural epiphytotics so that accurate distinction between resistant and susceptible types is impossible. Whenever accurate classifications cannot be made, repetition is necessary for dependable results, and valuable time and effort may consequently be wasted. For this reason, most modern programs of breeding for disease resistance are based on artificially induced epiphytotics. The ability to set up conditions insuring regular and heavy infection of susceptible materials often depends on specialized knowledge which falls in the province of the plant pathologist. Plant pathology, like plant breeding, has become increasingly specialized so that cooperative efforts between men schooled in the two disciplines is becoming a common practice in programs of breeding for disease resistance.

Stability of Resistance

Little stock was placed in disease resistance in many quarters half a century ago. The bridging-host hypothesis predicted that resistance would decline in new varieties as fast as they were produced. To those who accepted this hypothesis, resistance was a mirage, and it was natural that they should disparage the practicality of attempting to control diseases by breeding resistant varieties. These early ideas regarding the causes of changes in the resistance of host varieties were replaced when it was discovered that new pathogenic forms arise through hybridization, heterocaryosis, and mutation. Nevertheless, the new concepts have not entirely dispelled pessimism regarding the stability of resistant varieties, because it is now clear that pathogenic organisms have enormous potential in developing new virulent forms. Many investigators have come to regard the fungi and bacteria as shifty and treacherous opponents with weapons outmatching any the plant breeder can hope to muster in opposition. Other investigators have been more optimistic, and, as a result of their continued efforts, large numbers of resistant varieties have been released to farmers. The best evidence of the stability and value of disease resistance in agricultural practice comes from the long-term performance of these resistant varieties.

One of the first disease-resistant varieties developed in the present century was the Conqueror Watermelon, a wilt-resistant type bred by W. A. Orton about 1907 from a cross between the Eden Watermelon and the wilt-resistant African Citron. The Conqueror did not gain wide acceptance because of its horticultural deficiencies, but it served as a resistant parent to other varieties that have retained their resistance to the present day. A disease of cabbage caused by another soil-inhabiting fungus has also been kept under control by means of resistant varieties. Cabbage yellows was so destructive in many areas about 1910 that production had to be abandoned. However, occasional resistant plants occur in most commercial varieties, and by selecting these plants it was possible to develop varieties whose resistance has kept this disease under control. The situation is similar in the *Periconia* disease of sorghums. Although resistance is governed by a single gene and the resistant varieties have been grown over wide areas, the disease has not caused trouble since the release of resistant varieties in the decade 1930 to 1940. Still another example of stable resistance is provided by varieties of the Lima bean resistant to the root-knot nematode. The variety Hopi 5989, released to California farmers in 1932, was widely grown without decline in resistance for about a dozen years. Its popularity decreased when another resistant variety, Westan, with superior agronomic qualities became available in 1944. During the next 14-year period there was no apparent decline in the resistance of either of these varieties. These are but a few examples of varieties with lasting disease resistance.

The history of the wilt disease of flax is interesting for the light it throws on the ability of the flax plant to match new variability in the pathogen. Flax was long a migratory crop in the United States, moving continually westward to new lands as the soils in each locality became "flax sick." In 1900 it was shown that flax sickness was due to *Fusarium*, a fungus that accumulated and persisted in the soil after being introduced with flax seed. Usually a few plants survived in flax-sick soils, and propagation of seed from these survivors led to the development of several wilt-resistant varieties about 50 years ago. But the varieties lost their resistance after a few years in farmers' fields. The discovery of physiological races in fungi provided the clue necessary for developing a method of more permanent control. Flax-wilt plots were established in which the soil was inoculated with infective materials from many sources, and all new varieties were required to pass a test of survival in these plots. By continued testing and selection under these extreme conditions a succession of varieties has been produced that have kept flax wilt under control. This procedure appears to have been effective for two reasons: (1) a few resistant plants occur in most flax varieties; and (2) identification of these plants is easy since flax wilt selects rigidly, killing susceptible plants outright and leaving only the most resistant plants. Continuous effort to control flax wilt has made it possible to counteract new variability in the fungus and to maintain the productivity of flax during the last half century.

In other instances the stability of resistant varieties has been less impressive. In oats, for example, Stevens and Scott (1950) found that the average usefulness of varieties resistant to stem and crown rust was 5 years in the corn belt of the United States. Some varieties survived but a single year. They concluded that a new oat variety would be needed every 4 or 5 years to meet the threat of new races of stem and crown rust.

The reasons for these differences in stability of resistant varieties are by no means entirely clear at present. However, evidence from epidemiological studies with certain diseases have provided clues that promise to be helpful in solving the problem of stabilization of resistance.

Epidemiology of Physiological Races

Since 1916, when physiological races of black stem rust of wheat were discovered, an annual census of geographical distribution and population trends among races has been taken in North America. Following the severe epiphytotics of 1916, durum-wheat varieties were widely substituted for bread wheats in the hard-red-spring-wheat region in the United States and Canada because of their rust resistance. Within a few years, however, rust races appeared to which the durum wheats were susceptible. In 1926 the rust-resistant common-wheat variety Ceres was distributed to farmers. It soon became the most widely grown variety in this wheat area and seemed to have solved the stem-rust problem in spring wheat. Two years after the release of Ceres a new race of stem rust, race 56, to which Ceres was susceptible was isolated from barberries in Iowa to the south of the main spring-wheat area. This race gradually increased, and by 1934 it had become the most prevalent race in the United States. In 1935 race 56 dominated the rust population and ended the career of Ceres.

The resistant bread-wheat variety Thatcher had been released in 1934 (Hayes et al., 1936). This variety was resistant to race 56, and it escaped damage in the great epiphytotics of 1935 and 1937. Shortly after the release of Thatcher, a series of bread-wheat varieties carrying the resistance of Hope (a variety whose resistance comes from Yaroslav Emmer, McFadden, 1930), were released. Varieties of durum wheat carrying the resistance of Vernal Emmer also became available. These new varieties were highly resistant to all the North American races of rust. During this period the number of prevalent races decreased to three or four, and no major epiphytotics occurred. But many different races were being found on barberries, and among them was the highly virulent race 15B. The question whether this race would become widespread was answered dramatically in 1950 when it spread over most of North America in a single season. In 1950, 1953, and 1954, it caused extensive damage on both bread and durum wheats. Since 1950, race 15B itself appears to have increased in diversity, and it now includes a large number of subraces with different pathogenic capabilities. There are numerous sources of resistance to race 15B, and the process of adding this resistance to varieties with good agronomic characteristics is now under way.

A clear picture of the pattern of shifts and prevalence of races of stem rust has emerged from these studies. A great diversity of races arises as a result of hybridization and recombination on barberry plants. For example, in a study of four wheat fields next to barberry bushes in Virginia, twelve races not found elsewhere in the United States were isolated in the period 1950–1953. Certain of these races were prominent early in the season but declined in prominence as the season progressed. Other races were detected only late in the season. Race 56 tended to dominate at the end of the season. Thus, among the many races produced on barberries, some apparently cannot maintain themselves even on varieties they can infect, but others become predominant races. When environmental conditions are favorable, new races combining virulence and the necessary attributes of adaptation can spread very rapidly over large areas.

Once a race becomes established, its prevalence is determined by the varieties being grown. This is illustrated by shifts which occurred in the prevalent races found in two geographical areas of Mexico. In central Mexico race 56 was a dominant race until race 15B made its appearance. Two of the several varieties grown in the area proved to be resistant to race 15B, and as these two varieties were increased in acreage, there was a concomitant decrease in race 15B. However, race 49, a race that had formerly been prevalent in the United States and Mexico, but had virtually disappeared because of its avirulence on varieties deriving their resistance from Hope, made a spectacular comeback on these two varieties. In the Pacific Coastal plain, where a different set of varieties is grown, drastic shifts in races also occurred in the same period. These shifts were associated with the appearance of race 15B and also with the adoption of new varieties with different resistance from the previous varieties.

The epidemiology of stem-rust races in Australia has followed a somewhat different pattern. Waterhouse (1952) concluded from surveys of races conducted over a 30-year period that a single race, number 126, made up about 90 per cent of the inoculum prior to the release of the variety Eureka. Eureka was resistant to this race and, so far as was known, to all other races present. With the cultivation of Eureka, a new race was soon isolated that differed from the prevalent race only in its ability to attack Eureka. Since it is unlikely that barberries played any part in the origin of this new race, Waterhouse concluded that it had arisen by mutation. When Eureka was replaced by varieties having a different type of resistance, new races of the fungus were soon isolated from them. Again the reactions of the new races agreed with those of race 126 except for differentiation against the specific genes for resistance carried by each of the new varieties. Unlike the situation in North America, where new virulent races usually have very different pathogenic capabilities from previously prominent races, Australian races arising independently of barberries seem to originate by continuing step-by-step mutations for virulence. Thus, in Australia, not only is the prevalence of races associated with the screening effects of the varieties cultivated, but the origin of new races is presumed to be closely connected with the varieties used in breeding.

Watson (1956) has evidence for similar step-by-step changes in pathogenicity of leaf rust, *Puccinia triticina*. Only two races, 26 and 95, were isolated in Australia prior to 1930. However, with the use of new sources of resistance in breeding programs, at least sixteen races and biotypes of the leaf-rust organism can now be identified.

The studies of Flor on the rise and fall of races of flax rust are high points among investigations of the epidemiology of physiologic races. Flax rust has played an important role in the popularity of flax varieties in the north-central part of the United States, and, conversely, the prevalence of different races of rust has been largely determined by the varieties of flax that are grown. Since this rust attacks only flax, the survival of a race depends entirely on the continued cultivation of flax varieties on which it is virulent. Races unable to attack current commercial varieties disappear, whereas the races that can attack prominent varieties tend to increase.

Flor (1953) has been able to analyze these changes in the prevalence of races of flax rust in terms of genes for resistance in the host and for virulence or avirulence in the pathogen. In the period 1931 to 1940, when Bison was the leading flax variety, 92 per cent of the isolates of rust were races 1, 2, and 3. Race 1 carries no recessive (virulence) alleles, and races 2 and 3 each carry 1 pair. The variety Bison was released as a rust-resistant variety in 1926, but it has been susceptible to all collections of flax rust made in North America since 1931, when the census of physiological races began. It is known from studies in Australia that Bison carries an allele for resistance at the L locus. On this evidence Flor has suggested that when Bison became the dominant variety in North America, races unable to attack it died out, and those attacking it replaced them completely.

In the period 1942 to 1947, 77 per cent of the isolates were races 1, 2, 3, and 210. Race 210 differs from race 1 only in its ability to attack Koto, a variety developed as a rust-resistant replacement for Bison and widely grown during this period. Koto carries a pair of resistance alleles at the P locus. From 1948 to 1951, when both Koto and Dakota (Dakota carries a pair of resistance alleles at the M locus) were leading varieties, 68 per cent of the isolates were races 166, 180, and 210. These races differ from race 1 only in their ability to attack either Koto or Dakota, or both varieties. The dominant races during each of these three periods were races carrying the least number of recessive alleles compatible with survival on the currently dominant flax varieties. The proportion of isolates carrying genes for virulence that were not pertinent to ability to attack existing varieties decreased in each successive period. Thus the tendency has been toward loss of alleles unnecessary for virulence, and there has been no aggressive tendency on the part of the pathogen toward the development of races of wider virulence than required to attack cultivated varieties.

Maintenance of Resistance

It can be concluded from studies in epidemiology that plant breeders must be prepared to face increases in races to which their resistant varieties are susceptible. Races of concern might be either undetected ones already in existence or new ones arising by hybridization, heterocaryosis, parasexuality, or mutation. Efficient plant-breeding programs therefore include plans to deal with the threat of new races before they can cause excessive damage.

In the past, the usual procedure has been to replace a variety that has succumbed to a new race with a variety that is resistant to the new race. Most often it has been necessary to breed the replacement after the appearance of the new race, a procedure with obvious disadvantages. Frequently the replacement variety does not carry genes conditioning resistance to races suppressed by former varieties in the series, with the result that another shift of races occurs, and little if anything is gained in the long run.

There have been a number of suggestions to overcome these problems. In 1952 Jensen suggested the use of multi-line or composite varieties. Such varieties would consist of a blend of compatible lines, each selected for similarity of height, maturity, and other agronomic characters, but carrying different genes for resistance. Borlaug (1954) has advocated a similar, but somewhat more sophisticated approach based on lines developed by standard backcross breeding. As many such lines could be developed and held in reserve as there are different sources of resistance. The composition of the composite variety would then be modified from time to time on the basis of surveys of prevalent races.

Flor has made two suggestions toward the maintenance of varietal resistance. In 1947, he started a backcross program to develop lines of flax essentially alike except for a single rust-conditioning gene. Bison was selected as the recurrent parent because of its suitability for growing in the north-central part of the United States. Bison-like lines that are pure for rust-conditioning alleles derived from thirty-two different varieties have been developed. Fifteen of these thirty-two lines possess satisfactory resistance to North American races of flax rust; hence a wide choice of Bison-like resistant lines is available. Flor believes there is no real need to use these lines to develop composite varieties as Borlaug has suggested. Should the need arise, however, the Bison-like lines developed by backcrossing could be utilized in this way.

Bison-like lines possessing any desired combination of genes for rust resistance can be developed, within the limits of allelism, by using races of flax rust to identify resistance genes in the progeny of hybrids of the Bison backcrosses. Flor has used this technique to combine into single lines alleles at the L, M^3, and N loci that confer resistance to all North American races of flax rust. Since virulence is usually recessive, new races can be established only if they possess, as homozygous recessives, the genes for pathogenicity that complement each gene for resistance in a variety. A race can become homozygous for a mutant gene only if the identical mutation occurs in both dicaryotic nuclei in a single spore or if haploids carrying an identical mutation are combined during sexual reproduction. In either case the chance of securing two or three pairs of recessive genes in a single biotype of flax rust is much smaller than the chance of securing one pair. Plants combining three genes for resistance clearly present a formidable problem to the flax-rust organism, and on this basis it can reasonably be predicted that Flor's strain carrying multiple resistance will have a long and useful life.

REFERENCES

Ausemus, E. R. 1943. Breeding for disease resistance in wheat, oats, barley and flax. *Bot. Rev.* 9: 207–260.

Borlaug, N. E. 1954. Mexican wheat production and its role in the epidemiology of stem rust in North America. *Phytopath.* 44: 398–404.

Cobb, N. A. 1892. Contributions to an economic knowledge of the Australian rusts (Uredineae). *Agric. Gaz. N. S. Wales* 3: 44–68, 181–212.

Flor, H. H. 1953. Epidemiology of flax rust in the north central states *Phytopath.* 43: 624–628.

Flor, H. H. 1955. Host-parasite interaction in flax rust—its genetics and other implications. *Phytopath.* 45: 680–685.

Hayes, H. K., E. R. Ausemus, E. C. Stakman, C. H. Bailey, H. K. Wilson, R. H. Bamburg, M. C. Markley, R. F. Crim, and N. R. Levine. 1936. Thatcher Wheat. *Minn. Agric. Exp. Sta. Bull.* 325.

Hayes, H. K., F. R. Immer, and D. C. Smith. 1955. *Methods of plant breeding.* McGraw-Hill Co., New York.

Jensen, N. F. 1952. Intra-varietal diversification in oat breeding. *Agron. Jour.* 44: 30–34.

McFadden, E. S. 1930. A successful transfer of emmer characters to *vulgare* wheat. *Jour. Amer. Soc. Agron.* 22: 1020–1034.

Pugsley, A. T. 1949. Backcrossing for resistance to stem rust of wheat in South Australia. *Emp. Jour. Exp. Agric.* 17: 193–198.

Stakman, E. C., W. Q. Loegering, D. M. Stewart, W. M. Watson, and C. W. Roane. 1956. Physiologic races of *Puccinia graminis tritici* collected from wheat near barberry bushes. *Rept. Third Intern. Wheat Rust Conf.,* 112–113.

Stevens, N. E. and W. O. Scott. 1950. How long will present spring oat varieties last in the central corn belt? *Agron. Jour.* 42: 307–309.

Walker, J. C. 1951. Genetics and plant pathology. In *Genetics in the 20th century,* pp. 527–554. The Macmillan Co., New York.

Walker, J. C. 1953. Disease resistance in vegetable crops. *Bot. Rev.* 19: 606–643.

Waterhouse, W. L. 1952. Australian rust studies. IX. Physiologic race determinations and surveys of cereal rusts. *Proc. Linn. Soc. N. S. Wales* 77: 209–258.

Watson, I. A. 1956. The occurrences of three new wheat stem rusts in Australia. *Proc. Linn. Soc. N. S. Wales* 80: 186–190.

Watson, I. A. 1956. Epidemiology of wheat rusts in Australia. (abs.) *Phytopath.* 46: 30.

Watson, I. A., and D. Singh. 1952. The future for rust resistant wheat in Australia. *Jour. Australian Inst. Agric. Sci.* 18: 190–197.

Yearbook of Agriculture, U. S. Dept. Agric. 1936, 1937, 1943–1947, 1953.

28

General Features of Polyploidy

The rule that somatic nuclei contain two, and the gametes only one, of each kind of chromosome characteristic of the species is subject to a series of exceptions that have important effects on plant-breeding procedures. These exceptional cases can be classified into two very general groups. In the first group, the deviation takes the form of an unusual number of repetitions of one or a few particular chromosomes in the complement. In the second group, the organism is characterized by an unusual number of repetitions of full complements of chromosomes. This distinction is in a way an arbitrary one, because variations in the number of repetitions of individual chromosomes (or parts of chromosomes) or entire sets of chromosomes are continuous and overlapping in many organisms. The terminology that has been developed to describe these chromosomal repetitions has become too complex to be described here in detail (see Stebbins, 1949), but before proceeding to consideration of the impact of polyploidy on systems of variability in plants, its main features must be reviewed.

Terminology of Polyploidy

The term aneuploidy is a general term to describe organisms whose chromosome number is not a whole-number multiple of the basic number of the group. Thus if a basic number of chromosomes in a species is x and the normal somatic complement is $2x$, a deviate having $2x$ plus or minus one or more chromosomes is an aneuploid. Many aneuploid combinations are possible. If, for example, both members of one pair of chromosomes are missing from the normal complement of the somatic cells, the organism is termed a *nullisomic*. A *monosomic* lacks one chromosome of the normal complement. A *trisomic* has two complete sets of chromosomes plus a single extra chromosome—that is, one particular chromosome is present in triplicate. If

370

two extra chromosomes of one particular chromosome are present, the term *tetrasomic* is applied to the organism. If, however, an extra chromosome each of two different pairs is present, the organism is called a double trisomic. Similar terminology is applied to the many other combinations that are possible. The pattern of the terminology of aneuploidy is summarized in Table 28–1.

TABLE 28–1. Guide to the Terminology of Ploidy

Name	Formula	Somatic Chromosome Complement, Where A, B, and C Are Nonhomologous Chromosomes
Aneuploids		
nullisomic	$2x - 2$	(AB)(AB)
monosomic	$2x - 1$	(ABC)(AB)
double monosomic	$2x - 1 - 1$	(AB)(AC)
trisomic	$2x + 1$	(ABC)(ABC)(C)
double trisomic	$2x + 1 + 1$	(ABC)(ABC)(A)(B)
tetrasomic	$2x + 2$	(ABC)(ABC)(A)(A)
monosomic-trisomic	$2x - 1 + 1$	(ABC)(AB)(A)
Euploids		
monoploid	x	(ABC)
triploid	$3x$	(ABC)(ABC)(ABC)
autotetraploid	$4x$	(ABC)(ABC)(ABC)(ABC)
allotetraploid	$2x + 2x'$	(ABC)(ABC)(A'B'C')(A'B'C') or (ABC)(ABC)(DEF)(DEF)

The table is based on multiplications of the monoploid complement of a diploid species with $x = n = 3$ chromosomes. The somatic chromosome number of any form is here designated $2n$ and its gametic number n, regardless of its degree of polyploidy, unless it is an unbalanced type (aneuploid). The most probable basic number for any group of plants is designated by the letter x, and the somatic numbers of autopolyploids by $3x$, $5x$, and so forth.

Euploidy is a general term covering situations in which the issue is the number of repetitions of entire chromosome sets. A *monoploid* organism has only a single complement of a basic chromosome set of the species. A *triploid* has three full sets, a *tetraploid* four, a *pentaploid* five, a *hexaploid* six, and so on. A further distinction among euploids is made on the basis of the degree of differentiation of the sets of chromosomes that are repeated. In *autopolyploids* each of the repeated sets of chromosomes is considered to be identical, or at least very closely similar, to each other set. In *allopolyploids* the two or more basic sets of chromosomes (genomes) making up the multiple set of chromosomes are considered to be differentiated from each other. Thus if a hybrid is made between two basic diploid species whose chromosome sets are completely differentiated, the genomic formula of the F_1 is x_1x_2, and upon doubling the chromosome complement of this F_1, an allotetraploid $x_1x_1x_2x_2$ is produced.

It is clear that the distinction between auto- and allopolyploidy will not always be sharp because hybridization and subsequent doubling of chromosome number can involve species whose genomes are in various stages of differentiation. Furthermore, in some genera, aneuploidy, autopolyploidy, and allopolyploidy have been superimposed upon one another in very complex ways. The basic terminology of euploidy is given in Table 28–1.

In this table x has been used to designate the monoploid sets of chromosomes. The use of x in this fashion distinguishes the monoploid set from the haploid or gametic number of chromosomes (n), which in allopolyploids includes two or more monoploid (x) sets. The haploid (n) and monoploid (x) chromosome number of a basic diploid species are thus the same, but in an allopolyploid the gametic or haploid number (n) depends on the number of basic (x) sets of chromosomes whose evolutionary history has been merged by polyploidization.

Effects of Polyploidy on the Phenotype

The morphological and physiological effects of ploidy vary greatly in different materials. In general, however, deficiency or repetition of particular chromosomes results in unbalances in the genotype that cause different aneuploids to be morphologically distinct from each other and distinct from the diploid form. Aneuploids are usually less vigorous than their diploid progenitors. Nullisomics and monosomics are ordinarily viable only in species with polyploid ancestry where previous duplication of chromosomes appears to cover for missing chromosomal materials. Trisomics are not found in some species, presumably because the unbalance caused by even a single extra chromosome is lethal. In species where extra chromosomes are tolerated, trisomy usually exerts a profound effect on morphology, particularly in species that appear to be basic diploid types. In some polyploid species—wheat, for example—the degree of tolerance to repetitions of particular chromosomes is great, and even tetrasomics may be nearly indistinguishable from normal diploid plants.

A common effect of autopolyploidy is to increase the size of the vegetative portions of the plant, making autopolyploids huskier and rather more vigorous than the corresponding diploids. This effect is far from universal, however, and many autopolyploids are weak and lacking in vigor. Some investigators believe there is an optimum level of polyploidy in each group of plants. In most groups this optimum appears to be achieved at fairly low levels of autopolyploidy, but in some groups octoploids or even higher polyploids are still vigorous types.

Allopolyploidy is equally or even more unpredictable than autopolyploidy in its morphological and physiological effects on the phenotype. In general, however, allopolyploids combine in more or less blending fashion the characteristics of the species from which they are derived.

Origin of Polyploids

All types of ploidy apparently have their origin in cytological accidents of one sort or another. Monosomics and trisomics arise sporadically from gametes with $n - 1$ or $n + 1$ chromosomes. These gametes occur as a result of nondisjunction in one or another pair of chromosomes. Monoploids and triploids are another source of aneuploids. Aneuploids appear to arise most commonly, however, as a result of the troubles at meiosis that are caused by asynaptic genes.

Euploidy seems to arise most commonly through misadventures in meiosis that cause unreduced ($2n$) gametes to form. Polyploids with whole-number repetitions of basic chromosome sets are also formed from somatic cells in which a failure of mitosis has resulted in doubling of the chromosome complement. If the doubling occurs early in development of the plant, it may produce a sectoral chimera that ultimately bears flowers. The haploid gametes produced by such flowers have the same number of chromosomes ($2n$) as the original undoubled plant. Fertilization of unreduced gametes by normal gametes of the same species produces triploids with $3n$ chromosomes. (It should be noted that triploids are unstable cytologically, and they in turn often produce a great array of aneuploid progeny.) If the original plant was a diploid species, the result of self-fertilization in the polyploid sector is the production of an autotetraploid. If, however, the original plant was an F_1 interspecific hybrid, the result is a merging of the two species into a new amphidiploid species. The formation of unreduced gametes appears to be particularly frequent in wide crosses, apparently as a result of the disturbances in meiosis that are associated with lack of chromosome pairing.

Extent and Importance of Polyploidy in Cultivated Plants

Some authorities estimate that at least one third of the angiosperms are polyploids. Among cultivated species the proportion is equally great, if not even greater. This argues strongly that the cytological accidents discussed above leading to polyploidy are common and also argues for the evolutionary importance of polyploidy. An idea of the importance of polyploidy among crop plants can be obtained from the accompanying sample tabulation of chromosome numbers in cultivated and related wild species.

Species	Common Name	Gametic (n) Chromosome Number
Avena		
strigosa	sand oats	7
brevis	short oats	7
barbata	slender wild oats	14
fatua	wild oats	21
sativa	cultivated oats	21
byzantina	red cultivated oats	21
Bromus		
secalinus	chess bromegrass	7, 14
mollis	soft chess	14
inermis	smooth bromegrass	21, 28, 35
carinatus	mountain bromegrass	28
Festuca		
elatior	meadow fescue	7, 14, 21, 35
ovina	sheeps fescue	7, 14, 21, 70 (also $2n = 21, 49$)
rubra	red fescue	7, 21, 28, 35
Gossypium		
arborium	cultivated Asiatic cotton	13
thurberi	wild American cotton	13
barbadense	see island cotton	26
hirsutum	American upland cotton	26
Hordeum		
vulgare	cultivated barley	7
jubatum	squirrel tail barley (wild)	14
nodosum	foxtail barley (wild)	21

Species	Common Name	Gametic (n) Chromosome Number
Lotus		
uliginosus	big trefoil	6
corniculatus	birdsfoot trefoil	12
Medicago		
hispida	California burclover	7
lupulina	black medic	8, 16
falcata	yellow alfalfa	8, 16
sativa	common alfalfa	16
Nicotiana		
sylvestris	wild tobacco	12
tomentosa	wild tobacco	12
tabacum	cultivated tobacco	24
Prunus		
americana	American plum	8
avium	sweet cherry	8
persica	peach	8
cerasus	sour cherry	16
domestica	European plum	24
Sorghum		
versicolor	wild sorghum	5
vulgare	cultivated sorghums	10
halepense	Johnsongrass	20
Trifolium		
agrarium	hop clover	7
incarnatum	crimson clover	7
pratense	red clover	7
alexandrinum	berseem clover	8
fragiferum	strawberry clover	8
hybridum	alsike clover	8
subterraneum	subterranean clover	8
dubium	small hop clover	16
repens	white clover	16
medium	zigzag clover	40, 42, 48, 49

The origin of monoploid (x) sets of chromosomes is known for only a few polyploid species. The best-known cases of cultivated plants where particular monoploid sets have been identified with diploid species are those in *Brassica*, cotton, tobacco, and wheat.

Since polyploidy has been an important factor in the evolution of some of the most important cultivated species, knowledge of its mechanics is essential to an understanding of plant breeding. In Chapter 29 we shall consider the cytogenetics of aneuploids. Chapter 30 will be devoted to inheritance in autopolyploids and Chapter 31 to the cytogenetics of allopolyploids. Finally, in Chapter 32, we shall consider the role of induced polyploidy in practical plant breeding.

REFERENCES

Darlington, C. D., and E. K. Janaki Ammal. 1945. *Chromosome atlas of cultivated plants.* Allen and Unwin, London.

Goodspeed, T. H. 1953. Species origins and relationships in the genus *Nicotiana. Univ. Calif. Pub. in Bot.* 26: 391–400.

Goodspeed, T. H., and R. E. Clausen. 1928. Interspecific hybridization in *Nicotiana.* VIII. The sylvestris-tomentosa-tabacum hybrid triangle and its bearing on the origin of *tabacum. Univ. of Calif. Pub. in Bot.* 11: 245–256. 1928.

Muntzing, A. 1936. The evolutionary significance of autopolyploidy. *Hereditas* 21: 263–378.

Stebbins, G. L., Jr. 1949. Types of polyploids: Their classification and significance. *Adv. in Gen.* 1: 403–409.

Stebbins, G. L., Jr. 1950. *Variation and evolution in plants.* Columbia University Press, New York.

Stephens, S. G. 1947. Cytogenetics of *Gossypium* and the problem of the origin of new world cottons. *Adv. in Gen.* 1: 431–442.

U, N. 1935. Genome analysis in *Brassica* with special reference to the experimental formation of *B. napus* and peculiar mode of fertilization. *Jap. Jour. Bot.* 7: 389–425.

29

Cytogenetics of Aneuploids

Aneuploids are for the most part less vigorous than normal plants, presumably because of physiological disturbances that are associated with unbalanced numbers of chromosomes. Aneuploids also tend to be irregular meiotically, and as a result they are likely to be partly or even highly sterile. This sterility, combined with the genetic instability and physiological unbalance that characterize aneuploids, is the very antithesis of progress in plant breeding, and as a result aneuploids have found little if any direct use as market varieties.

Although aneuploids rarely have agricultural advantages over normal plants, nevertheless certain types of aneuploids, particularly nullisomics, monosomics, and trisomics, are finding a place in plant breeding because of their usefulness in locating genes on particular chromosomes. This method is being used by a number of plant breeders, and a good start has been made toward determining linkage groups in certain crop species—for example, tobacco (Clausen and Cameron, 1944), wheat (Sears, 1953), and tomatoes (Rick and Barton, 1954). The method is particularly useful in polyploids, the very species in which nullisomics and monosomics are most likely to be viable, because it overcomes some of the difficulties in genetic analysis caused by large chromosome numbers and the frequency of duplicated genes in polyploid species. In nonpolyploid species, also, aneuploid analysis performs useful services. In the tomato, for example, trisomics have been used to identify chromosomes with their linkage groups.

Another use that has been found for aneuploids, particularly nullisomics, is in the transfer of particular chromosomes with desirable genes from one variety to another, or from one species to another.

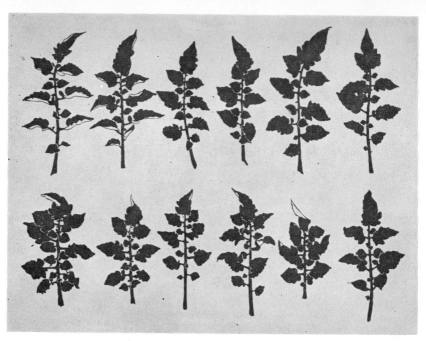

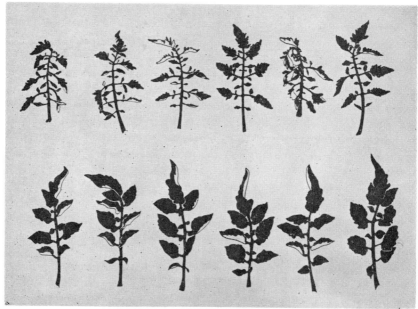

FIGURE 29-1. Phenotypic effects of aneuploidy in the tomato. Top row, variation in leaf form in a normal 2N variety. Second, third, and bottom rows, trisomics 10, 4, and 6 respectively. How many chromosomes do trisomic tomatoes have? How many primary trisomics are possible in tomatoes? (Photograph courtesy Dr. C. M. Rick, California Agricultural Experiment Station.)

Cytogenetics of Aneuploids

Before turning to the uses of aneuploidy in plant breeding, let us take up the general effects of different types of aneuploidy on meiosis and stability in sexual reproduction. In aneuploids with an uneven number of some particular chromosome (monosomics, trisomics), there is clearly no way for an equal number of each chromosome to end at the two poles in the division figure. Monosomics, for example, produce two kinds of gametes, n and $n - 1$. The odd chromosome passes at random to either pole in meiosis, but frequently it will lag at anaphase and not be included in either daughter nucleus. For this reason gametes with $n - 1$ chromosomes are more frequent than gametes with n chromosomes. This bias toward $n - 1$ gametes is usually not reflected as strikingly in the zygotic chromosome numbers, however, because gametes with $n - 1$ chromosomes often do not function, and, furthermore, zygotes with $2n - 2$ chromosomes (nullisomics) are inviable except in a few polyploid species. Thus most of the progeny of monosomics are either normal diploids or monosomics.

Trisomics tend to be somewhat more stable genetically than monosomics. Nevertheless, there is no assurance that two chromosomes will always go to one pole and one to the other pole at meiosis, and we must be careful about making generalizations concerning patterns of inheritance in trisomics.

Tetrasomics often behave more regularly than the aneuploids with odd numbers of chromosomes. The four homologs tend to form a quadrivalent at meiosis, and disjunction often proceeds fairly regularly, two by two. Nevertheless, quadrivalents are not always formed, nor is disjunction always regular, so that tetrasomics must also be classified as genetically unstable.

Use of Aneuploidy in Detecting Linkage

The use of aneuploids in detecting linkage depends basically on identifying aberrant genetic ratios caused by deficient or extra chromosomes in the complement or, alternatively, by associating certain chromosome constitutions with particular phenotypes. Sears (1953) has used this latter method for locating certain dominant genes in the wheat variety Chinese Spring, in which the complete set of twenty-one nullisomics is available. Thus Chinese Spring has a dominant gene for red seeds, and this gene was located on chromosome XVI by the direct observation that the seeds produced by nullisomic XVI are

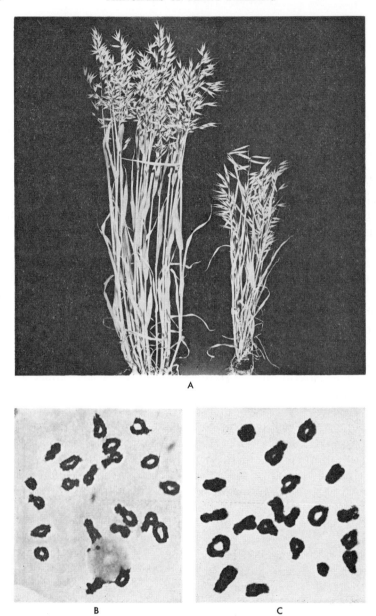

FIGURE 29-2. *A*, normal oat (*Avena byzantina*) left, and nullisomic, right. *B*, diakinesis in normal oats. *C*, meiotic metaphase in nullisomic oats. (After Ramage and Suneson, *Agron. Jour.* 50:52–53. Photograph courtesy United States Department of Agriculture.)

white. Similarly, chromosomes VIII and X each carry a dominant inhibitor of awns, as shown by the increased awn development in nullisomics VIII and X.

Genes in varieties other than Chinese Spring can be located by crossing to each of the twenty-one nullisomics and observing the ratios obtained in the F_2 populations. In actual practice, however, monosomics are used instead of nullisomics, because the greatly reduced vigor and fertility of the nullisomics make them difficult to handle experimentally. The pioneer work on the monosomic technique was done in tobacco by Clausen and has been used with conspicuous success by him and his colleagues in the genetic analysis of that species. Briefly, the method consists of crossing a line carrying a particular gene with each of the twenty-four different tobacco monosomics. Some of the F_1 plants are normal (24 pairs), and some are monosomic. The monosomics, of course, result from the union of a gamete with $n - 1$ chromosomes with a gamete with n chromosomes from the line under test. If the gene being tested is recessive, the monosomic plants of the critical F_1 family will all be recessive in phenotype. If the gene is dominant, F_2 populations are grown from monosomic F_1 plants. These populations yield two types of offspring, monosomic and disomic (nullisomics are inviable in tobacco). In the critical F_2 family neither monosomic nor disomic individuals can have received the chromosome carrying the gene under test from the monosomic parent but only from the normal parent. Hence the critical family can be distinguished from all others by the fact that it does not segregate for the gene under test.

An Example of Monosomic Analysis

An example of the use of the monosomic technique to locate a pair of linked duplicate genes governing stem-rust resistance in wheat is given in Table 29–1. The resistant parent was C.I. 12633, a common-wheat variety ($n = 21$) deriving its stem-rust resistance from *Triticum timopheevi,* a wheat species with 14 pairs of chromosomes. Since duplicate genes govern stem-rust resistance and they are linked and located about 21 crossover units apart, a ratio of about 5.4 resistant to 1 susceptible plant was expected in the F_2 generation. From Table 29–1 it can be seen that the expected ratio occurred in all F_2 families except the one derived from monosomic XIII, in which there was a deficiency of susceptible plants. The three susceptible plants observed were nullisomic, as expected if the resistance genes were

TABLE 29–1. Stem-Rust Reaction of F_2 Plants in Crosses between Monosomic Lines and C.I. 12633

(After Nyquist, 1957)

Chromosome	Resistant	Susceptible	χ^2	P
I	68	15	0.232	.70–.50
II	16	3	0.083	.80–.70
III	43	8	0.028	.90–.80
IV	75	14	0.010	.95–.90
V	57	6	1.314	.30–.20
VI	73	15	0.057	.90–.80
VII	109	28	2.133	.20–.10
VIII	78	17	0.239	.70–.50
IX	83	14	0.027	.90–.80
X	86	12	0.584	.50–.30
XI	86	7	3.970	.05–.02
XII	69	12	0.001	.99–.98
XIII*	131	3	17.084	.01
XIII†	58	20	5.301	.05–.02
XIV	31	6	0.013	.95–.90
XV	80	16	0.026	.90–.80
XVI	43	8	0.028	.90–.80
XVII	74	15	0.037	.90–.80
XVIII	70	9	0.748	.50–.30
XIX	94	20	0.209	.70–.50
XX	65	10	0.138	.80–.70
XXI	54	5	1.774	.20–.10

Stem-rust resistance is governed by duplicate linked genes 21 crossover units apart. Hence the expected ratio is 5.43 resistant : 1 susceptible in F_2.

* Monosomic F_1 plants

† Normal F_1 plants

located on chromosome XIII. Furthermore, only 6 susceptible plants occurred among 627 F_3 plants derived from 24 different resistant F_2 plants. Again, all the susceptible plants were nullisomics. These observations, combined with the normal segregation observed in all other families (and families derived from normal F_1 plants obtained in the hybrid with monosomic XIII), allow the inference that the duplicate dominant genes for stem-rust resistance are located on chromosome XIII.

To date, genes for stem-rust resistance in nine different varieties have been associated with particular chromosomes as shown in Table 29–2. This information has cleared up some of the puzzling results

TABLE 29–2. Association of Genes for Stem-Rust Resistance with Specific Chromosomes in Wheat

(After Unrau, 1958)

Line	Chromosome Carrying Gene(s) for Resistance								
	III	VI	VIII	X	XIII	XVI	XVII	XIX	XX
Timstein				2					
Hope			1			1			
Thatcher	1				1		1		
McMurachy									1
Red Egyptian		1			1			1	
Redman	1	1		1					
Kenya Farmer		1			1	1			
C.I. 12633					2				
Synthetic Thatcher	1			2	1			1	1
Synthetic Lemhi				2				1	

obtained in breeding for stem-rust resistance. But more important, this information is helpful in combining chromosomes carrying known genes for resistance. For example, Kuspira and Unrau have developed a synthetic Thatcher, which in addition to the resistance of Thatcher also has the resistance of Timstein and McMurachy or a total of six genes for resistance (Unrau, 1958).

Chromosome Substitution Using Aneuploids

The use of nullisomics and monosomics for the substitution of whole chromosomes and parts of chromosomes so they can be studied for gene content has been described by Unrau (1950). Later, Kuspira and Unrau (1957) found that the substitution of particular homologous chromosomes from one variety to another caused significant modifications of yield, earliness, lodging resistance, and other characters. It is possible that aneuploids will become valuable in wheat breeding by allowing particular characters in outstanding varieties to be modified by replacing specific chromosomes with their homologues from other varieties.

REFERENCES

Clausen, R. E., and D. R. Cameron. 1944. Inheritance in *Nicotiana tabacum* XVIII. Monosomic analysis. *Genetics* 29: 447–477.

Kuspira, J., and J. Unrau. 1957. Genetic analyses of certain characters in common wheat using whole chromosome substitution lines. *Can. Jour. Plant Sci.* 37: 300–326.

Nyquist, W. E. 1957. Monosomic analysis of stem rust resistance of a common wheat strain derived from *Triticum timopheevi*. *Agron. Jour.* 49: 222–223.

Rick, C. M., and D. W. Barton. 1954. Cytological and genetical identification of the primary trisomics of the tomato. *Genetics* 39: 640–666.

Sears, E. R. 1953. Nullisomic analysis in common wheat. *Amer. Nat.* 87: 245–252.

Sears, E. R. 1954. The aneuploids of common wheat. *Missouri Agric. Exp. Sta. Res. Bull.* 572.

Unrau, J. 1950. The use of monosomics and nullisomics in cytogenetic studies of common wheat. *Sci. Agric.* 30: 66–89.

Unrau, J. 1958. Cytogenetics and wheat-breeding. *Proc. Tenth Intern. Congr. Gen.* 1: 129–141.

30
Inheritance in Autopolyploids

A number of agricultural crops are autopolyploids or have been found to behave cytologically as autopolyploids. Among these cultivated autopolyploids, monoploids as a group are of negligible importance. Monoploid plants, when viable, are typically smaller and less vigorous than diploids. The chromosomes have no regular partner with which to pair during meiosis. At telophase some chromosomes go to one pole and some to the other. This chaotic meiosis in monoploids leads to daughter nuclei that are nearly always deficient for one or more chromosomes and, as a result, the great majority of gametes are nonfunctional. Nevertheless, monoploids are of some interest to plant breeders because doubling the chromosome number of a monoploid gives rise to a diploid pure line, homozygous at all loci. Chase (1952) has discussed the use of monoploidy as a short-cut method in establishing homozygous lines in corn.

Triploids, like monoploids, are usually highly sterile. Unlike monoploids, however, they occupy a fairly important position among cultivated plants, because favorable morphological and physiological effects on plant and fruit characters are sometimes associated with the 3n condition. Triploids are fairly common among crop plants in which the sterility barrier can be overcome by vegetative propagation, for example, among ornamentals. Among fruits there are a number of important triploid varieties, such as the Baldwin, Stayman Winesap, and Gravenstein apples and many European varieties of pears. The triploids are generally vigorous growers that produce large fruit. When planted alone or in blocks with other triploids, they tend to give sparse crops because of their poor pollen-producing ability. Consequently, in commercial practice, triploids are generally grown in combination with diploid varieties that serve as pollinators.

The banana is perhaps the outstanding example of a triploid crop. In this species maximum vigor is associated with the triploid state.

Moreover, since diploid bananas have hard seeds that make their fruit unacceptable commercially, the sterility that accompanies triploidy has the important commercial advantage of genetically deseeding the banana in an efficient and dependable way.

Autotetraploidy

If we judge by the frequency with which the different levels of autopolyploidy appear among crop plants, we must consider autotetraploidy to be more important to agriculture than the other levels. Four of the major crop species may be autotetraploids. These are alfalfa (Ledingham, 1940), the potato (Lamm, 1945), coffee (Krug and Carvalho, 1951), and the peanut (Husted, 1936). In addition several forage grasses and many ornamentals appear to be autotetraploids. Critical studies will probably reveal additional cases of autotetraploidy among cultivated plants and also plants in which chromosomal behavior indicates a constitution intermediate between auto- and allotetraploidy. Birdsfoot trefoil, for example, has been tentatively assigned to this latter category.

Clear-cut cases of higher levels of autopolyploidy among crop plants are not frequent. The sweet potato, *Ipomoea batatas*, appears to be an autohexaploid, and some forage grasses and ornamentals also appear to be high-level autopolyploids.

The Cytology of Autotetraploids

Before we attempt to discuss the genetics of autotetraploids, it is advisable to recapitulate the main features of their cytological behavior at meiosis: In leptotene, the four homologous chromosomes of each type emerge from the resting stage. In zygotene, they become associated in pairs. Association by pairs begins at a number of different points along the chromosomes, so that different chromosomes pair with different partners at different places. When pachytene association has been completed, a chromosome may thus have been paired first with one chromosome and then with another—that is, a number of changes of partner may have occurred (Figure 30–1). By the end of pachytene, each chromosome has divided into two chromatids, and chiasmata have formed as a result of crossing over between paired chromosomes. This, of course, causes portions of different chromatids to become attached to the same kinetochore and leads to *equational* separation at anaphase whenever an odd number of chiasmata form between a locus and the kinetochore. Separation at the kinetochore is always *reductional* at first anaphase.

Variations in the number and position of chiasmata lead to a

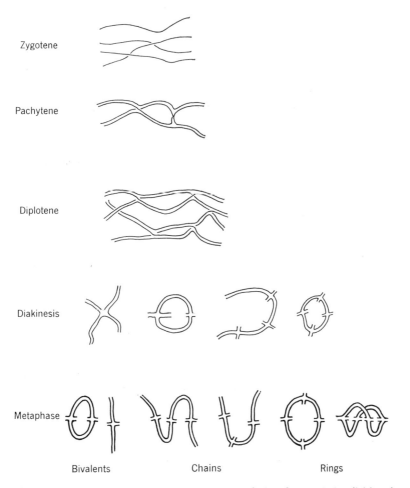

Zygotene

Pachytene

Diplotene

Diakinesis

Metaphase

Bivalents   Chains   Rings

FIGURE 30-1. Diagrammatic representation of the first meiotic division in a quadrivalent. (See text.)

number of different configurations (e.g., bivalents, chains, rings) at diplotene. At metaphase, a ring may disjoin so that adjacent chromosomes go to either the same or to different poles. The same two possibilities exist for chains. But when no two chiasmata straddle an exchange between partner chromatids, two bivalents form, and the two chromosomes of each bivalent must go to opposite poles at anaphase. Disjunction of quadrivalents at first anaphase is usually two by two, but numerical aberrations sometimes occur. Unbalanced gametes are often poorly competitive, however, so that zygotic frequencies may be little effected by these numerical aberrations of disjunction. At the second division of meiosis, the two chromatids attached to the same kinetochore at the first division always go to opposite poles, and so into different gametes.

The essential feature of the cytology of tetraploids is the partition of the eight chromatids of the first meiotic division into four pairs, each pair corresponding to one of the four gametes produced by each sporocyte. This meiotic process is much more complicated than that of diploids, which involves the partition of only four chromatids, one to each of the four gametes. In an autotetraploid five genotypes, $AAAA$ (quadruplex), $AAAa$ (triplex), $AAaa$ (duplex), $Aaaa$ (simplex), and $aaaa$ (nulliplex), are possible at each locus. These zygotic combinations arise from the fusion of three different types of gametes, AA, Aa, and aa, whose relative frequency, for a given genotype, depends on the cytological events of meiosis. The real problem in calculating expected segregations in tetrasomic inheritance is to determine the ways in which cytological events affect the partition of eight chromatids, two at a time, to the four cells of the tetrad formed by each sporocyte.

Random Chromosome Assortment in Autotetraploids

Aside from the genotype of the sporophyte, factors affecting gametic output in autotetraploids are (1) the regularity with which quadrivalents are formed, and (2) the randomness of disjunction from quadrivalents, which depends primarily on the distance between the kinetochore and the locus in question.

Let us first consider gamete formation in a simplex ($Aaaa$) individual in which only bivalents are formed. The two chromatids carrying the dominant alleles are always joined to the same kinetochore. Consequently they always go to the same pole in the first anaphase and always separate in the second anaphase. The two dominant alleles therefore never appear in the same gamete. The three pairs of chromatids bearing recessive alleles behave similarly. In other words, sister chromatids—chromatids derived from the same chromosome—never end in the same gamete. Thus, if we represent the eight chromatids as (A_1, A_2), (a_3, a_4), (a_5, a_6), and (a_7, a_8), A_1 and A_2 cannot appear in the same gamete, but each has an equal chance of appearing in the same gamete with a_3, a_4, a_5, a_6, a_7, or a_8. Similarly a_3 and a_4 cannot appear in the same gamete, but each has an equal chance of appearing with A_1, A_2, a_5, a_6, a_7, or a_8, and so on for the other pairs of chromatids. This type of partition of chromatids is called *random chromosome assortment*. The cytological features of random chromosome assortment are illustrated diagrammatically in Figure 30–2.

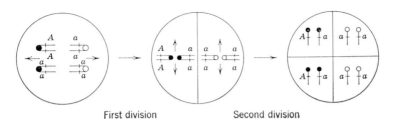

First division   Second division

FIGURE 30-2. Gamete formation in a simplex quadrivalent in which the first division is reductional, that is, the A chromatids are joined to the same kinetochore at first division, since crossing over has not taken place between the locus and the kinetochore.

Gametic output under random chromosome assortment for a simplex individual is as follows:

	A_1	A_2	a_3	a_4	a_5	a_6	a_7	a_8
A_1			A_1a_3	A_1a_4	A_1a_5	A_1a_6	A_1a_7	A_1a_8
A_2			A_2a_3	A_2a_4	A_2a_5	A_2a_6	A_2a_7	A_2a_8
a_3					a_3a_5	a_3a_6	a_3a_7	a_3a_8
a_4					a_4a_5	a_4a_6	a_4a_7	a_4a_8
a_5							a_5a_7	a_5a_8
a_6							a_6a_7	a_6a_8
a_7								
a_8								

The gametic ratio is therefore $1Aa:1aa$. By similar reasoning, the gametic ratios for duplex and triplex individuals are found to be $1AA:4Aa:1aa$ and $1AA:1Aa$, respectively.

Once the gametic frequencies have been established, the prediction of zygotic frequencies for various types of matings follows as a matter of routine. For example, the zygotic expectation on selfing a simplex plant is as follows:

		♂ gametes	
		1Aa	1aa
♀ gametes	1Aa	AAaa	Aaaa
	1aa	Aaaa	aaaa

If one A in the genotype produces the dominant phenotype, the expected phenotypic ratio in F_2 is $3A:1a$, exactly as in disomic inheritance. Expected phenotypic ratios for some other types of matings, assuming random chromosome assortment, are given in Table 30–1.

In the example we have just considered, the allocation of chromatids (and hence genes) to gametes by the process of random chromosome assortment was brought about by failure of quadrivalent formation in the first meiotic division. At this point we should note that exactly the same results obtain, even when quadrivalents always form, when the locus in question is inseparably linked to the kinetochore. The effect of the linkage is to cause the locus to separate reductionally at the first anaphase and equationally at second anaphase, exactly as when quadrivalent formation fails in the first division.

Random Chromatid Assortment

When quadrivalents form and the locus in question is far enough removed from the kinetochore to permit chiasmata to form between it and the kinetochore, sister chromatids at this locus can end up attached to different kinetochores. Consequently sister alleles may be included in the same gamete or in different gametes, depending on the distribution of the chromatids at the two meiotic anaphases (Figure 30–3). When quadrivalent formation is complete and 50 per cent crossing over occurs between kinetochore and locus, the partition of chromatids to gametes will be at random. Gametic output for a simplex individual can then be predicted as follows:

	A_1	A_2	a_3	a_4	a_5	a_6	a_7	a_8
A_1		A_1A_2	A_1a_3	A_1a_4	A_1a_5	A_1a_6	A_1a_7	A_1a_8
A_2			A_2a_3	A_2a_4	A_2a_5	A_2a_6	A_2a_7	A_2a_8
a_3				a_3a_4	a_3a_5	a_3a_6	a_3a_7	a_3a_8
a_4					a_4a_5	a_4a_6	a_4a_7	a_4a_8
a_5						a_5a_6	a_5a_7	a_5a_8
a_6							a_6a_7	a_6a_8
a_7								a_7a_8
a_8								

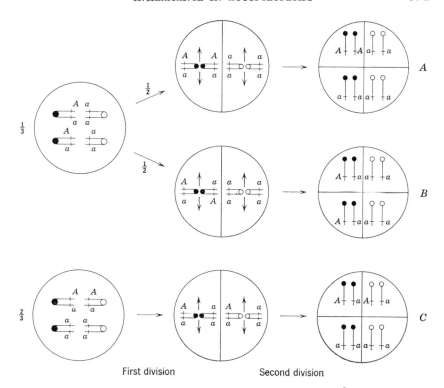

First division   Second division

FIGURE 30-3. Gamete formation in a simplex quadrivalent in which the chromosome carrying the dominant allele crosses over with another chromosome. The results is two *Aa* and two *aa* chromosomes. The two *Aa* chromosomes go to the same pole in the first division one-third of the time (top) and to opposite poles two-thirds of the time (bottom). When they pass to the same pole in the first division, and also in the second division (sequence *A* above), some homozygous gametes are formed. This process is called "double reduction." Gamete formation in duplex quadrivalents is much more complicated because of the many different combinations that are possible.

The gametic ratio for a simplex plant is thus $1AA : 12Aa : 15aa$. For duplex and triplex quadrivalents the gametic ratios are $3AA : 8AA : 3aa$ and $15AA : 12Aa : 1aa$, respectively.

Upon selfing a simplex plant, the zygotic genotypic ratio expected is:

Quadruplex	1
Triplex	24
Duplex	174
Simplex	360
Nulliplex	225
Total	784

If we assume that one A allele in the genotype produces the dominant phenotype, we can see that the F_2 phenotypic ratio of a simplex plant is $559A : 225a$ or approximately $2.48A : 1a$. Some of the other phenotypic ratios expected under random chromatid assortment are summarized in Table 30–1.

TABLE 30–1. Phenotypic Ratios Expected on the Basis of Two Hypotheses Concerning Gamete Formation in Autotetraploids

Mating	Random Chromosome Assortment	Random Chromatid Assortment
$AAAA$ selfed	All A	All A
$AAAa$ selfed	All A	$783A : 1a$
$AAaa$ selfed	$35A : 1a$	$20.8A : 1a$
$Aaaa$ selfed	$3A : 1a$	$2.5A : 1a$
$aaaa$ selfed	All a	All a
$AAAa \times AAaa$	All A	$130A : 1a$
$AAAa \times Aaaa$	All A	$51.3A : 1a$
$AAAa \times aaaa$	All A	$27A : 1a$
$AAaa \times Aaaa$	$11A : 1a$	$7.7A : 1a$
$AAaa \times aaaa$	$5A : 1a$	$3.7A : 1a$
$Aaaa \times aaaa$	$1A : 1a$	$0.87A : 1a$

One A allele is assumed to produce the dominant phenotype.

The Parameter Alpha

Random chromosome assortment, with its characteristic gametic output, occurs when separation is reductional at the first meiotic division, owing to failure of quadrivalent formation or to complete linkage between kinetochore and locus. Random chromatid assortment, with its different gametic output, occurs when quadrivalents form and 50 per cent crossing over takes place between kinetochore and locus. Random chromosome and random chromatid assortment therefore represent the two extremes of gametic output in tetrasomic inheritance. We recognize immediately that gametic output will be intermediate between these extremes when (1) quadrivalents sometimes form and sometimes do not, and (2) kinetochore and locus are partly linked.

The net effect of partial quadrivalent formation can be summarized in a parameter, a. This parameter takes its minimum value of zero when quadrivalents never form. It reaches its maximum value of one when quadrivalents always form and disjoin randomly two by two.

The net effect of the second factor, variable chiasma number, can be summarized by a parameter, e, which reflects the mean frequency with which sister chromatids arrive in the same gamete. When the two chromosomes in a gamete are derived from sister chromatids, *double reduction* is said to have occurred. The parameter e takes its minimum value of zero when linkage between kinetochore and locus is complete, because when this is true all separations are reductional in the first meiodic division, and sister chromatids cannot end in the same gamete (Figure 30–2).

In deducing the maximum value of e, we must first identify the sequence of events by which sister chromatids end in the same gamete. For double reduction to occur, it is necessary that (1) a single crossover (or any odd number of crossovers) occur between kinetochore and locus, so that sister chromatids are attached to two different kinetochores; (2) kinetochores bearing sister chromatids go to the same pole in first anaphase; and (3) these sister chromatids also go to the same pole at second anaphase. This sequence of events is illustrated diagrammatically in Figure 30–3.

The probability of each of these events is as follows. Assuming 50 per cent crossing over between kinetochore and locus in a simplex individual, any one chromatid, say A_1, has an equal chance of being attached to the same kinetochore as A_2, a_3, a_4, a_5, a_6, a_7, or a_8. Among these 7 equally likely possibilities, only one, attachment to the same kinetochore as A_2, excludes the possibility of double reduction. In the other 6 possibilities sister chromatids are attached to different kinetochores and hence can, but do not necessarily, end in the same gamete. The probability of the first event stated above is therefore 6/7.

Now, if sister chromatids are attached to different kinetochores, random orientation on the metaphase plate will cause them to go to the same pole at first anaphase in one third of cases and to the same pole at second anaphase in one half of cases. Hence the probabilities of events 2 and 3 above are one third and one half respectively, and the maximum value of e is

$$\tfrac{6}{7} \times \tfrac{1}{3} \times \tfrac{1}{2} = \tfrac{1}{7} \quad \text{or} \quad .1429.$$

Gametic output in tetrasomic inheritance is therefore a function of the parameters a and e, whose joint effect can be summarized in a third parameter, alpha, which is the product of $a \times e$. Alpha takes its minimum value of zero when either a or e is zero. It takes a maximum value of one seventh when $a = 1$ and $e = \frac{1}{7}$.*

The significance of alpha in determining gametic output can be illustrated by a hypothetical example similar to many encountered in actual breeding materials. Suppose that quadrivalents form in 50 per cent of cases and that the locus in question is independent of the kinetochore. Then $a = .5$, $e = \frac{6}{7} \times \frac{1}{3} \times \frac{1}{2} = \frac{1}{7} = .1429$, and $\alpha = .5 \times .1429 = .07145$. For a simplex organism, it can be shown that gametic expectancies are

$$AA = \frac{\alpha}{4}, \qquad Aa = \frac{2 - 2\alpha}{4}, \qquad \text{and} \qquad aa = \frac{2 + \alpha}{4}.$$

Gametic output for our hypothetical example can be compared with that for random chromosome and random chromatid segregation, as shown in the following table. Gametic output for the hypothetical example is, as expected, intermediate between that of random

		Theoretical Proportions		
Type of Gamete	General	Random Chromosome Assortment $\alpha = 0$	Hypothetical Example $\alpha = 0.07145$	Random Chromatid Assortment $\alpha = 0.1429$
AA	$\dfrac{\alpha}{4}$	0.0000	0.0179	0.0357
Aa	$\dfrac{2 - 2\alpha}{4}$	0.5000	0.4643	0.4286
aa	$\dfrac{2 + \alpha}{4}$	0.5000	0.5179	0.5357
Total	1	1	1	1

Gametic ratios, AA : Aa : aa are for:
(1) Random chromosome assortment, 0 : 1 : 1
(2) Hypothetical example,   1 : 13.4 : 28.9
(3) Random chromatid assortment,   1 : 12 : 15

chromosome and random chromatid assortment. Hence phenotypic ratios for various types of matings are also intermediate between the extremes represented by random chromosome and random chro-

* The maximum value of alpha has been erroneously reported in the literature to be $\frac{1}{6}$. In the event that one and only one chiasma always forms between kinetochore and locus, $e = \frac{5}{8} \times \frac{1}{3} \times \frac{1}{2} = \frac{5}{32} = 0.1563$, and alpha also takes a maximum value of 0.1563. There are, however, no good biological reasons to assume that one and only one chiasma will always form in a particular chromosome segment, so for all practical purposes the maximum value of alpha is $\frac{1}{7}$.

matid assortment. The equations expressing gametic output in terms of alpha are useful, as we shall see later, in estimating alpha from observed segregations in various types of matings.

Selective Pairing in Quadrivalents

Pairing in a tetraploid can be either random or selective for any given group of four homologous chromosomes. With random pairing, any one of the four homologous chromosomes pairs equally frequently with any of the other chromosomes. Selective pairing occurs when the four homologs are not equally homologous, but display differential affinity. Homology is a relative term and can vary from complete identity of chromosomes, expected when a homozygous diploid is doubled, to very weak homology, expected within certain quadrivalents when the F_1 hybrid between sharply differentiated species is doubled.

To illustrate the effect of selective pairing on segregation, let us assume we have a duplex hybrid $A_1A_1a_2a_2$, where similar subscripts designate similar pairs of chromosomes. If selective pairing is complete, only gametes of the constitution A_1a_2 can be formed, and all the progeny will be of the genotype $A_1A_1a_2a_2$. Hence there will be no segregation. If selective pairing is incomplete so that A_1 occasionally pairs with a_2, A_1A_1 or a_2a_2 gametes will occasionally be formed, and the latter can unite to give homozygous recessive progeny. Tetraploids with the constitution $A_1a_1A_2a_2$ arise under certain conditions. When this happens, complete selective pairing produces gametes with the constitution A_1A_2, A_1a_2, a_1A_2, and a_1a_2 in equal proportions. Upon selfing such plants, the phenotypic ratio is the familiar $15A:1a$. Moreover, two types of duplex hybrids will be produced: (1) types like the parents and (2) nonsegregating hybrids genotypically $A_1A_1a_2a_2$ or $a_1a_1A_2A_2$. Newly arisen amphidiploids are almost certain to display considerable selective pairing. On the other hand, long-established autotetraploids such as the potato and alfalfa may also show a great deal of selective pairing, presumably as a result of the accumulation, over long periods of time, of small changes in the chromosomes. Stated in another way, there is a tendency in autopolyploids toward "diploidization" as a result of gradual changes in originally identical chromosomes.

Application of Principles of Polyploid Genetics

Because many important crop plants are polyploids, plant breeders frequently must face the difficulties associated with the genetics of polyploids. As an example let us consider an experiment in alfalfa designed to determine the inheritance of resistance to the root-knot nematode (*Meloidogyne hapla*), a soil-infesting eel worm that causes severe damage to alfalfa in mild climates.

Among a number of resistant plants of alfalfa selected from commercial fields in California by B. P. Goplen, one, designated M–9, was immune to all races of the organism. When clonal material of this plant was selfed and the resulting S_1 population scored for resistance, 319 resistant and 119 susceptible plants were observed. In a test cross of the M–9 clone to a susceptible clone, there were 46 resistant and 55 susceptible plants. These results can be interpreted as indicating that clone M–9 is heterozygous (Aa) for a single gene segregating in a disomic manner or, alternatively, that it is a tetrasomic simplex ($Aaaa$). Differentiation between these hypotheses is not possible with these data alone, because the expectations are nearly identical for the disomic and the various possible tetrasomic hypotheses (Table 30–1).

However, a simplex plant giving a 3:1 ratio in the S_1 generation is expected to have quite a different pattern of segregation in the S_2 generation than a disomic heterozygote. The pattern expected under the hypothesis of tetrasomic inheritance differs from disomic segregation as shown in Table 30–2. The actual numbers of the different types of S_2 families obtained by Goplen are shown in Table 30–3. The distribution of the S_2 families observed clearly excludes disomic inheritance; the fact that one S_2 family did not segregate and was

TABLE 30–2.   Expected Segregation in S_2 of a Disomic Heterozygote and a Tetrasomic Simplex

	Expected Ratio under		
Type of S_2 Family	Disomic Inheritance	Random Chromosome Assortment	Random Chromatid Assortment
Homozygous susceptible	$\frac{1}{4}$	$\frac{1}{4}$	225/784
Segregating approximately 3 resistant: 1 susceptible	$\frac{1}{2}$	$\frac{1}{2}$	360/784
Segregating 21 resistant : 1 susceptible			174/784
Segregating 35 resistant : 1 susceptible		$\frac{1}{4}$	
Segregating 783 resistant : 1 susceptible			24/784
Homozygous resistant	$\frac{1}{4}$		1/784

TABLE 30–3.   Segregation of S_2 Families of the Resistant Selection M–9 and Goodness of Fit to Expectations under Disomic Inheritance and Two Tetrasomic Hypotheses

		Expected		
			Random	Random
		Disomic	Chromosome	Chromatid
Type of Family	Observed	Inheritance	Assortment	Assortment
Homozygous susceptible	12	9	9	10.33
Segregating approximately 3 : 1	20	18	18	16.53
Segregating 21 : 1 to 35 : 1	3		9	7.99
Nonsegregating resistant	1	9		1.15
Total	36	36	36	36.00

Random chromosome assortment $\chi^2_{[2]} = 4.00$, $P = 0.10$–0.20.
Random chromatid assortment $\chi^2_{[3]} = 4.146$, $P = 0.20$–0.30.

presumably triplex suggests that double reduction occurs in the quadrivalent in question.

When alpha was calculated from combined S_1, test-cross, and S_2 data, it was found to have a value of .01 ± .04, which is not significantly different from zero, the value alpha takes for random chromosome segregation. This suggests either that quadrivalents are rarely formed or that the locus is tightly linked to the kinetochore. The former explanation seems more likely for alfalfa because, in meiosis, usually only two quadrivalents are observed of the eight that are possible.

This example illustrates the difficulties of the genetics of tetraploids and points to some problems of considerable practical importance in polyploid genetics. Consider first the problem of obtaining pure lines in tetraploid species. In diploids this is relatively simple because selection of an F_2 or later-generation plant that produces uniform progeny is a virtual guarantee of homozygosity, and the breeder feels safe in bulking the seed and using it for an increase. Such a procedure might not be safe in a tetraploid because the selected plant might be triplex. The progeny of a triplex would be highly uniform phenotypically (Table 30–1), but genotypically it would still be segregating, producing duplex and simplex plants.

Another problem that arises is the difficulty of obtaining multiple recessive types from hybrids. Suppose the dihybrid $AAaaBBbb$ is selfed. The phenotypic ratio expected on the basis of random chromo-

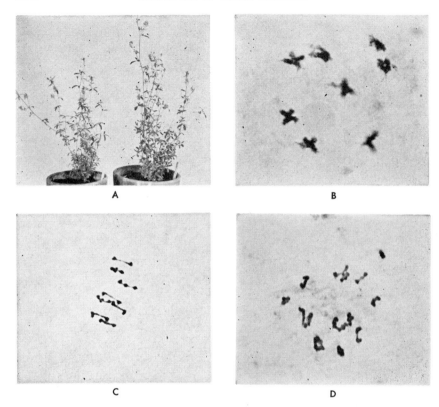

FIGURE 30-4. *A*, left, dihaploid plant in *Medicago sativa*. Right, colchicine induced tetraploid from the dihaploid. *B*, diakinesis in the dihaploid plant. *C*, metaphase in the dihaploid plant. *D*, metaphase in the induced tetraploid showing twelve or fourteen bivalents and one or two quadrivalents. What is a dihaploid? What significance does the fact that the dihaploid forms eight bivalents have in clarifying the evolution of alfalfa? Metaphase in normal alfalfa $(2n = 32)$ is the same as in the induced tetraploid. What significance does this have concerning expected patterns of inheritance in alfalfa? (Photograph courtesy of Dr. W. M. Clement, Jr., California Agricultural Experiment Station.)

some assortment is $1225AB : 35Ab : 35aB : 1ab$. The corresponding diploid ratio from selfing a disomic $AaBb$ plant is $9AB : 3Ab : 3aB : 1ab$. In an autotetraploid trihybrid the frequency of triple recessives varies from approximately 1 in 8000 to 1 in 47,000, depending on the value of a and e. It is obvious that one must either grow very large populations in tetraploids or devise some step-by-step scheme for obtaining desired recessive types.

The situation becomes tremendously more complicated in autohexaploids and higher-level autopolyploids. The student who is willing to brave the complexities of hexasomic inheritance is referred to the papers of Sir Ronald Fisher and his collaborators.

REFERENCES

Chase, S. S. 1952. Monoploids in maize. In *Heterosis,* pp. 389–399. Iowa State College Press.

Dawson, C. R. D. 1941. Tetrasomic inheritance in *Lotus corniculatus, L. Jour. Gen.* 42: 49–72.

Fisher, R. A. 1944. Allowance for double reduction in the calculation of genotypic frequencies with polysomic inheritance. *Am. Eug.* 12: 169–171.

Husted, L. 1936. Cytological studies on the peanut, *Arachis.* II. Chromosome number, morphology and behavior, and the application to the problem of the origin of cultivated forms. *Cytologia* 7: 399–423.

King, J. R., and R. Bamford. 1937. The chromosome number of *Ipomoea* and related genera. *Jour. Hered.* 28: 279–282.

Krug, C. A., and A. Carvalho. 1951. The genetics of *Coffea. Adv. in Gen.* 4: 127–158.

Lamm, R. 1945. Cytogenetic studies in *Solanum* sect. *Tuberarium. Hereditas* 31: 1–128.

Ledingham, G. F. 1940. Cytological and developmental studies of hybrids between *Medicago sativa* and a diploid form of *M. falcata. Genetics* 25: 1–15.

Little, T. M. 1945. Gene segregation in autotetraploids. *Bot. Rev.* 11: 60–85.

Little, T. M. 1958. Gene segregation in autotetraploids. II. *Bot. Rev.* 24: 318–339.

Mather, K. 1935. Reductional and equational separation of the chromosomes in bivalents and multivalents. *Jour. Gen.* 30: 53–78.

Mather, K. 1936. Segregation and linkage in autotetraploids. *Jour. Gen.* 32: 287–314.

Simmonds, N. W. 1949. Genetical and cytological studies in *Musa.* X. Stomatal size and plant vigor in relation to polyploidy. *Jour. Genetics* 49: 57–68.

Stanford, E. H. 1951. Tetrasomic inheritance in alfalfa. *Agron. Jour.* 43: 222–225.

31

Cytogenetics of Allopolyploids

Among the various types of polyploidy, allopolyploidy has been the most important in the ancestry of cultivated species. This is true whether we judge from the numbers of species or from the economic value of allopolyploids compared with other types of polyploids. Sometimes the hybridization preceding polyploidization was between interfertile diploid subspecies, as in *Dactylis,* and in such cases there is some question whether the polyploidy should be described as auto- or allopolyploidy. Often, however, distinctly different species were involved, and the most important polyploids among cultivated species, for example, wheat, cotton, oats, species of *Brassica,* tobacco, and sugar cane, are unambiguously amphidiploid in their origin.

The Cytology of Amphidiploids

The most precise information on the cytology of amphidiploids has come from the many amphidiploids that have been produced by artificially doubling the chromosome number of F_1 interspecific hybrids. One of the best guides to the fertility and stability of these induced amphidiploids has been the extent of chromosome homology between the genomes of the constituent species. When the two genomes are highly divergent as indicated by little pairing in the $2n$ hybrid, there are for practical purposes only two chromosomes of each kind in the amphidiploid. These pairs of chromosomes tend to form bivalents fairly regularly at meiosis, and disjunction is often normal with a full haploid $(n = x_1 + x_2)$ complement of chromosomes being allocated to each gamete. Such amphidiploids are often fertile and reasonably stable, both cytologically and genetically. There are, however, many exceptions to this generalization, and it is now clear that factors other than chromosome homology can influence the fertility and genetic stability of cytologically regular amphidiploids.

400

When an amphidiploid is made between species whose genomes are partly divergent, as judged from considerable pairing in the undoubled hybrid, the fertility and the stability of the 4n type have frequently been rather low. This condition can be ascribed in part to irregular pairing among incompletely homologous chromosomes, leading to unequal partition of the chromatin to the gametes. Again, however, few investigators are willing to charge the entire difficulty to cytological disturbances because some amphidiploids with quite regular cytological behavior are sterile, while others with irregular pairing are reasonably fertile. The opinion has been expressed that the sterility of polyploids is physiological in nature and is frequently associated with genically controlled unbalance, rather than with irregular chromosome behavior alone (Randolph, 1941; Stebbins, 1950).

The most common defects of artificially induced amphidiploids have been slower growth rate and low fertility. These difficulties also were presumably faced by the raw amphidiploids that were ancestral to present highly successful polyploid crop species. Since the successful allopolyploids now behave much like diploids, it seems evident that natural selection has led to an increase in preferential pairing by selecting toward increased differentiation among partly homologous chromosomes. It also seems likely that the great store of potential genetic variability associated with the increased number of genes in polyploids provides an opportunity for radically new genotypes to develop. Some of these different genotypes might adapt their bearers to new habitats, for example, the new habitats associated with cultivation. Amphidiploids thus appear to have the potential to overcome an initial disadvantage in sterility by means of progressive diploidization. They should also be able to overcome an initial disadvantage in physiological balance by means of recombinations of the great genetic variability that is inherent in the 4n condition. Apparently, however, long-continued selection is necessary to convert "raw" amphidiploids into smoothly functioning types such as *Triticum vulgare* or *Gossypium hirsutum*.

Polyploidy in Wheat and Its Relatives

The genus *Triticum* provides the classical example of a polyploid series. There are three groups of species in the genus, with chromosome numbers $n = 7$, 14, and 21. Moreover, every species in the five genera of the subtribe *Triticinae*—namely, *Triticum, Aegilops, Agropyron, Secale, Haynaldia*—has either seven pairs of chromosomes or some multiple of seven pairs.   Cytological data from hybrids indicate that the chromosome complement of each polyploid species consists of groups (genomes) of seven pairs of chromosomes each.   Each of the genomes of the polyploids is considered to be a more or less distinct entity, homologous or homoeologous with the seven chromosomes of a particular diploid species.

Assessment of the degree of relationship among species in the *Triticinae* has been based largely on the amount of pairing at meiosis in F_1 interspecific hybrids.   This measure of relationship is subject to errors caused by translocations between chromosomes of different genomes and from nonhomologous associations that tend to reduce genomic individuality.   Genomic designations are therefore undoubtedly oversimplifications.   Nevertheless, they seem to come close to the true picture, and they serve a useful purpose in reconstructing the pattern of polyploidy in this group.

In the genus *Triticum* itself, the species within each of the three groups have homologous chromosomes, with some possible exceptions to be discussed below.   The species of wheat and their genomic formulas are given in Table 31–1.   Crosses between Einkorn wheats ($n = 7$) with members of the Emmer and Vulgare groups discloses that the Einkorn genome of seven chromosomes is common to all three groups.   This genome, designated A, does not correspond perfectly with the A genome of the tetraploid (or hexaploid) wheats, but there seems little doubt that the tetraploid group arose as an amphidiploid hybrid of either *T. aegilopoides* or *T. monococcum* with some other species with $n = 7$.   McFadden and Sears (1947) suggested on cytological evidence that the B genome of the Emmer wheats was derived from *Agropyron triticeum*, and Sarkar and Stebbins (1956) have postulated on morphological grounds that the B genome evolved from the S genome of *Aegilops speltoides*.

There is general agreement that the A and B genomes of the Emmer wheats are closely related to the A and B genomes of the Vulgare wheats, because most hybrids between Emmer and Vulgare types regularly have fourteen pairs of chromosomes.   However, in some hybrids fewer than fourteen pairs are formed, and in a few Emmer-Vulgare combinations striking abnormalities in pairing occur.   The variety Khapli, for example, although morphologically a good Emmer type and interesting to wheat breeders because of its high degree of rust resistance, is notoriously infertile in hybrids with bread wheats.

Two species with $n = 14$ chromosomes, *T. timopheevi* and *T. armeni-*

TABLE 31–1.   The Species of *Triticum* and the Genomic Formulas

Species	Chromosome Number (n)	Formula
T. aegilopoides	7	A
T. monococcum	7	A
T. dicoccoides	14	AB
T. dicoccum	14	AB
T. turgidum	14	AB
T. persicum	14	AB
T. polonicum	14	AB
T. durum	14	AB
T. timopheevi	14	AG
T. armeniacum	14	AG
T. spelta	21	ABD*
T. vulgare	21	ABD
T. mocha	21	ABD
T. vavilovi	21	ABD
T. compactum	21	ABD
T. sphaerococcum	21	ABD

*American and British workers commonly use C instead of D, but the symbol D has priority since much of the work in identifying the genomes in wheat has been done by Kihara and his coworkers, who have based their designations on D.

acum, have been singled out on both morphological and cytological grounds as distinct from the species of the Emmer group of wheats. Although the evidence is not conclusive, it seems possible that the second genome of these species was derived from a different diploid species from the one contributing the B genome of the true Emmers. $T.$ $timopheevi$ and $T.$ $armeniacum$ are consequently given the genomic formula AG to emphasize the distinctiveness of their second genome, whether it be merely a highly differentiated form of the B genome or whether it had a completely different origin.

After much speculation by many investigators, the third chromosome set of the Vulgare wheat was finally identified as homologous to that of the diploid $(2x)$ species $Aegilops$ $squarrosa.$ The amphidiploid of $T.$ $dicoccoides$ × $Ae.$ $squarrosa,$ produced by McFadden and Sears, resembles $T.$ $spelta$ closely and forms fertile, meiotically regular hybrids with $T.$ $vulgare$ and $T.$ $spelta.$

McFadden and Sears have postulated that wheats have arisen through the following series of evolutionary events:

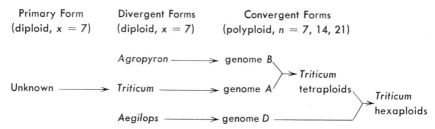

| Primary Form (diploid, x = 7) | Divergent Forms (diploid, x = 7) | Convergent Forms (polyploid, n = 7, 14, 21) |

Chromosome Behavior in Pentaploid Wheats

The cytological and genetic behavior of hybrids between members of the Emmer and the Vulgare series are particularly interesting to plant breeders because of the possibility of transferring desirable genes from one group to the other. The chromosome behavior of a pentaploid cross is as follows:

Parents	Emmer ($A_eA_eB_eB_e$)	$\times$	Vulgare ($A_vA_vB_vB_vD_vD_v$)
	14II		21II
Parental gametes	14E (A_eB_e)	$\times$	21V ($A_vB_vD_v$)
F_1 zygotes		$A_eA_vB_eB_vD_v$	
F_1 meiosis		14II ($A_eA_vB_eB_v$) + 7I (D_v)	
F_1 gametes		14–21 chromosomes (14AB + 0 to 7D_v)	
F_2 meiosis		14II + 0 to 7I made up of: ($14_e + 0_v$) to ($0_e + 14_v$) + ($0 - 7D_v$)	

In the F_1 the 14 Emmer chromosomes pair with the 14AB Vulgare chromosomes. Since these chromosomes usually pair and disjoin normally, each of the gametes receives 14 of these chromosomes. Crossing over occurs so that the chromatids attached to each kinetochore are likely to carry a mixture of genes derived from the two parents. The 7 univalent chromosomes are, however, distributed at random to the gametes. Thus the gametes produced by the F_1 have 14 AB chromosomes and 0 to 7D chromosomes. Since the distribution of the chromosomes of the D genome is expected to conform to the binomial expansion $(\frac{1}{2} + \frac{1}{2})^7$, only a small portion of the gametes should have $14 + 0$ or $14 + 7$ chromosomes, and the great majority should have $14 + 3$ or $14 + 4$ chromosomes. It was shown by Thompson and Armstrong in 1932 that the chromosome numbers of the gametes conform fairly closely to this expectation. The zygotic frequencies of plants formed by fusion of these gametes do not, however, conform to the expectation that most plants will have 35 or 36 chromosomes. Apparently pollen grains with intermediate numbers of chromosomes frequently fail to function, so that plants with 35 to 39 chromosomes occur less frequently than expected, and plants with 28, 29, 30, and 31 or 40, 41, and 42 chromosomes occur in excess of expectation.

The F_1 hybrid between Emmer and Vulgare, when related to the Vulgare series, can be regarded as monosomic for all 7 of the D chromosomes. On the same basis, the F_2 generation from a pentaploid hybrid is made up of a large number of different types of nullisomic and monosomic aneuploids. Thus a plant with 14 pairs of chromosomes is nullisomic for all 7 of the D chromosomes, and one with 41 chromosomes (20II + 1I) is monosomic for only 1 of the D chromosomes. Kihara has shown that types with chromosome numbers intermediate between 28 and 42 are unstable. Plants with 29 to 34 chromosomes thus ultimately produce 28-chromosome types, and plants with 35 to 41 chromosomes revert, but somewhat less rapidly, to produce 42-chromosome types.

It might therefore be anticipated that the ease of gene exchange between Emmer and Vulgare types depends on the genomes involved. There should be no difficulty in transferring genes from Emmer to common wheats, and this has been found to be true (excluding Khapli Emmer and a few other exceptional varieties). The main limitation of such gene transfers is associated with the infrequency with which 42-chromosome plants occur in a selfing series from the F_1 hybrid. This difficulty can be overcome by backcrossing the F_1 hybrid to the Vulgare parent. The resulting zygotes have 35 to 42 chromosomes and are at least monosomic for all of the 21 Vulgare chromosomes. Hence Kihara's ascending series takes over on subsequent selfing, resulting ultimately in 42-chromosome plants.

Allopolyploidy in Other Crops

Although wheat and its relatives provide the classical example of an allopolyploid series and illustrate the essential features of naturally occurring amphidiploidy, there is decisive evidence concerning the parentage of a number of other important amphidiploids. The first case to be described was in *Primula*, an important genus of ornamentals. In 1912, Digby discovered that a spontaneous fertile type in the sterile interspecific hybrid *P. verticillata* × *P. floribunda* contained a doubled number of chromosomes. However, the significance of this cytological observation was not recognized until after 1917, when Winge in Denmark, unaware of Digby's work, formulated his hypothesis that hybridization followed by polyploidy might be a

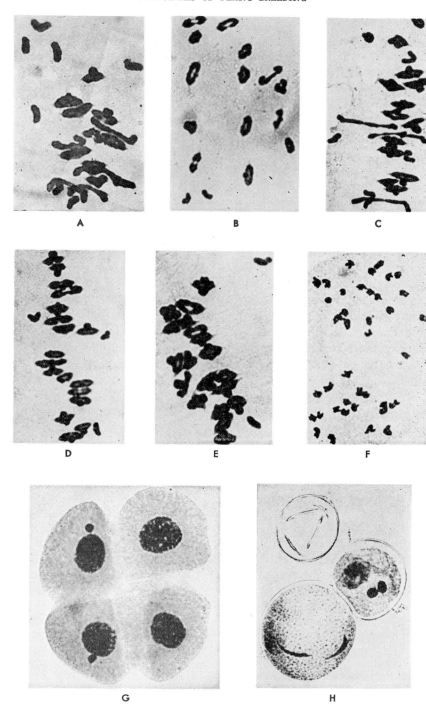

A   B   C

D   E   F

G   H

method of speciation. The first experimental verification of Winge's hypothesis was made in 1925 by Clausen and Goodspeed, who synthesized *Nicotiana digluta* from *N. tabacum* × *N. glutinosa*. These same workers (Goodspeed and Clausen, 1928) soon established that cultivated tobacco ($n = 24$) is an amphidiploid of *N. sylvestris* ($n = 12$) and *N. tomentosa* ($n = 12$).

In the next two decades, decisive evidence was found concerning the parents of various other amphidiploids among cultivated plants. Two of the most interesting cases are those of cotton and the Brassicas.

Among the 20 species in the genus *Gossypium*, about half are Old World species (Asia, Africa, Australia) and the remainder are American species. All 9 Old World species have $n = 13$ chromosomes. Eight of the New World species also have $n = 13$ chromosomes, but their chromosomes are distinctly smaller than those of the Old World species. The 3 remaining New World species, *G. hirsutum*, *G. barbadense*, and the wild *G. tomentosum* of Hawaii, have $n = 26$ chromosomes of which 13 are large and 13 are small. This suggests that the tetraploid New World species are amphidiploids between Old and New World types. The amphidiploid nature of the tetraploid New World types has now been proved as conclusively as such things can be proved, but how this amphidiploidy occurred between diploids confined to the Pacific slope of the Americas and Afro-Asiatic diploids remains an open question.

Some investigators favor the idea of an ancient origin in North America through hybridization between one group of diploids now restricted to South America and another group of diploids now confined to the Old World. Others favor a more recent origin which assumes that the cultivated Asiatic cottons (*G. arbaceum* or *G. arboreum*) were carried by an early civilization across the Pacific to the New World and that hybridization of the cultivated crop with a neighboring wild American species gave rise to the first amphidiploid. Regardless of the time and place where the Old and New World

FIGURE 31-1. Meiosis in a pentaploid wheat hybrid. *A,* metaphase in the F_1 hybrid showing 14 bivalents and 7 univalents. *B–E,* metaphase in F_2 plants. *B,* 14 bivalents and 2 univalents. *C,* 15 bivalents and 6 univalents. *D,* 18 bivalents and 3 univalents. *E,* 19 bivalents and 2 univalents. *F,* First anaphase in an F_2 plant with $2n = 39$ chromosomes showing an 18:21 separation into the daughter nuclei. *G,* tetrad stage in an F_2 plant showing micronuclei. *H,* pollen of an F_2 plant showing normal, retarded, and empty grains. All of the above plants can be regarded as aneuploids. Classify them as to type of aneuploidy with respect to the Emmer wheat parent; Vulgare wheat parent. What relation does the irregular chromosome pairing have to formation of micronuclei? To pollen abortion? To fertility?

species came together, there is strong evidence that the cultivated tetraploids arose from the amalgamation of the Old World *G. aboreum* and an American diploid similar or identical to *G. raimondii,* a wild cotton of Peru (Stephens, 1947).

Almost all of the cotton grown in the United States is annual American Upland Cotton, *G. hirsutum,* developed by selection from perennial shrubs native to Guatemala and Mexico. The other important cultivated species, *G. barbadense,* is a perennial whose center of variability is in Peru-Bolivia.

The vegetable crops furnish us with another well-documented case of an amphidiploid series which is made the more interesting because it also illustrates another and entirely different type of change in chromosome number, one in which the basic haploid number increases

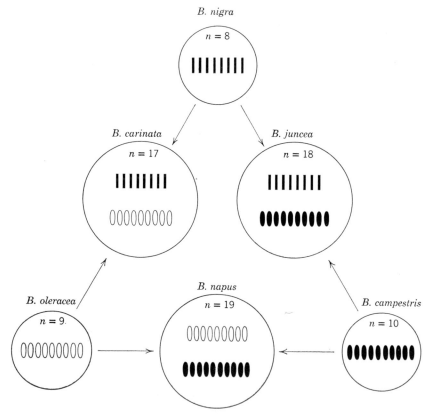

FIGURE 31-2. The triangle of *U.* The basic diploid species have chromosome numbers of 8, 9, and 10. These species have combined as shown in the diagram to form a series of amphidiploids. Gametic chromosome (*n*) numbers are given.

or decreases by one chromosome at a time. Such changes in chromosome number appear to be brought about by duplication or loss of a kinetochore, combined with a series of translocations. Three common diploid species in the genus *Brassica, B. oleracea, B. nigra,* and *B. campestris,* have haploid numbers of 8, 9, and 10, respectively, which probably represent a phylogenetically ascending series (Manton, 1932). *B. juncea* is a natural amphidiploid (*AABB*), combining the basic sets of *B. campestris* (*AA*) and *B. nigra* (*BB*). *B. napus* (*AACC*) and *B. carinata* (*BBCC*) are also natural amphidiploids which combine the basic sets of *B. campestris* (*AA*) and *B. oleracea* (*CC*), and *B. nigra* (*BB*) and *B. oleracea* (*CC*), respectively. These relationships are conveniently represented by arranging the basic diploid species in a triangle as shown in Figure 31–1. This triangle is known as the triangle of U, after Nagaharu U (1935). The amphidiploid combinations possible are $n = 16, 17, 18, 19, 20, 24, 25, 26, 27, 28, 29, 30$, and so on, and of these numbers $n = 17, 18, 19, 27$, and 29 are known in either natural or artificially produced amphidiploids. Establishment of these relationships among species cleared up many of the puzzling features of the genetics of the genus *Brassica* and were helpful in providing a foundation for applied breeding in the many morphologically distinct types in the genus that are important commercially.

REFERENCES

Clausen, R. E., and T. H. Goodspeed. 1925. Interspecific hybridization in *Nicotiana*. II. A tetraploid *glutinosa-Tabacum* hybrid, an experimental verification of Winge's hypothesis. *Genetics* 10: 279–284.

Digby, L. 1912. The cytology of *Primula kewensis* and of other related *Primula* hybrids. *Ann. Bot.* 26: 357–388.

Goodspeed, T. H., and R. E. Clausen. 1928. Interspecific hybridization in *Nicotiana*. VIII. The *sylvestris-tomentosa-tabacum* triangle and its bearing on the origin of *Tabacum. Univ. of Calif. Pub. in Bot.* 11: 245–256.

Manton, I. 1932. Introduction to the general cytology of the *Cruciferae. Ann. Bot.* 46: 509–556.

McFadden, E. S., and E. R. Sears. 1947. The genome approach to radical wheat breeding. *Jour. Amer. Soc. Agron.* 39: 1011–1026.

Randolph, L. F. 1941. An evaluation of induced polyploidy as a method of breeding crop plants. *Amer. Nat.* 75: 347–363.

Sarkar, P., and G. L. Stebbins, Jr. 1956. Morphological evidence concerning the origin of the *B* genome in wheat. *Amer. Jour. Bot.* 43: 297–304.

Sears, E. R. 1941. Chromosome pairing and fertility in hybrids and amphidiploids in the *Triticinae. Missouri Agric. Exp. Sta. Res. Bull.* 337.

Stebbins, G. L., Jr. 1950. *Variation and evolution in plants.* Columbia University Press, New York.

Stephens, S. G. 1947. Cytogenetics of *Gossypium* and the problem of the origin of New World cottons. *Adv. in Gen.* **1**: 431–442.

Thompson, W. P., and J. M. Armstrong. 1932. Studies on the failure of hybrid germ cells to function in wheat species crosses. *Can. Jour. Res.* **6**: 362–373.

U, N. 1935. Genome analysis in *Brassica* with special reference to the experimental formation of *B. napus* and peculiar mode of fertilization. *Jap. Jour. Bot.* **7**: 389–452.

32

Induced Polyploidy in Plant Breeding

When it was discovered in 1937 that polyploidy can be induced more or less at will in plants through the action of the drug colchicine, many geneticists and plant breeders held high hopes that a way had been opened for rapid development of novel and superior types that would revolutionize agriculture. The reasoning was as follows. First, many of the most important agricultural species are polyploids. Hence, the doubling of chromosome numbers under the guiding hand of man should lead to further valuable polyploids that had been missed by the random methods of nature. Second, polyploids have larger cells than their diploid counterparts, and it seems reasonable to expect this larger cell size to contribute to larger plant size and higher yields. On these grounds large numbers of both auto- and allopolyploids were produced in the first flush of enthusiasm for the new method. To other plant breeders, however, it was difficult to imagine that doubling of chromosome numbers could accomplish results in any short period that had required hundreds of years in the established polyploid crops. They recognized that successful varieties must conform to particular specifications in most crops and that any radical departure from traditional types is unlikely to receive the approval of either farmers or consumers.

In reviewing the situation after 20 years of attempts to make practical use of polyploidy in plant breeding, it appears that the skeptics were more nearly right than the proponents of polyploid breeding. The sanguine hope that polyploidy would revolutionize breeding methods has been thoroughly dispelled, because no case has yet been reported among crop plants where polyploidization has suddenly produced a conspicuously superior type. Most, if not all, newly induced polyploids have been infertile and genetically unstable to the point where they cannot satisfy the specialized standards set by modern agriculture. Moreover, it has become increasingly clear that doubling chromosome

411

numbers only rarely endows crop plants with superiority and in fact is much more likely to produce types that are inferior to their ancestral diploids.

It must be remembered, however, that induced polyploidy was added to plant-breeding methods less than a quarter of a century ago and that few of the established methods of plant breeding are capable of producing a new variety in less than ten years. The attitude toward polyploidy has gradually changed with increasing experience until it is now regarded by most plant breeders as capable of furnishing raw materials for breeding programs, but seldom, if ever, capable of producing strains ready for immediate release. With this change in attitude came the realization that progress depends upon provision of substantial variability at the polyploid level to allow for extensive programs of hybridization and selection among the polyploid types. It was also realized, taking into account the complexities of tetrasomic inheritance, that selection of desirable recombinants from populations created by hybridization at the polyploid level is likely to be a slow and difficult process. Nevertheless, it cannot be overlooked that so many important agricultural plants are polyploids, and this fact provides hope that persistence may ultimately be rewarded.

Induced polyploidy has another and distinct area of usefulness in plant breeding in addition to providing raw materials for the selection of novel types. In those groups of plants characterized by differences in level of polyploidy, artificial polyploidy can be used as a step in the process of transferring single valuable characteristics, such as disease resistance, from one species to another. The value of polyploidy in overcoming barriers to interspecific transfers of genes may, in the long run, prove to be its most important contribution to practical plant breeding.

Experimental Techniques for Inducing Polyploidy

The induction of polyploidy involves upsetting the normal sequence of events in nuclear division. This can be accomplished in a number of ways. For example, cold or heat shocks not infrequently lead to doubled-chromosome constitution; shoots that arise near the union of stock and scion in grafts are sometimes polyploid, as are new shoots that arise from decapitated plants. A number of chemical agents also induce polyploidy, including acenapthene, chloral hydrate, ethylmercury-chloride, and sulfanilamide. The drug colchicine, which is derived from the corms of the autumn crocus, is, however, by far the most effective agent. This drug is easily handled, because it is water soluble and produces a high proportion of polyploid cells at concentrations that are nontoxic to a wide variety of plant species. When colchicine is applied to plants in a lanolin paste, dripped onto a cotton pad held against meristematic tissue, or applied to plants in several other ways, spindle fibers fail to form in many cells. The chromosomes do not line up on the equatorial plate but divide without moving to the poles. The duplicated chromosomes then go through a regular telophase, and a membrane forms around the nucleus with a doubled chromosome number. So long as a critical concentration of colchicine is maintained in the cell, doubling is repeated again and again until, after 3 or 4 days, several hundred chromosomes may occur in some cells. If, however, the colchicine is applied for only a short time, the spindle reforms, and the polyploid cells produce daughter nuclei like themselves. As a rule only cells with tetraploid, or sometimes octoploid, numbers are able to reproduce themselves and give rise to sectors of tissue from which the plant breeder can perpetuate the polyploid strain.

Induced Autopolyploidy

Perhaps the most successful case of induced polyploidy among crop plants is autotetraploid rye with $n = 14$ instead of $n = 7$ chromosomes. The advantages of the tetraploid over its diploid counterpart are larger kernel size, superior ability to emerge under adverse conditions, and higher protein content. The tetraploid also has several disadvantages, perhaps the most serious being the necessity for isolating it from diploid rye to avoid formation of sterile triploids in the tetraploid variety. Among the other disadvantages of the tetraploid type are reduced tillering capacity (and hence poorer ability to adjust to inadequacies in stand), taller straw, making combine harvesting more troublesome, and the objection of millers to the larger kernels. The last objection arises because milling equipment must be adjusted one way for small-seeded $2n$ types and another way for large-seeded $4n$ types, thereby increasing the labor requirement in the milling operation.

Many different diploid varieties of rye have been made into tetraploids, and differences have been noted in the value of the different tetraploids relative to their diploid progenitors (Muntzing, 1951). Tetraploids equal to diploids in yielding ability were obtained only when a number of tetraploids had become available and the most promising among them subjected to selection against undesirable characteristics. The tetraploid derived from the Stub variety was the most successful. In eighteen different trials its yield varied from 97 to 103 per cent of its diploid counterpart. Different tetraploid varieties have now been intercrossed in the Swedish rye-breeding program to provide favorable material for selection against the weakness of the present tetraploids (Muntzing, 1954).

Seedless Watermelons

It might be anticipated that crops in which the seed is not the commercial product would produce better autotetraploid varieties than seed crops such as rye. In the watermelon, for example, seedlessness would be an advantage, provided some suitable way could be found to propagate the seedless strain. Diploid watermelons ($n = 11$) are fully fertile types that produce large numbers of seeds per fruit. Although tetraploids ($n = 22$) are reduced in fertility, it is possible to propagate them from seed, and they are reasonably stable genetically (Kihara, 1951). Triploids can be produced season after season by crossing tetraploids with diploids. The cross $4n\ ♀ \times 2n\ ♂$ produces viable triploid seeds, but the reciprocal cross is unsuccessful. In the triploid, normal viable gametes are produced only when one or more complete sets of 11 chromosomes are present. Disjunction at the first meiotic division in the triploid usually results in 2 chromosomes going to one pole and 1 to the other pole, so that gametes with 11 to 22 chromosomes are formed. The frequency of these gametes is given by the binomial expansion $(\frac{1}{2} + \frac{1}{2})^{11}$. Thus $(\frac{1}{2})^{11}$ of the gametes are expected to have either 11 chromosomes or 22 chromosomes, and $1 - (\frac{1}{2})^{10}$ of the gametes are expected to have 12 to 21 chromosomes. Less than 1 per cent of the gametes are therefore expected to be viable, and the triploid should be greatly reduced in fertility. This expectation is fully borne out. As a rule the triploids produce no true seeds, but only small, white, rudimentary structures similar to unripe cucumber seeds. Occasionally seeds of normal size and color, though empty, are found in the triploid fruits, rendering not strictly accurate the name "seedless" applied to these types.

Since fruit setting depends on pollination and the triploids are pollen sterile, it is necessary to interplant diploid plants among the triploids as pollinators in commercial fields. According to Kihara (1951) and Stevenson (1958), 1 diploid to about 5 triploids will suffice. The yield of the triploids has been high in Japan, and consumer acceptance has been good both there and in the United States (Stevenson, 1958). There are, however, many problems to be overcome. These include finding better ways of maintaining the $4n$ lines, improving the irregular fruit shape and tendency toward hollowness in the fruit, eliminating the empty seeds, and finding $4n \times 2n$ combinations that show greater heterosis than present combinations.   The most serious problem in triploid-watermelon culture is the large amount of hand labor required to make crosses between $4n$ and $2n$ types.   The resulting high cost of seed prohibits direct field seeding and greatly increases the cost of production by requiring that the plants be started in beds and transplanted to the field.

Another fruit that may benefit from induced autopolyploidy is the grape (Olmo, 1952).   Spontaneous autotetraploids appear not uncommonly in grapes, and some of them have been used as commercial varieties in Japan.   In California the Pierce variety, a tetraploid, has been grown successfully in field plantings.   Many tetraploids have also been induced in grapes using colchicine.   The advantages of tetraploid grapes are large berry size and fewer seeds per berry.   The disadvantages are poorly filled clusters, brittle canes and clusters, too few canes to protect the fruit from sunburn, and low total yield.   Olmo believes these disadvantages can be reduced by a program of hybridization and selection at the tetraploid level.

Autopolyploid Forage Species

The forage crops might also be expected to produce better autopolyploids than the seed crops because, in these crops, seeds are necessary only to establish new stands, and hence the problem of reduced fertility is less acute.   Alsike and red clover appear to have produced the most successful autotetraploids among the forage species.   The Swedish tetraploid strains are higher in hay yield than corresponding diploids largely because of better recovery after clipping or grazing.   Fertility, while not equal to that of diploids, is good in some strains.   Market varieties have nevertheless not been released, primarily because bees, the chief pollinators of these species, discriminate against the tetraploids.   The style of the tetraploids is long, and the bees cannot reach the nectar in the flower.   Consequently they do not revisit the tetraploids, and as a result large-scale seed production has been unsuccessful.   In addition, admixtures of diploid types tend to increase rapidly, making for difficulty in maintaining the tetraploid stocks.   As in other species, the best diploid varieties do not necessarily produce the best tetraploids.   Experience such as that gained from rye and red clover suggests that not too much should be expected of autotetraploidy in the short term, but that it may nevertheless make real contributions in the long term.

Triploid Sugar Beets

Among the root crops, triploid sugar beets appear to be farthest along the road toward successful commercial exploitation. In this species tetraploid roots tend to be smaller than the roots of diploids. The tetraploids also produce lower yields per acre. Triploidy apparently represents the optimum level of polyploidy because $3n$ plants have larger roots than diploids and also yield more sugar per unit area. Because of this triploid advantage a number of plant breeders in northern Europe and Japan have been seeking ways of producing hybrid triploid seed on a commercial scale. The sugar beet flower is small, and any practical scheme requires that the seed be produced from crossing blocks in which diploids and tetraploids have been interplanted. There are two difficulties: (1) total seed yield is high on diploid plants, but only a small proportion of the seeds are triploid; and (2) the proportion of triploid seeds is high on tetraploids, but the total seed yield is small on the low-yielding tetraploid plants. Despite these difficulties, commercial varieties of sugar beets that include a substantial proportion of triploid plants now occupy about 20 per cent of the acreage in Germany. They are also being grown in some other countries in northern Europe and in Japan. Both male sterility and genetic incompatibility are being investigated as a means of improving the efficiency of production of triploid sugar-beet seed.

Autopolyploid Ornamental Species

It is among ornamental plants that induced autopolyploidy has been most successful. Autopolyploids often have qualities different from those of diploids, and in garden plants novelty itself is often a virtue. Many autopolyploid ornamentals, however, have had valuable properties beyond distinctiveness. In some, the effect of autopolyploidy has been to increase flower size; in others, to improve keeping qualities or lengthen the blooming period. Reduced seed-setting ability and other disabilities that would be fatal to varieties of crop plants assume lesser importance in ornamentals because of the willingness of gardeners to take special pains with varieties they fancy.

Allopolyploidy

Induced allopolyploidy has found an even smaller place in practical plant breeding than induced autopolyploidy. *Triticum-Secale* and

Triticum-Agrophyron amphidiploids have received greater and more sustained attention than any other allopolyploids, especially in Sweden. Interest in *Triticale* (*T. vulgare* × *Secale cereale*) has been promoted by the possibility of producing a cereal combining the baking qualities of wheat with the hardiness of rye. This amphidiploid with $2n = 56$ chromosomes (42 from wheat and 14 from rye) has occurred naturally on a number of occasions and has also been produced artificially by a number of plant breeders. According to Muntzing (1951) *Triticale* yields about half as much as standard varieties of wheat in the early generations following the production of the amphidiploid. Fifteen generations of selection improved the yield of *Triticale* to about 90 per cent of the wheat standards. *Triticale* is not yet grown commercially in Sweden, since even the best strains are not able to compete with wheat on wheat soils or with rye on rye soils. Apparently there is no place where *Triticale* is able to compete successfully with one or the other of its parents. The main difficulty is the infertility associated with cytological and genetic instability. Also, none of the *Triticale* strains combines all the good features that occur separately in the numerous types now available. Muntzing has embarked on a program of extensive hybridization and selection at the polyploid level in an attempt to combine the desirable features in a single strain.

The amphidiploids of *Triticum* and *Agropyron* have received even more attention recently than *Triticale*. The original objective in producing these amphidiploids was to develop perennial wheats that would not need to be established every year. Although some progress may have been made in this direction, the chief contribution of this work has been the selection of wheat-like types with the excellent resistance to rusts and other diseases that is contributed by the *Agropyron* parent. The difficulties with *Triticum-Agropyron* amphidiploids are the usual ones, cytological and genetic instability, inability to stabilize chromosome numbers, and low fertility (Pope and Love, 1952).

Stebbins (1949) has produced numerous artificial amphidiploids in the genera *Agropyron, Bromus, Elymus,* and *Sitanion* and established them in various habitats of the California range. The great diversity of climatic, topographic, and edaphic features of the area provide a wide range of habitats in which these amphidiploids and their segregation products can express ecological preferences. Some of the amphidiploids have disappeared, but others promise to become established and provide new sources of forage in ecological niches where they are adapted.

Alien Substitution Races

Induced polyploidy may possibly find its greatest value in practical plant breeding as a means of overcoming interspecific sterility barriers. In some genera, transfers of desirable genes from one species to another is hindered or prevented by differences in chromosome number between species. In such instances the transfer can be made by first producing a polyploid with a lower-numbered species in the series. The best-known example is tobacco. The cultivated species, *Nicotiana tabacum* ($n = 24$), is believed to have arisen as an amphidiploid from the hybrid *N. sylvestris* ($n = 12$) $\times$ *N. tomentosa* ($n = 12$). *N. tabacum* is susceptible to tobacco mosaic, but *N. glutinosa* ($n = 12$) is resistant. *N. glutinosa* carries a different genome from either of those of *N. tabacum*, and its F_1 hybrid with the cultivated species is sterile. However, the amphidiploid *N. digluta* (*N. tabacum* $\times$ *N. glutinosa*) is reasonably fertile when crossed with *N. tabacum*, and, by backcrossing this hybrid repeatedly to *N. tabacum*, always using resistant plants, Holmes (1938) developed a resistant tobacco-like line with 24 pairs of chromosomes. The resistance of this line appeared to be governed by a single gene. Its cytological behavior was unusual in that its F_1 hybrid with varieties of *N. tabacum* formed 23 pairs and 2 univalents. Gerstel (1945) showed by monosomic analysis that the transfer of resistance had been accomplished by substituting the pair of *N. glutinosa* chromosomes carrying resistance for the H chromosome pair of *N. tabacum*. This process, by which a nonhomologous chromosome is substituted into the genome of another species, was designated by Gerstel as *alien-chromosome substitution*.

The transfer of an entire chromosome has limited the usefulness of this source of mosaic resistance. Although the H chromosome of *N. glutinosa* can substitute for the H chromosome of *N. tabacum* in function, the H chromosomes from the two sources do not pair (possibly because of structural differences). Hence all genes on the *glutinosa* chromosome are linked with mosaic resistance, and strains carrying this chromosome retain undesirable features such as the slow growth rate, low yield, and reduced leaf size that characterize *N. glutinosa*.

An example of alien-chromosome substitution with somewhat different features occurs in cotton. Beasley (1942) induced fertility in the hybrid *Gossypium thurberi*, Arizona wild cotton ($n = 13$) $\times$ *G. arboreum*, an Asiatic species ($n = 13$) by artificially doubling the chromosome number of the sterile F_1 hybrid. The genomic formulas of *G. thurberi* and *G. arboreum* are *AA* and *DD*, respectively, and that of the amphidiploid is *AADD*, the same as the genomic formula of *G. hirsutum*, American upland cotton. Pairing between the genomes derived from the different sources is far from complete, however, indicating that considerable differentiation has occurred in the long period that the amphidiploid, *G. hirsutum*, has been separated in evolution from its parental diploids. Although the induced amphidiploid is only partly fertile with *G. hirsutum*, it is possible to develop fertile lines by backcrossing the F_1 hybrid to upland cotton. One of the interesting features of this hybrid is the appearance in the backcross generations of individuals with greater fiber strength than any previously known in cultivated upland types. *G. arboreum* is not outstanding in fiber strength, and Arizona wild cotton is a lintless species. This again illustrates the fact that the results of hybrid polyploidy are unpredictable. Fiber strength is apparently controlled by a number of genes, and some of these genes are apparently located in chromosome segments that are differentiated in the parental types. Resulting reduction in crossing over seems to be the explanation for the difficulty in separating the fiber-strength genes from linked deleterious genes and also for the limited progress that has been made in utilizing the improved fiber strength in a practical way.

Alien-Addition Races

The addition of an alien chromosome to the normal complement of the host species is an alternative to alien-chromosome substitution. This technique was used by Gerstel (1945) in developing a mosaic-resistant race of *N. tabacum* by adding to the complement of the cultivated species the H chromosome pair from *N. glutinosa*. The addition race was stocky and had short broad leaves. In general, however, its agricultural features were less disturbed than those of the comparable alien-substitution race. Alien-addition races have also been developed in wheat by the addition of rye chromosomes to the wheat complement and have been produced in cotton by adding chromosomes from diploid species to the complement of *G. hirsutum*.

Transfer of Small Chromosome Segments

Whole-chromosome substitutions or additions have usually been found to be unsatisfactory from a practical point of view, apparently because so much genetic material is introduced from the donor species that undesirable characters are brought along with the wanted characteristic. In the transfer of resistance to leaf rust from *Aegilops umbellulata* to common wheat, Sears (1956) has demonstrated a method for introducing small segments of chromosome from the foreign species. *Ae. umbellulata* ($n = 7$) is a wild grass from the Mediterranean region whose genomic designation is CC. The genomic formula of wheat is $AABBDD$. It has not been possible to obtain germinable seeds when this species is crossed as either male or female parent with common wheat. An amphidiploid of *T. dicoccoides* ($n = 14$, genomes A and B) with *Ae. umbellulata* was used as a bridge for combining the genomes of *Ae. umbellulata* and common wheat. Two backcrosses of the F_1 hybrid to common wheat produced a plant that was resistant to leaf rust and carried a single added chromosome from *Aegilops*. This added chromosome had a deleterious effect both on the plant and on viability of pollen. Some of the progeny of this plant were X-rayed prior to meiosis, and the irradiated pollen was used to pollinate wheat plants. Of 6091 offspring, 132 were resistant to leaf rust, including 40 in which part of the *Aegilops* chromosome was translocated to a wheat chromosome. At least 17 different translocations were involved among these 40 plants. Data from one of these translocations provided evidence indicating that the gene for leaf-rust resistance was located near the kinetochore. Another translocation was an intercalary type that showed essentially normal pollen transmission and was cytologically undetectable. Since homozygous plants were distinguishable from normal ones only by their rust resistance and slightly later maturity, the translocated segment from *Aegilops* was apparently a small one that included little genetically active material besides the gene for leaf-rust resistance.

Résumé

Several points having a bearing on the future use of polyploidy emerge from this brief review of the most successful examples of induced polyploidy in practical plant breeding. First, it is now clear that only a small proportion of diploids produce successful autopolyploids. Success cannot be predicted in advance because poor diploids sometimes make better autopolyploids than good diploids. The most common defects in induced autopolyploids are slower growth rate, genetic instability, and reduced fertility.

Second, induced allopolyploids have been even less successful as new crop plants than induced autopolyploids because of sterility and other disabilities resulting from interactions between the genetic materials of the parental species.

Third, new polyploids can rarely be expected to achieve immediate success (except possibly with ornamentals in which large flower size, improved texture of petals, or novelty itself are able to overbalance the disadvantages accompanying polyploidy). In crop plants, the breeding of polyploids should therefore provide for hybridization and selection at the polyploid level. If this is done, it may be possible to assemble, in one or a few types, desirable genes dispersed in many different raw polyploids derived from different diploid strains.

Fourth, although allopolyploidy cannot be expected to produce striking improvements in established, highly selected crops, it does provide novel genetic variability from which it may be possible to develop entirely new crops.

Finally, it seems possible that induced polyploidy may find its greatest usefulness in overcoming interspecific sterility barriers and hence in aiding the transfer of desirable genes from one species to another.

REFERENCES

Beasley, J. L. 1942. Meiotic chromosome behavior in species, species hybrids and induced polyploids of *Gossypium*. *Genetics* 27: 25–54.

Blakeslee, A. F., and A. G. Avery. 1937. Methods of inducing chromosome doubling in plants. *Jour. Hered.* 28: 393–411.

Gerstel, D. U. 1945. Inheritance in *Nicotiana tabacum*. XIX. Identification of the *tabacum* chromosome replaced by one from *N. glutinosa* in mosaic-resistant Holmes Samsoun tobacco. *Genetics* 30: 448–454.

Gerstel, D. U. 1945. Inheritance in *Nicotiana tabacum*. XX. The addition of *Nicotiana glutinosa* chromosomes to tobacco. *Jour. Hered.* 36: 197–206.

Holmes, F. O. 1938. Inheritance of resistance to tobacco-mosaic disease in tobacco. *Phytopath.* 28: 553–561.

Kihara, H. 1951. Triploid watermelons. *Proc. Amer. Soc. Hort. Sci.* 58: 217–230.

Muntzing, A. 1951. Cytogenetic properties and practical value of tetraploid rye. *Hereditas* 37: 2–84.

Muntzing, A. 1954. Genetics in relation to plant breeding. *Proc. Indian Acad. Sci.* 34: 227–241.

Olmo, H. P. 1952. Breeding tetraploid grapes. *Proc. Amer. Soc. Hort. Sci.* 59: 285–290.

Pope, W. K., and R. M. Love. 1952. Comparative cytology of colchicine-induced amphidiploids of interspecific hybrids. *Hilgardia* 21: 411–423.

Randolph, L. F. 1941. An evaluation of induced polyploidy as a method of breeding crop plants. *Amer. Nat.* 75: 347–363.

Richmond, T. R. 1951. Procedures and methods of cotton breeding with special reference to American cultivated species. *Adv. in Gen.* 4: 213–245.

Sears, E. R. 1956. The transfer of leaf-rust resistance from *Aegilops umbellulata* to wheat. *Brookhaven Symposia in Biology* No. 9: 1–22.

Stebbins, G. L., Jr. 1949. The evolutionary significance of natural and artificial polyploids in the family *Gramineae*. *Proc. Eighth Intern. Congr. Gen.*, 461–485.

Stebbins, G. L., Jr. 1956. Artificial polyploidy as a tool in plant breeding. *Brookhaven Symposia in Biology* No. 9: 37–52.

Stevenson, E. C. 1958. Seedless watermelons. *Amer. Vegetable Grower* 6(3): 11.

Unrau, J., C. Peterson, and J. Kuspira. 1956. Chromosome substitution in hexaploid wheats. *Can. Jour. Bot.* 34: 629–640.

33

Cytology and Genetics of Interspecific Hybrids

From a genetic point of view, the union of any two gametes differing in allelic constitution at one or more loci produces a hybrid. Thus many degrees of hybridity are possible and, as we shall see, many different kinds of hybrids. The most common type of hybridity is that present in individuals of sexually reproducing cross-pollinated species that share the same gene pool. Such individuals are almost certain to be heterozygous or "hybrid" at many loci, and upon mating they produce variable offspring as a result of segregation and recombination. Whenever there are restrictions on free exchange of genes between subpopulations, genetic differentiation must sooner or later take place. In general, therefore, crosses between individuals from populations that have been isolated from one another in reproduction for a period of time display greater hybridity than crosses between individuals that share the same gene pool.

Genetic Disability and Reproductive Incapacity

When two gene pools have been separated in reproduction for many generations and have become differentiated in gene frequency to the point where the two populations have been assigned subspecific rank, hybridization between members of the two groups often has two properties in addition to ordinary segregation and recombination. The first of these properties is *genetic disability* in the F_2 and later generations. The genetic consequences of segregation in crosses between subspecies are segregation and recombination of the alleles responsible for important differences in morphological appearance, physiological responses, and adaptation that separate the subspecies. If the parental gene pools represent balanced combinations of genes

423

(Chapter 16), the chances are high that these new combinations of genes will be inferior to the original ones, and the hybrid swarm will be on the average less fit than either of the parental types.

When genetic differentiation has proceeded still further and the two populations have achieved specific rank, the hybrid usually acquires still another property, the property of *reproductive incapacity*, or sterility, in the F_1 generation. If the differences between the two species are entirely ones of morphological differentiation and ecological preference, the F_1 hybrid may be fully fertile and different from sub-species hybrids only in the proportion of disharmonious types appearing in the segregating generations. More often, however, interspecific hybrids suffer a loss in reproductive capacity, with both F_1 and later generations showing a greater or lesser degree of hybrid sterility. In addition, the segregating generations usually display weaknesses associated with genically controlled physiological disturbances (genetic disability). Hybrid sterility is not, however, confined to interspecific crosses. Sterility in various degrees occurs in intraspecific crosses in certain genera, although the sterility is usually not as pronounced as in crosses between species. Interspecific hybrids therefore differ quantitatively rather than qualitatively from hybrids within species even in sterility, which is the most significant characteristic of species crosses.

Interspecific hybrids run the gamut from complete fertility to complete sterility. In general, however, they can be divided into two groups: (1) those capable of producing some viable eggs or pollen, and consequently capable of being propagated by selfing, intercrossing, or backcrossing to the parental species; and (2) those that cannot be propagated except by vegetative means or by making them into allopolyploids. The possibilities in the practical breeding of hybrids that produce even occasional offspring are of course quite different from the possibilities of the completely sterile hybrids.

Barriers to Interspecific Hybridization

The first concern in breeding programs involving interspecific hybridization is whether or not a viable F_1 hybrid plant can be obtained. No certain guide aside from actual trial has been discovered. In general, however, close relationship as indicated by taxonomic status is the most useful criterion of crossability. This correlation is far from complete. For example, the common cabbage, *Brassica oleracea*, will cross with the radish, *Raphanus sativus*, but not with the more closely related turnip, *Brassica rapa*, or with rape, *Brassica napus*.

Failure to obtain thrifty F_1 hybrids can result from genetic or from cytoplasmic incompatibilities that are expressed either in failure of fertilization or in death of the zygote at any stage between early cleavage divisions and maturity. In the genus *Datura*, pollen tubes sometimes burst in the styles of foreign species. In *Nicotiana* and in hybrids between *Zea mays* and *Tripsicum* species, failure of fertilization sometimes results from the fact that the style of the maternal species is much longer than the style of the pollen parent. Pollen tubes not adapted to traversing the great length between stigma and ovary therefore fail to effect fertilization. In addition, the thick pollen tubes of polyploid species sometimes have difficulty growing in the slender styles of diploids or lower polyploids. This type of barrier to fertilization can sometimes be overcome by making the cross in the reciprocal direction or by shortening the style artificially.

Barriers to interspecific hybridization that act by preventing the growth of the F_1 hybrid are common in plants. Hybrid inviability can be caused by single genes, by general incompatibility of the genotypes of the parents, or by disharmonies between the developing embryo and the endosperm. A few examples of these types of barriers follow.

In 1930, Hollingshead reported that hybrids between *Crepis tectorum* and *C. capillaris* died in the cotyledonary stage or reached maturity, depending on whether the *C. tectorum* parent carried the lethal gene *l* or its nonlethal allele *L*. The gene *l* also caused the early death of *C. tectorum* hybrids with *C. bursifolia* and *C. leantodontoides*. The gene *l* had no discernible effect on *C. tectorum* itself. Similar situations have been described in many other interspecific crosses and also in both intervarietal crosses (e.g., in wheat) and intergeneric crosses. In the cross between Einkorn wheat and *Aegilops umbellulata*, Sears (1944) distinguished three alleles, L^e for early lethality, L^l for late lethality, and *l* for viability.

Lethality is common in hybrids between species of *Gossypium*. In an extensive study of the wild American species, *G. gossypioides*, Brown and Menzel (1952) found that all but one of the interspecific hybrids that could be obtained with this species died in some stage ranging from early embryogeny until after meiosis had taken place. Hybrids of another wild American species, *G. davidsoni*, with *G. barbadense*, *G. hirsutum*, *G. stocksi*, and *G. thurberi* die in the cotyledonary stage (some other hybrids with *G. davidsoni* fail in earlier stages). This behavior was attributed by Silow (1941) to general genotypic disharmony rather than to the action of specific lethal genes. Stephens (1950) and Gerstel (1954), however, have been able to demonstrate that hybrid inviability is controlled by simple genetic mechanisms in some of the other interspecific hybrids in cotton.

Differences between reciprocal crosses are often observed when the species that are hybridized do not have the same number of chromosomes. Greater success usually attends hybridizations in which the species with the larger number of chromosomes is the seed parent. In the cross of hexaploid × tetraploid wheats, the combination $(4x)$ ♀ × $(6x)$ ♂ often produces badly shriveled seeds which fail to germinate. The seeds from the combination $(6x)$ ♀ × $(4x)$ ♂ are reduced in size but nearly normal in germination. Thompson (1930) attributed the difference to the chromosome balance of the endosperm. The endosperm in the combination $6x$ ♀ by $4x$ ♂ has $21 + 21 + 14$ chromosomes, or **88** per cent of the chromosome complement of the hexaploid endosperm. In the reciprocal cross, the endosperm has $14 + 14 + 21$ chromosomes, or only **78** per cent of the hexaploid complement. The abnormal endosperm may be due in part to quantitative relationships between the chromosome numbers of the endosperm and embryo, but it seems likely that unbalance of gene dosage may be responsible for some or perhaps even all the difficulty in some crosses.

If the growth of the embryo is inhibited by the endosperm, the hybrid can sometimes be obtained by dissecting out the young embryo and growing it in nutrient culture. Success was obtained in this way by Brink, Cooper, and Ausherman (1944) in the hybrid between *Secale cereale* and *Hordeum jubatum*. It has also been used with success by Blake in making peach hybrids, by Laibach in flax, and by various investigators working with nonagricultural species.

Certain interspecific hybrids in *Melilotus* fail because the F_1 hybrid seedlings are albinistic defectives. Smith (1945) overcame this barrier by grafting the albino seedlings to one of the parental species, whereupon they grew to maturity and produced viable seeds.

In a number of genera there are barriers of hybrid weakness that result from the interaction of certain hybrid genotypes and the cytoplasm of one of the parental species. This subject has been reviewed in detail by Caspari (1948). Good reviews of the general subject of hybrid inviability and its genetic basis are those of Stebbins (1950, 1958).

Cytological and Genetic Basis of Hybrid Sterility

The most characteristic feature of interspecific hybrids is sterility in greater or lesser degree. Although the causes of hybrid sterility are probably numerous, all seem to have their basis in the same kind of genetic unbalance that is responsible for hybrid inviability. If the unbalance affects metabolism in early stages of development or affects later vegetative stages of development, the result is hybrid inviabilty. If, however, the unbalance is not expressed until gametogenesis, the result is hybrid sterility.

Sterility that becomes apparent only at the time of the formation of the gametophytes or gametes can be divided into two types, chromosomal and genic. There is a long list of species hybrids in which reduced pairing, presence of ring and chain configurations, bridge-fragment configurations, and other cytological abnormalities indicate lack of structural homology between the chromosomes of the parental species. These cytological configurations can be interpreted as indicating heterozygosity for translocations, inversions, and differences in the position of the centromere. The disorganized disjunction in structurally heterozygous hybrids usually causes less than a full complement of chromosomes to be partitioned to each gamete in the meiotic process, and, as a result, many or all gametes are nonfunctional. The demonstration that amphidiploids derived from certain hybrids become fertile once each chromosome has a fully homologous mate with which to pair shows that the sterility of the F_1 is sometimes due entirely to chromosomal sterility and not to genic unbalance.

Structural differentiation of chromosomes has been observed in all degrees in sterile interspecific hybrids. In some hybrids the differentiation has proceeded to the point where pairing fails completely. At the other extreme, the F_1 hybrid may be highly sterile, even though bivalents are formed regularly at meiosis and disjunction is normal. Sterility associated with completely normal pairing of chromosomes is usually called *genic* sterility. Stebbins (1950, 1958) has emphasized the difficulties in distinguishing between genic and chromosomal sterility. In most interspecific hybrids, chromosome pairing has been analyzed at the first meiotic metaphase, when the chromosomes are strongly contracted and are associated at only a few points along their length. Even quite large structural differences may not be detected at this stage. Stebbins has presented evidence for the widespread existence of chromosomal sterility due to heterozygosity for structural differences too small to affect pairing materially and has proposed a name for this phenomenon: *cryptic structural hybridity*.

Perhaps the best evidence for genic sterility has been presented by Oka (1955). In cultivated rice, hybrids between different varieties, and especially hybrids between varieties of the subspecies *indica* and *japonica,* show varying degrees of sterility. There is no difficulty in hybridizing different subspecies. Although the F_1 subspecific hybrids show no disturbances in chromosome pairing, both megaspores and microspores deteriorate shortly after meiosis in some hybrids. Oka studied segregation ratios in a number of hybrids and concluded that a series of genes affecting development of the gametophyte are responsible for the sterility. These genes appeared to act in a complementary fashion so that the presence of two recessive alleles in one megaspore or microspore inhibits full development. Oka found certain of the sterility genes to be linked with genes governing known morphological characters. Mashima and Uchiyamada (1955), however, have reported that induced intervarietal tetraploids have a smaller number of quadrivalents and a higher frequency of bivalents than autotetraploids derived from a single variety. This was taken as an indication that structural differentiation has occurred between chromosomes of *japonica* and *indica,* and that Oka's gametophyte development genes may be rearrangements of very small chromosome segments.

Clearly it is difficult to distinguish between genic and chromosomal sterility. For the present, it seems safe to conclude only that the disharmonious effects of gene recombination and the effects of deficiencies, duplications, and other alterations of chromosomal segments operate singly or jointly to produce hybrid sterility. For a comprehensive discussion of the subject of hybrid sterility, the reader is referred to the review of Stebbins (1958).

Segregation in Interspecific Hybrids

Before we discuss specific examples of the role of interspecific hybridization in plant breeding, some generalizations about segregation in wide crosses must be introduced. The first of these generalizations relates to the tremendous diversity of types that appear in the F_2 and later generations as a result of the extreme heterozygosity of interspecific F_1 hybrids. Each individual in an F_2 progeny is likely to be different from each other individual in a large number of characters. Moreover, many segregants are likely to have entirely new characteristics, different from those of either parent, that could not have been predicted from a study of the morphology or physiology of the parental species.

The second of these generalizations pertains to the difficulty in accounting in precise Mendelian terms for inheritance in species crosses. The meiotic process rarely functions with complete normality in interspecific hybrids. Thus the mechanism causing genes to assort to the gametes with great precision in intervarietal crosses breaks down; and even when pairing appears to be fairly regular, segregation often does not fit classical Mendelian patterns.

Third, although segregation in the F_2 or later generations produces extremely heterogeneous recombination types, the recombinations that actually appear are by no means a random sample of the total possible recombinations of parental characteristics. This has been emphasized by Anderson (1939) in his analysis of recombination products produced by the hybrid between *Nicotiana alata* $\times$ *N. langsdorffi*. These two species differ in many visible characteristics, for example, in size and shape of the flowers and flower parts (Figure 33–1).

If the recombinations of any two characters are considered, the ones actually observed form, more or less, an ellipse running diagonally across the correlation rectangle from one parental combination to the other (Figure 33–2). The distribution of various combinations of characters studied by Anderson suggested that even in an infinitely large F_2 the combinations would fall far short of reaching the upper left-hand and lower right-hand corners of the correlation diagrams. The actual recombinations in F_2, though extremely diverse, in reality represent only a narrow segment of the total imaginable recombinations that would occur with complete shuffling of the traits of the two species. In Figure 33–3, data for three characters have been idealized and combined into a three-way correlation diagram showing the relation between the total possible recombinations of the three characters and actual recombinations obtained in experiment. The variability that seemed impressive in the F_2 generation becomes surprisingly small when compared to the recombinations that might have occurred (Figure 33–1). This suggests that powerful restrictions to recombination are operating in interspecific hybrids. Anderson believes these restrictions are caused by gametic elimination, zygotic elimination, pleiotropy, and linkage.

The fourth and final generalization is based on the fact that male gametogenesis is more easily upset by chromosomal or genic disharmonies than female gametogenesis. For this reason, the propagation of hybrids often depends on backcrossing the F_1 as the seed parent to one or the other of the parental species. Moreover, when the F_1 is pollen fertile and can be self-pollinated, the offspring may still be more like backcross types than true F_2 segregates. There are two reasons for this situation. First, functional male gametes often have genotypes very similar to the genotypes of one or the other parental species. Since male gametes with genes contributed by both parents tend to be

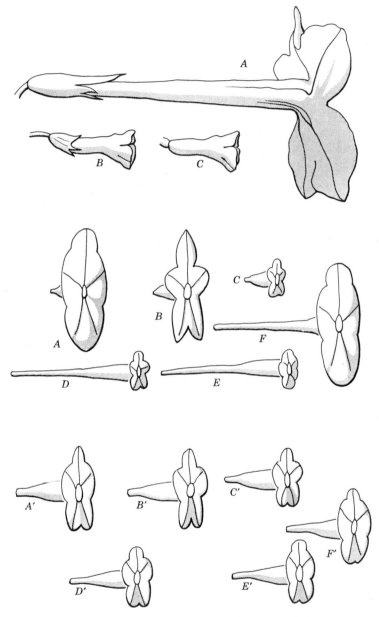

FIGURE 33-1. Top, representative flowers of (A) *Nicotiana alata* and (B) *Nicotiana Langsdorffii*; (C) corolla of *N. Langsdorffii*. Middle, extreme recombinations to be expected with complete recombination of tube length, limb width and lobing. Bottom, actual extremes obtained among 347 F₂ plants. (After Anderson, 1939.)

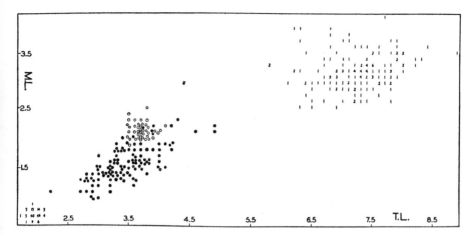

FIGURE 33-2. Correlation between limb width (maximum lobe) and tube length in *Nicotiana Langsdorffii, N. alata,* one F₁ family of 41 plants and one F₂ family of 118 plants. Open circles represent the F₁, solid circles the F₂; numbers represent *N. Langsdorffii* (lower left) and *N. alata* (upper right). (After Anderson, 1939.)

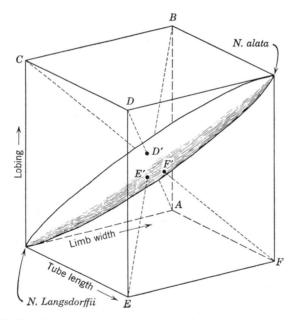

FIGURE 33-3. Diagram of correlation cube for three characters, showing the relation between complete recombination (total cube) and actual recombination (central spindle). Letters mark the positions of the ideal (*A* to *F*) and actual (*A′* to *F′*) extreme recombinations illustrated in Figure 33-1. Three recombinations (*A′, B′,* and *C′*) are supposed to be on the far side of the solid spindle. (After Anderson, 1939.)

inviable, self-fertilization thus has much the same effect as backcrossing to one of the parents. This situation has been observed in extreme form in the hybrid of *T. vulgare* × *T. timopheevi* and in certain crosses in *Collinsia,* where the only functional male gametes appeared to be nearly identical to those of one of the parental species. Of course, any tendency toward inviability of intermediate types of gametes, whether on the male or the female side, leads to an excess of parental zygotes and a deficiency of recombination zygotes. The second reason relates to the effect of natural selection on viable recombination zygotes. The combinations of characters represented by parental species are likely to be adaptive ones, and the chances that any new combinations will be equally adaptive are small. Hence the nearer a segregant is to one or the other parental species in characters of adaptive value, the better its chance of perpetuation, whether in nature or in a plant breeder's nursery.

REFERENCES

Anderson, Edgar. 1939. Recombination in species crosses. *Genetics* 24: 668–698.
Brink, R. A., D. C. Cooper, and L. E. Ausherman. 1944. A hybrid between *Hordeum jubatum* and *Secale cereale* reared from an artificially cultivated embryo. *Jour. Hered.* 35: 67–75.
Brown, M. S., and M. Y. Menzel. 1952. Additional evidence on the crossing behavior of *Gossypium gossypioides. Bull. Torrey Bot. Club* 79: 285–292.
Caspari, E. 1948. Cytoplasmic inheritance. *Adv. in Gen.* 2: 1–66.
Dobzhansky, Th. 1941. *Genetics and the origin of species.* Columbia University Press, New York.
Gerstel, D. V. 1954. A new lethal combination in interspecific cotton hybrids. *Genetics* 39: 628–639.
Hollingshead, L. 1930. A lethal factor in *Crepis* effective only in an interspecific hybrid. *Genetics* 15: 114–140.
Mashima, I., and Uchiyamada, H. 1955. Studies on the breeding of fertile tetraploid plant of rice. *Bull. Nat. Inst. Agric. Sci.* (Japan), Ser. D, 104–136.
Oka, H. 1955. Population genetics of rice and barley. *Ann. Rept. Nat. Inst. Genetics* (Japan), 1954 (5): 40–50.
Sears, E. R. 1944. Inviability of intergeneric hybrids involving *Triticum monococcum* and *T. aegilopoides. Genetics* 29: 113–127.
Silow, R. A. 1941. The comparative genetics of *Gossypium anamalum* and the cultivated Asiatic cottons. *Jour. Gen.* 42: 259–358.
Smith, W. K. 1948. Transfer from *Melilotus dentata* to *M. alba* of the genes for reduction in coumarin content. *Genetics* 33: 124–125.
Stebbins, G. L., Jr. 1950. *Variation and evolution in plants.* Columbia University Press, New York.

Stebbins, G. L., Jr. 1958. The inviability, weakness, and sterility of interspecific hybrids. *Adv. in Gen.* 9: 147–215.

Stephens, S. G. 1950. The genetics of "corky." II. Further studies on its genetic basis in relation to the general problem of interspecific isolating mechanisms. *Jour. Gen.* 50: 9–20.

Thompson, W. P. 1930. Causes of differences in success of reciprocal interspecific crosses. *Amer. Nat.* 64: 407–421.

34

Interspecific Hybridization in Plant Breeding

Early records of the use of interspecific hybridization in practical plant breeding are meager, although there are indications that some work had been done by the beginning of the eighteenth century. The first apparently authentic record appeared in 1717 when Thomas Fairchild reported a cross between the carnation and the sweetwilliam. This was a period when considerable interest had been aroused by the demonstration by Camerarius that plants are sexual organisms. However, the chief interest of the botanists of this period was not so much the production of better plants as the accumulation of evidence on sexuality in the plant kingdom.

Fairchild's hybrid received scant attention from the scientists of the day, but it aroused considerable interest among gardeners. The vast increase in new strains of garden plants that appeared in the eighteenth century and the lack of written records about these strains suggest that countless amateurs carried out breeding work, especially with flowers, once the existence of sex in plants had become widely known and accepted. The following résumé of the number of listed varieties in a few of the more common flowers gives some idea of the scope of this and later work: roses, 15,000; tulips, 8000; dahlias, 7000; gladioli, 2500; irises, 4000. Many of these varieties originated in backyard gardens. Consequently the hybrids from which they were selected were not always recorded. It has been possible to deduce from modern studies, however, that interspecific hybridization played a major role in the origin of modern types in some plant groups, although its role has been slight in others. Perhaps the best guide to future successful use of interspecific hybridization is information on the part it has played in the past and the factors that have influenced its usefulness in different types of plants. This subject has been considered in detail by Stebbins (1950), whose arguments are the basis of the following discussion.

Interspecific Hybridization in Ornamental, Orchard, Field, and Vegetable Crops

Among ornamental, orchard, field, and vegetable species, interspecific hybridization has been by far the most important in the ornamentals. A high proportion of the most popular perennial herbs and shrubs—for example, rhododendrons, irises, orchids, cannas, dahlias, gladioli, roses, poppies, and violets—are hybrid in origin. Wide crosses, including intergeneric hybrids, have been frequent in these ornamentals, especially among roses, orchids, and lilies (Emsweller et al., 1937).

Hybridization has been next in importance in the orchard and arbor crops (apples, plums, cherries, filberts, grapes, and various kinds of berries). It has been less important in the forage crops, and still less important in the cereal, fiber, and oil crops. Many of the field crops, such as wheat, oats, cotton, tobacco, and sugar cane, are very old allopolyploids originally derived from hybrids between highly differentiated species. The recent role of hybridization in the improvement of these old, established polyploids has been through the transfer of individual characters from one species to another and not through the selection of strikingly different types in the segregating generations. It is in the vegetable crops that hybridization between distinct species has been least important of all. Asparagus, beans, beets, carrots, celery, lettuce, Lima beans, and tomatoes are each derived from a single diploid species, and interspecific hybridization has played little if any role in their improvement. The exceptional vegetables are the potato and the sweet potato.

Stebbins believes that the differences between the classes of crops can be explained on the basis of two assumptions: first, that hybridization is more valuable in vegetatively propagated species than in those that reproduce by seed; second, that the value of hybridization is less when particular attributes of quality are more important than total quantity of yield. He believes the first factor explains the greater importance of interspecific hybridization in ornamentals and orchard crops contrasted to field and vegetable crops. Ornamental and orchard types are often propagated entirely by vegetative means, and hybrids are valuable because once a superior hybrid is obtained it can be perpetuated indefinitely. This is usually not possible in field or vegetable crops.

He believes differences in demands of quality explain the greater importance of hybridization in ornamentals contrasted to orchard crops and in field crops contrasted with vegetable crops. In garden flowers and shrubs, the demand is not for particular qualities but for values that can be achieved in various ways. As was noted previously, novelty itself is often desirable in ornamentals, and this is the very attribute of quality most likely to be achieved by interspecific or intergeneric hybridization. The demands of quality in fruits, on the other hand, are generally conservative in that the traditional flavors usually must be retained. Moreover, shipping and canning characteristics are exacting and particular. Hence wide crosses have not been so important in orchard species, because it is not so easy to obtain individual genotypes possessing the necessary attributes from interspecific as from intervarietal crosses.

In many field crops, especially the forage species, the demands of quality can be met in many ways. Here quantity of yield is often the most important characteristic, since animals can be raised successfully on pasturage or hay from various legumes, grasses, or forbs. In cotton, wheat, and oil crops and in vegetable crops, on the other hand, the requirements of quality are specific. The breeder, therefore, usually cannot destroy existing genic systems by introducing large numbers of genes from different species. In such crops, interspecific hybridization has been used largely to transfer specific genes from one species to another by a combination of hybridization, backcrossing, and selection. By these processes the characteristics of the economically important parent can be combined with one particular character—for example, disease resistance—or at most with a few characters from the other parent. Let us now turn to some specific examples from practical plant breeding to illustrate these general features of interspecific hybridization.

Interspecific Hybridization in Tomatoes

Interspecific hybridization in the genus *Lycopersicon* is particularly interesting because many different hybrids can be secured, and most of the hybrids that can be made are sufficiently fertile to permit extensive studies of their progeny. The genus *Lycopersicon* is naturally divided into two subgenera on the basis of color of fruit, position of the stigma, and some other characters. The red-fruited subgenus *Eulycopersicon* includes two species, *L. esculentum* and *L. pimpinellifolium;* the green-fruited subgenus *Eriopersicon* includes the species *L. hir-*

sutum, L. peruvianum, L. chilense, and two or three other species of less interest in interspecific hybridization.

The most extensively studied hybrid is that between the two species in *Eulycopersicon, L. esculentum* and *L. pimpinellifolium.* According to Rick and Butler (1956), no barriers of any consequence exist to the hybridization of these species or, in fact, to the production of progeny by the hybrid or to gene recombination in the hybrid. Although chiasma frequency and pollen fertility are very slightly reduced compared with most intervarietal crosses within *L. esculentum,* genes can be exchanged so freely between these species that most workers are inclined to view *L. pimpinellifolium* as a variety of *L. esculentum.* The attention directed to this hybrid can be attributed in part to the practical objective of transferring desired traits, particularly disease resistance, from *L. pimpinellifolium* to horticultural varieties of *L. esculentum.* Monogenic resistance to *Cladosporium fulvum, Fusarium oxysporium,* and *Stemphylium solani* have been transferred, and there have also been reports of transfers of disease resistance of more complex or unknown genetic determination.

Hybrids between the two members of the subgenus *Eulycopersicon* and the several members of *Eriopersicon* have several features in common. The crosses succeed only when species of *Eriopersicon* are used as the male parent. A similar limitation also applies to the F_1 hybrids in that they are successfully mated to *L. esculentum* only as the male and to the wild parent only as the female. A complicating factor in these interspecific hybrids is the self-incompatibility of most forms in *Eriopersicon* and the transmission of this incompatibility to all hybrids as if it were a dominant trait. All known hybrids between the two subgenera are characterized by remarkable regularity of chromosome behavior. Deviations from normal bivalent formation are scarcely any greater than those of the hybrid between *L. esculentum* and *L. pimpinellifolium.*

Hybrids of *L. esculentum* with *L. hirsutum* are highly fertile. Although transmission of certain marker genes shows small deviations from expected ratios in test crosses and indicates some type of upset in normal segregation, there is little difficulty in transferring genes from *L. hirsutum* to garden tomatoes. Among the economically interesting genes that have been transferred are a dominant gene governing resistance to *Septoria lycopersici* and a gene that causes a sharp change in content of carotenoid pigment.

Hybrids of *L. esculentum* with *L. peruvianum* and the closely related species *L. chilense* hold the greatest promise of contributing useful traits to horticultural varieties of tomatoes. These two species, according to Rick and Butler, offer promising resistance to numerous diseases and root-knot nematode, as well as a source of genes for high vitamin C content. In hybridizing *L. peruvianium* with *L. esculentum*, fruits are obtained only if *L. esculentum* is the pistillate parent, and even then it is necessary to resort to embryo culture to establish seedlings. It was first believed that the fertility of F_1 hybrid plants was very low. McGuire and Rick (1954) demonstrated, however, that the high level of incompatibility of the *L. peruvianum* parent also exists in F_1 hybrids and that, if two F_1 plants of different genotypes are crossed, seed yield is at least 10,000 greater than if single F_1 plants are selfed. Sib-mated F_1 hybrids are in fact reasonably fertile, yielding about 10 per cent as much seed as either parent. Backcrosses to the *L. esculentum* parent are, however, impeded by a barrier of incompatibility almost as severe as that separating the two parents. Some workers have obtained a few seeds of this backcross, but others have had to resort to embryo culture or have failed completely. Despite these difficulties, genes conferring resistance to *Septoria lycopersici* and to root-knot nematode have been transferred to *esculentum* types from *L. peruvianum*. Resistance to nematode is associated with late maturity and fruit cracking, presumably because of tight linkages of unfavorable genes with the resistance-conferring gene *Mi*.

F_1 hybrids of *L. esculentum* × *L. chilense* are more easily obtained than those with *L. peruvianum*. Later generations from the *L. chilense* cross are also higher in fertility. Otherwise the *esculentum-chilense* hybrid behaves much like the *esculentum-peruvianum hybrid*.

Interspecific Hybridization in Wheat

Further consideration of the wheat species will illustrate some of the features of interspecific transfers of genes between species with different chromosome numbers. It might be anticipated that the morphological characters distinguishing the Emmer wheats from Einkorn group are carried in the *B* genome and hence are untransferable from Emmers ($n = 14$) to Einkorns ($n = 7$). In a similar way it might be expected that the *D* genome would carry many of the genes which distinguish bread wheats ($N = 21$) from the Emmer group ($n = 14$), and that such genes also cannot be transferred to the group with the lower chromosome number. Thompson et al. (1935) have shown that this does represent the situation in a broad way. Thus it is relatively simple to transfer genes to species at the next higher level of polyploidy, but the reverse transfer can be carried out only for genes carried in genomes common to both species. Fortunately, in genera represented by different levels of polyploidy, the important cultivated species are generally the higher polyploids.

Interspecific hybridization in wheat has almost always had as its goal the transfer of genes for disease resistance. Some of the best stem-rust resistance is found in the Emmer series of wheats, and the transfer of this resistance has claimed the attention of many different wheat breeders. The transfer of the excellent stem-rust resistance of the durum wheat variety Iumillo to common wheat was an early and important step in the breeding of rust-resistant wheat varieties. Hayes et al. (1920) crossed Iumillo with Marquis wheat and from the progeny selected a stem-rust-resistant bread wheat they named Marquillo. Marquillo, although never a successful variety in its own right, was one of the parents of Thatcher, a variety released in 1934 and still widely grown in the spring-wheat belt of the United States and Canada.

Although a number of wheat breeders have made crosses between Iumillo and different varieties of *T. vulgare,* none of the *vulgare*-like derivatives have possessed the full resistance of Iumillo. This may be because stem-rust resistance is governed by many genes, and no one has succeeded in transferring all the genes, major and minor, that condition stem-rust resistance. Peterson and Love (1940) suggested an alternative explanation, that the difficulty in obtaining a fully resistant *vulgare* type is due to genes in the D genome that inhibit or reduce the effect of the genes for resistance. If this is true, it would be possible to capture the full resistance of Iumillo only if crossing over occurs between chromosomes in the A or B genomes with chromosomes of the D genome. In support of this possibility it should be recorded that a little homology remains between chromosomes of the different genomes. This evidence comes from the observation that some pairing occurs in haploid common-wheat plants.

The first transfer of resistance genes from *T. dicoccum* to *T. vulgare* was by McFadden in 1930, who selected Hope and H-44 from a hybrid between Yaroslav Emmer and Marquis. Although Hope and H-44 never achieved prominence on their own because of agronomic deficiencies, they have probably been used more widely as parents in crosses designed to produce stem-rust-resistant varieties than any other resistant type. In addition to their excellent stem-rust resistance, these varieties are also resistant to many races of leaf-rust, loose-smut, and bunt diseases. In general, transfer of genes is more difficult from *T. dicoccum* to *T. vulgare* than transfers from *T. durum* to *T. vulgare* because of the lesser chromosome homology and lower fertility of the former hybrid.

The Emmer-like wheat species *Triticum timopheevi* offers an additional source of disease resistance to wheat breeders. It will be recalled that its genomic formula is *AAGG*, rather than *AABB* as in the typical Emmers. Shands in 1941 succeeded in transferring the stem-rust resistance of this species to common winter-wheat types. The usefulness of the resistant lines was restricted, however, by tight linkage between the genes for stem-rust resistance and certain genes for late maturity. An additional transfer of disease resistance from *T. timopheevi* was made in 1949 by Allard, who paid particular attention to the efficiency of different breeding procedures in making the transfers.

Backcrosses made by using pollen of the F_1 on *vulgare* were nearly indistinguishable from the F_1 in morphological appearance, cytological behavior, and fertility. It was concluded that the only functional male gametes produced by the F_1 had almost exactly the same chromosome complement as *T. timopheevi*. Backcross plants obtained using pollen of *vulgare* on the F_1 were, however, intermediate between the F_1 and *vulgare*, and the resemblance to *vulgare* increased through four backcrosses made in this same direction. When, however, the second backcross was made using first-backcross plants as the pollen parent and *vulgare* as the seed parent, the return toward *vulgare* type was much more rapid than for the reciprocal backcross series. This was attributed to the greater sensitivity of male than female gametes to cytological and genetic unbalances, which leads to selection in favor of gametes carrying nearly a full complement of *vulgare* chromosomes.

It was found that resistance to stem-rust, leaf-rust, and mildew diseases could be maintained by selection without difficulty through several backcross generations while other *timopheevi* characters were being eliminated. However, despite selection of bunt-free plants during a backcross series, resistance to bunt disease was lost as the population became more *vulgare*-like cytologically and morphologically. This was interpreted as indicating that the genes governing stem-rust, leaf-rust, and mildew resistance are located in chromosomes in the *A* or *G* genomes that have homologs in the *vulgare* parent, but that genes for bunt resistance are located in chromosomes with lesser homology. It was later shown by Allard and Shands (1954) that stem-rust resistance is governed by duplicate linked genes and that these same two genes, or very closely linked ones, also govern mildew resistance. Nyquist, using monosomic analysis, subsequently identified chromosome XIII as the site of these genes (Chapter 29). Apparently chromosome XIII is one of the chromosomes of *T. timopheevi* that has retained considerable homology with its corresponding type in *T. vulgare*. It should be noted that, in the nullisomic series, the chromosomes of the A and B genomes are numbered I through XIV. It was therefore not surprising that the chromosome carrying the stem-rust resistance genes was found to be one of the chromosomes in the A or B ($=G$) genomes and not one in the D genome.

Interspecific Hybridization in Forest Species

The breeding of forest trees is influenced by at least two factors that tend to distinguish it from the breeding of traditional agricultural plants. First, forest trees cannot be provided by man with optimum conditions for growth, but they must be able to shift for themselves under more or less natural conditions. Genetic uniformity is thus not necessarily a desirable quality in stands of forest trees, which occupy a heterogeneous environment and in which the final stand of mature trees is usually much reduced from the initial stand. This can be illustrated by the hybrid between the Monterey and Knobcone pines, produced at the Institute of Forest Genetics at Placerville, California. The first-generation hybrids represent a distinct improvement over the parental species when grown in certain environments. The F_2 generation derived from the F_1 trees is highly variable. This variability is, however, no real disadvantage to the timberman, because many of the F_2 trees, perhaps as many as 50 per cent, are as good as the F_1 trees. A few are even better than the F_1 hybrid trees. Thus, in a stand produced by natural reseeding from F_1 trees, there are enough good trees in the second generation to repopulate the area and to crowd out the inferior trees before the stand reaches maturity. The timberman is accustomed to using trees from forests of natural origin, and the variability of the F_2 hybrids does not disturb him as it would the farmer.

The second factor is the amount of time required to grow a crop to maturity; it is much greater for forest trees than for most agricultural plants. This puts a high premium on the quality of seed and makes it economically possible to go to considerable expense to obtain the best possible strains. Thus even though F_1 hybrid seed may be more expensive than open-pollinated seed, its use can often be justified economically, because the seed cost is but a small fraction of the total expenditure over the many years required to bring a forest to maturity.

Hybrid trees are usually intermediate between the parents in growth rate and other quantitative characters. For this reason, hybrids are most often used to advantage in the area where the less desirable parent occurs. An example of an intermediate hybrid is the one obtained by crossing the poorly formed and not very desirable jack pine of the Lake States with the straight-growing lodgepole pine of the Sierra Nevada. At three years of age the hybrid trees approximate or slightly exceed the height of jack pines and are 179 per cent of the height of the lodgepole pine. The logical locality in which to use this hybrid is the one in which the jack pine is now planted, because the hybrid has the straight, erect growth habit of the lodgepole pine.

Although the intermediate type of hybrid has a definite field of usefulness, hybrids that show heterosis have greater potential value. Heterosis is rare in some genera of trees; in others, such as the poplars and pines, it is frequent. The hybrid between the eastern and western white pines, as an example, shows greater growth capacity than either parent. At Placerville, California, the two parents grew at about the same rate, but the average hybrid was 232 per cent of the height of the seed parent, the western white pine, at three years of age. The difference in volume, or weight, was even greater.

Few hybrid trees are used in forest plantings at the present time. However, indications that seed of some fast-growing hybrids in pines can be produced in abundance, together with the possibility that financial risk can be reduced during the necessary trial period by inter-planting hybrids with standard types, have caused a great increase in interest in reforestation using seedlings obtained directly from crossing.

In some genera of forest trees, hybrid seed is too difficult to obtain to use it directly for reforestation. Moreover, the manifestation of heterosis may be irregular in the F_1 generation. In poplars, for example, the variability of F_1 hybrid seedlings of *Populus alba* crossed with either *P. grandidendata* or *P. tremuloides* was very large (Johnson, 1942). Since vegetative propagation through cuttings is a practical procedure in poplars, the best way to improve yield in this genus may be to select parental trees that give vigorous hybrids and obtain the necessary planting stock by making cuttings from only the most vigorous F_1 genotypes.

REFERENCES

Allard, R. W. 1949. A cytogenetic study dealing with the transfer of genes from *Triticum timopheevi* to common wheat by backcrossing. *Jour. Agric. Res.* 78: 33–64.

Allard, R. W., and R. G. Shands. 1954. Inheritance of resistance to stem rust and powdery mildew in cytologically stable spring wheats derived from *Triticum timopheevi*. *Phytopath.* 44: 266–274.

Emsweller, S. L., P. Brierley, D. V. Lumsden, and F. L. Mulford. 1937. Improvement of flowers by breeding. *U. S. Dept. Agric. Yearbook* 890–998.

Hayes, H. K., J. H. Parker, and C. Kurtzwell. 1920. Genetics of rust resistance in crosses of varieties of *Triticum vulgare* with varieties of *T. durum* and *T. dicoccum. Jour. Agric. Res.* 9: 523–542.

Johnson, L. P. V. 1942. Studies on the relation of growth rate to wood quality in *Populus* hybrids. *Can. Jour. Res.*, C, 20: 28–40.

McFadden, E. S. 1930. A successful transfer of emmer characters to vulgare wheat. *Jour. Agric. Res.* 22: 1020–1034.

McGuire, D. C., and C. M. Rick. 1954. Self-incompatibility in species of *Lycopersicon* sect. *Eriopersicon* and hybrids with *L. esculentum. Hilgardia* 23: 101–124.

Peterson, R. F., and R. M. Love. 1940. A study of the transfer of immunity to stem rust from *Triticum durum* Var. Iumillo, to *T. vulgare* by hybridization. *Sci. Agric.* 20: 608–623.

Rick, C. M., and L. Butler. 1956. Cytogenetics of the tomato. *Adv. in Gen.* 8: 267–382.

Shands, R. G. 1941. Disease resistance of *Triticum timopheevi* transferred to common winter wheat. *Jour. Amer. Soc. Agron.* 33: 709–712.

Stebbins, G. L., Jr. 1950. *Variation and evolution in plants.* Columbia University Press, New York.

Thompson, W. P., T. J. Arnason, and R. M. Love. 1935. Some factors in the different chromosome sets of common wheats. *Can. Jour. Res.* 12: 335–345.

35

Mutation Breeding

Study of induced mutations commenced in 1927, when Muller showed in *Drosophila* that X-rays are able to induce genetic deviants indistinguishable from naturally occurring ones. The notable part of this discovery was the very great increase in mutation rate caused by irradiation treatment. Shortly afterwards, Stadler, who had been working on the effects of X-rays in plants, announced that it was possible through this agency to obtain very high mutation rates in maize and barley. The possibilities inherent in the ability to produce mutations artificially were grasped at once by many plant breeders, and a period followed in which considerable effort was directed toward utilizing the new discovery in practical breeding. Results were discouraging, however; and except for sustained effort on the part of a group of plant breeders in Sweden, attempts to use induced mutations in practical breeding gradually slowed to a virtual standstill.

There were good reasons for the rapid loss of interest on the part of plant breeders. First, it soon became clear that induced mutations, like spontaneous ones, were almost always deleterious in their effects on the phenotype. Second, many of the genes governing characteristics in which commercial varieties required improvement occurred in one of another known stock, and large numbers of other desirable characteristics were known to occur in the rapidly expanding world collections of germplasm. Plant breeders had more variability at their disposal than they could accommodate, and too many obvious tasks to perform in assembling the known desirable genes into new commercial varieties to be diverted very long by a procedure that apparently produced little constructive variability. Thus most plant breeders chose to make use of existing variability, rather than depend on haphazard mutations produced by radiation.

444

Mutation Breeding in Sweden

Soon after Muller's discovery of the mutagenic action of X-rays, Herman Nilsson-Ehle of the Swedish Seed Association at Svalof and one of his students, Åke Gustafsson, began experiments in mutation breeding that have been continued to the present. It was soon established that Stadler's results with barley and maize had general validity with diploid species. Chlorophyll mutations arose in great numbers in two-rowed barley following X-ray dosages to seeds. When the dosage approached the lethal point, mutation rates increased to about one thousand times the spontaneous rate. In the middle 1930's some mutants characterized by compact spikes and very stiff straw, which were designated erectoides, were observed. A great range of other mutants have since been reported in barley, including changes affecting height, date of maturity, seed size, leaf width, color, malting quality, and yield. In some cases, the mutants appeared to affect only one characteristic of the plant; in others, conspicuous pleiotropic effects were observed.

Some of the mutant types in barley, particularly the erectoides mutants, seemed to have useful agricultural properties. When the more promising of these mutants were tested in field trials (Gustafsson and Tedin, 1954; Froier, 1954), they generally produced yields about the same as the mother variety, but a few appeared to be significantly superior. Despite over 25 years of intensive work with barley by a large and enthusiastic group of investigators, and the favorable results they reported, no commercial variety of barley has as yet originated from the program. One of the difficulties seems to be the greater variation in different years displayed by the mutants compared to existing commercial varieties. Granting the practical value of certain distinctly better properties in the more promising mutations— high straw strength, earliness, and, apparently in several cases, increased yielding ability—the demand in modern agriculture for the smallest possible variation in yield and quality in different seasons has, according to Froier (1954), tended to restrict the use of mutant types.

In addition to barley, the Swedish workers have used mutation breeding with many other crops (wheat, oats, peas, vetches, soybeans, lupines, flax, oil turnips, oil rape, white mustard, sugar beets, potatoes, Kentucky bluegrass, timothy, red fescue, apples, pears, plums, cherries, ornamentals, and forest trees). These programs have resulted in two new varieties: Svalof Primex White Mustard, released for commercial production in 1950, and Regina II Summer Oil Rape, released in 1953. Primex Mustard was selected from a population that was irradiated in 1941. It is reported to exceed the parent population by 4 per cent in yield and 2 per cent in oil content. However, mustard is a highly heterozygous cross-pollinated species. It is therefore not definite whether the improvement in yield and oil content is related to the irradiation treatment, since it could have been due to effective selection for genetic variability that existed in the original population. Andersson and Olsson (1954) believe the increase in yield and oil content was greater in the irradiated population than that obtained by selection in nontreated materials, and hence they credit induced mutations with part of the superiority of Primex. Rape is about one third cross-pollinated, and hence it also does not demonstrate conclusively the benefit of induced mutations.

Mutation Breeding in other Countries

Similar but less extensive mutation-breeding programs have been conducted at a number of breeding institutes in Europe. Mutation-breeding work on a large scale started in Germany about 1940 (barley, wheat, oats, lupine, flax, hemp, tomatoes, currants, fruit trees). This program has apparently produced one variety, a variety of common bean that was released for commercial production in 1950. Mutation breeding has also been investigated in France (wheat), Finland (barley, wheat, oats, peas, red clover), England (barley, wheat, sugar beets, Brussels sprouts, fruit trees), Holland (tulips, gladioli), and Norway (barley, oats, tomatoes).

Experiments in mutation breeding were renewed in the United States and Canada following World War II, partly as a result of reports of successes in Sweden. The interest stimulated in radiation research by the arrival of the atomic age and the financial support that became abundantly available as a result were no doubt also important in its revival. For the most part, the mutations observed in the North American programs have been much the same as the ones reported in Sweden. The one important difference in the American programs has been the greater emphasis placed on search for mutations conditioning disease resistance, a subject to be covered in more detail later. Apparently only one commercial variety originating from irradiated materials has been released for commercial production in North America, namely Sanilac, a variety of common bean released by Down and Andersen in Michigan in 1956. Michelite, a vine (indeterminate) variety of pea bean had been radiated in 1941, and one of the mutants that resulted was a bush (determinate) type about twelve days earlier in maturity than Michelite. Resistance to the anthracnose disease was added to the mutant strain by backcross breeding. As released, Sanilac was resistant to anthracnose, six days earlier, superior in yield, and adapted to direct combine harvesting because of its upright growth habit. It should be noted that early-maturing bush types are abundant in the common-bean species; hence the fact that the gene for bush habit of growth carried by Sanilac was from an X-ray-induced mutation is only incidental to the breeding of this variety.

Unsolved Problems in Mutation Breeding

There have now been enough indications of positive results with mutation breeding to attest to the potential value of artificially induced mutations in practical plant breeding. However, many questions remain to be answered before it can be decided whether mutation breeding will assume a place as a major plant-breeding method or whether it will fit in as a minor adjunct to other methods. The most important questions remaining to be answered are: (1) Do artificially induced mutations differ in any way from natural mutations, or do mutagenic agents merely reproduce the same spectrum of variability that occurs naturally? Induced mutations are the plant breeder's one hope for freedom from complete dependence on nature as the only source of the genetic variants necessary in plant improvement. Hence the answer to this question is potentially an important one in the long-range future of plant improvement. (2) Do mutations with phenotypically constructive expressions occur often enough to make the search for them profitable and their incorporation into commercially acceptable varieties competitive with other methods of breeding? If not, can the mutation process be brought under experimental control so as to increase the proportion of constructive changes? The key to these problems appears to lie in understanding of the mutation process itself.

The Nature of Induced Mutations

In artificial mutagenesis, as in nature, it has been postulated that two basic mutation processes are operating. In the first process mutations are assumed to occur as a result of mechanical disruption of the hereditary materials in accordance with the principles of the target theory (Lea, 1946). The second process is assumed to induce mutations more or less indirectly by chemical alteration of the hereditary materials, according to the indirect activation theory. The attitude at present is to accept the simultaneous operation of both processes. Thus, among the mutations produced by any mutagenic agent, some may result from purely mechanical forces, some from purely chemical forces, and others from a combination of the two forces.

The densely ionizing radiations (alpha particles, neutrons) are believed to act primarily in a mechanical way, because a high proportion of the changes they induce are chromosomal alterations (segmental rearrangements, losses, and so on). However, the densely ionizing radiations also produce some point mutations. Some of these point mutations may be produced by direct mechanical action of the radiation. Others, however, may be caused by secondary and indirect action of chemicals produced by the action of the ionization on the protoplasm.

The more sparsely ionizing radiations (X-rays, gamma-rays) have a less pronounced tendency to produce chromosomal alterations, and a greater proportion of the genetic changes they induce are point mutations. Ultraviolet light leads to excitation of electrons, but not ionization. It has an even lesser tendency to induce chromosomal alterations and a still higher proportion of its effects are point mutations.

The alkylating chemical mutagens, such as mustard gas and its derivatives, also act in an intermediate way, producing some chromosomal alterations and some point mutations. The mutagenic nucleosides and other chemical mutagens with special effects on the nucleic acids appear to produce mutations by special chemical affinities. Thus, if mutagenic agents are arranged in decreasing order of the energy they dissipate, there appears to be a direct relation between the proportion of chromosomal disturbances and point mutations produced. In other words, one particular mutagen tends to produce only part of the entire spectrum of possible genetic changes, and the type it produces is a reflection of its energy potential.

Other factors also influence the genetically visible spectrum of mutations produced by particular mutagens. For example, a change in the metabolism of the treated organism, such as pretreatment with colchicine or even a change in mineral nutrition, can alter the genetic response to a particular mutagen. The amount of control possible is, however, disappointingly small. Although it has been possible to shift the proportions of certain mutants, for example, to increase one type of chlorophyll deficiency at the expense of another, mutagens have not yet been discovered that increase the proportion of constructive mutation much above the ratio of 800 unfavorable mutations to 1 favorable one that characterizes X-rays. Intensive work on the detection of more active chemical mutagens should increase the possibilities of producing more point mutations; and, with the discovery of more specific mutagens, the possibilities of directing the mutation process should also increase.

The range of genotypic variability produced by artificial means has been impressive. In barley, one of the most intensively studied species, more than 200 different kinds of vital, visible mutations have been described (Nybom, 1954), and new types are being added continually. Nevertheless, the natural variability of barley is so extensive (Harlan, 1956) that it would be idle to pretend that the range of artificially induced mutants produced thus far duplicates it even in small part. Moreover, since the Herculean task of classifying the enormous naturally occurring variability has hardly started, it would be unwise to state that any particular induced mutation does not already occur naturally. There has, in fact, been little incentive to postulate that entirely new types have been induced, because the variants that have appeared thus far in mutation experiments all seem to have natural counterparts. For the present at least, the fact that mutation breeding may free the plant breeder from dependence on natural variability hardly makes a compelling case for adopting mutation breeding as a general method for plant improvement.

Methods of Mutation Breeding

Even though the great stores of unexploited natural variability remaining in most agricultural species would seem to obviate any immediate need for production of additional variability by artificial means, excluding a few special cases, mutation breeding still has a place in the planned improvement of agricultural plants, provided constructive changes can be induced frequently enough to make the search for them economically competitive with other methods of breeding. Two facts must be taken into account in judging the economic feasibility of searching for favorable mutants: (1) deleterious changes outnumber constructive changes by a factor of several hundred to one; and (2) most mutants are recessive. It follows, therefore, that very large second-generation populations are necessary if the desired improvements are to be obtained and detected. It also follows that mutation breeding is most likely to be successful when applied to the improvement of characteristics for which efficient screening techniques have been developed.

An example of an efficient screening technique is the one described by Konzak (1956) for screening irradiated populations of oats for resistance to the Victoria blight disease. Susceptible reaction to this root-rotting disease is caused by sensitivity to a powerful toxin produced by the causal organism, *Helminthosporium victoriae*. Sensitivity to the toxin is inherited as a simple dominant. In Konzak's test, seeds of a selected variety of oats are irradiated, sown, and the resulting M_1 plants grown to maturity. (The letter M is used as a general term for mutagenic treatment. M_1 refers to the generation in which treatment was performed, M_2 to the generation following the treatment, and so on.) Each M_1 plant is harvested individually, and an M_2 progeny is grown from it by rolling the seeds in a moistened paper towel, which is placed in a moist chamber for two days. The seeds are then sprayed with a solution of the toxin. After toxin treatment, the seeds are allowed to continue germination for another three or four days. The toxin kills the roots of sensitive plants, but does not affect the roots of resistant homozygous recessive plants from the same lot of seeds (Figure 35–1). Resistant mutants are carried to maturity by transferring them to soil. An indication of the frequency of resistant mutants is given in Table 35–1.

The same sort of mutation occurs spontaneously. Luke and Wheeler (1956), using a mass-screening technique, were able to isolate blight-

A   B   C

FIGURE 35-1. Screening test for isolation of blight-resistant mutants in oats. A and B show the appearance of progenies two days after exposure to toxin from a culture of *Helminthosporium victoriae*. One resistant seedling was present in progeny B. C shows the appearance of progeny without toxin treatment. (After Konzak, 1956. Photograph courtesy Brookhaven National Laboratory.)

TABLE 35–1. Frequency of Victoria Blight-Resistance Mutations Induced in Oats by Radiation

(After Konzak, 1956)

Seeds Exposed to	Number of Progenies	
	Tested	Showing Mutants
Thermal neutrons ($1.5 \times 10^{13} n_{th}/\text{cm.}^2$)	542	8
Thermal neutrons ($7.6 \times 10^{12} n_{th}/\text{cm.}^2$)	363	4
X-rays 25,000–30,000 r	330	5
Control (no irradiation)	403	0

resistant mutants also carrying crown-rust resistance by screening large populations of nonirradiated varieties of Victoria parentage.

There seems to be little doubt that mutation breeding or use of mass screening to identify naturally occurring blight-resistant mutants is at least as economical as standard hybridization techniques in breeding blight-resistant types with Victoria crown-rust resistance.

Some other examples of disease resistance induced by mutagenic agents include resistance to stripe rust in wheat, crown rust in oats, mildew in barley, flax rust, and leaf-spot and stem-rot resistance in peanuts.

Morphological and Physiological Mutants

Effectiveness in breeding for constructive morphological and physiological mutations is limited by the efficiency with which such changes can be identified. Chlorophyll mutations have been reported in great numbers, very likely because they are easily detected in the seedling stage, when large numbers of plants are easily examined. Probably for similar reasons, striking morphological alterations such as dwarf and tall mutants, deviants with different size or shape of inflorescence, and the like have frequently been reported. Early or late deviants are also common, but physiological alterations with low heritability (for example, cold or heat tolerance) are rarely reported. Refinements in experimental methods may ultimately make it profitable to search for the less easily identified characters.

Quantitative Characters

If it is assumed that induced mutations are polydirectional, the expected consequences of mutagenic treatment on quantitative characters are as follows: (1) In a population treated with a mutagenic agent, it should be possible to make genetic advance under selection; (2) such quantitative characters should become fixed at random under inbreeding; (3) normal-appearing but mutant-inbred lines should show heterosis when hybridized; and (4) quantitative characters should be represented in a series of intensities in independent mutant families.

Gregory (1956) has shown that mutant progenies with significantly higher yield than the control variety from which they arose occurred in a population of irradiated peanuts (a self-pollinating species). Some of Gregory's data are summarized in Table 35–2. The entries

TABLE 35–2. Mutant versus Control Lines of Peanuts under the Same Selection Pressure Over a Three-Year Period

(After Gregory, 1956)

Control Lines	Yield of Fruit in Lbs. Per Plot
C-1–46	20.0
C-3–43	21.8
C-8–50	20.6
C-2–41	21.3
C-7–42	21.4
Average, control lines	21.0
Selected mutant lines	
YT 36	22.4
YT 32	22.4
YT 12	23.1
YT 13	23.0
YT 9	21.3
Average, selected mutant lines	22.4

in the table represent the 5 best selections from an original sample of 192 randomly chosen M_3 plants. The mutants were first tested in the M_4 at which time the 10 best lines were selected. In the M_5 generations, the 5 best lines among the 10 were selected. The same procedure was used for the control. The selected mutant lines maintained a small but measurable superiority over the control lines over a 3-year test period.

Gregory (1956) has obtained evidence from his work with peanuts to indicate that the other postulates stated above are also fulfilled in irradiated populations.

Special Merits and Disadvantages of Mutation Breeding

Mutation breeding seems to be especially useful in changing single simply inherited characteristics in highly developed genic systems. When dealing with a highly developed variety, the breeder is reluctant to use standard hybridization methods because they may disrupt a superior combination of genes. This situation is often encountered when some outstanding variety succumbs to a new race of a disease or is inferior in some specific morphological or physiological attribute. Whether mutation breeding or the standard backcross technique should be used depends on two factors: (1) the ease with which the desired improvement can be induced, and (2) the number of deleterious mutants that accompany the specific mutation for which the breeding program is undertaken.

The point is sometimes made that undesirable alterations in other characters are easily handled in mutation-breeding programs because the mutant lines are so similar to the parent variety that a few backcrosses will restore the desired background genotype. However, marker genes brought in by a genetically dissimilar parent often allow very effective selection toward the genotype of the recurrent parent in standard backcross programs and more rapid return to the type of the recurrent parent than otherwise would be possible. It is therefore doubtful whether the number of backcrosses required to guarantee recovery of the genotype of the outstanding variety is less in mutation breeding than in standard backcross breeding.

Mutation breeding appears to have special advantages in adding specific characteristics to fruit trees and other vegetatively propagated crops. Varieties in these crops are usually highly heterozygous clones not especially suited to improvement by the selection of recombinant types in successive generations following hybridization or amenable to improvement by backcross breeding.

The disadvantages of mutation breeding are largely associated with the necessity for testing large second-generation populations. The field work required to achieve some particular improvement is often substantially greater with mutation breeding than that required in conventional methods of breeding. As a result the practical use of the method is now limited to the improvement of a small number of characteristics for which efficient screening methods have been developed (Konzak, 1956). In a review on the Swedish work in mutation breeding, MacKey (1956) concluded: "Considering the low fraction of progressive mutations and considering at least the present limitation in selective mutagenesis, it is rather obvious that mutation breeding cannot be considered such a revolutionary tool in crop improvement that it will replace old methods. It means a definite contribution to our plant breeding methods, but it should not be over- or under-estimated just because it is new."

REFERENCES

Andersson, G., and G. Olsson. 1954. Svalof's Primex white mustard—a market variety selected in x-ray treated material. *Acta Agric. Scand.* 4: 574–577.

Auerbach, C. 1949. Chemical mutagenesis. *Biol. Rev.* 24: 355–391.

Catcheside, D. G. 1948. Genetic effects of radiations. *Adv. in Gen.* 2: 271–358.

Down, E. E., and A. L. Andersen. 1956. Agronomic use of an x-ray-induced mutant. *Science* 124: 223–224.

Froier, K. 1954. Aspects of the agricultural value of certain barley x-ray mutations produced and tested at the Swedish Seed Association, Svalof, and its branch stations. *Acta Agric. Scand.* 4: 515–548.

Granhall, I. 1954. Spontaneous and induced bud mutations in fruit trees. *Acta Agric. Scand.* 4: 594–600.

Gregory, W. C. 1956. Induction of useful mutations in the peanut. *Brookhaven Symposia in Biology* No. 9: 177–190.

Gustafsson, A., and J. MacKey. 1948. Mutation work at Svalof. Svalof, 1886–1946: 338–355.

Gustafsson, A., and O. Tedin. 1954. Plant breeding and mutations. *Acta Agric. Scand.* 4: 633–639.

Harlan, J. R. 1956. Distribution and utilization of natural variability in cultivated plants. *Brookhaven Symposia in Biology* No. 9: 191–208.

Konzak, C. F. 1956. Induction of mutations for disease resistance in cereals. *Brookhaven Symposia in Biology* No. 9: 157–176.

Lea, D. E. 1946. *Actions of radiations on living cells.* Cambridge University Press.

Luke, H. H., and H. L. Wheeler. 1956. Disease-resistant oat selections obtained from susceptible varieties. *Phytopath.* 46: 18.

MacKey, J. 1956. Mutation breeding in Europe. *Brookhaven Symposia in Biology* No. 9: 141–156.

Nybom, N. 1954. Mutation types in barley. *Acta Agric. Scand.* 4: 430–456.

Singleton, W. R. 1955. The contribution of radiation genetics to agriculture. *Agron. Jour.* 47: 113–117.

Stadler, L. J. 1954. The gene. *Science* 120: 811–819.

Wettstein, D. 1954. The pleiotropic effects of erectoides factors and their bearing on the property of straw-stiffness. *Acta Agric. Scand.* 4: 491–506.

36
Distribution and Maintenance of Improved Varieties

The purpose of applied plant breeding is to develop better varieties. But the benefits of improved varieties cannot be realized until enough seed has been produced to allow the new variety to be grown on a commercial scale over the entire area to which it is adapted. Nor do problems in the utilization of new varieties end with the initial distribution of improved types to farmers. Unless provision is made to maintain varietal purity, much of the effort expended in developing improved types may come to naught.

In response to these problems, most countries of the world have developed procedures for the orderly increase, distribution, and maintenance of the products of plant breeding. In general these procedures recognize three overlapping areas of responsibility in the process by which new varieties are developed and utilized. These areas are breeding, certification, and commercial seed production. In this division of responsibility the primary function of the plant breeder is to develop new varieties and carry out initial small-scale seed increases. The certification agency is concerned with the steps by which the seed is turned over to the seed grower, and it also administers regulations of production and marketing that are designed to assure varietal purity and reasonable standards of seed quality. Commercial seed production is the responsibility of seedsmen and selected farmers who have the equipment and experience to grow, clean, and market large quantities of pure seed. Usually these three stages in variety development and distribution are entrusted to entirely separate groups of people, but occasionally two or even all three stages are handled by the same people.

455

Plant-Breeding Agencies

In most countries, plant breeding is conducted both by publicly supported agencies and by private individuals or companies. Tax-supported agencies which sponsor plant-breeding projects in the United States are the United States Department of Agriculture and the colleges of agriculture and agricultural experiment stations of the various states. The situation is almost exactly parallel in Canada. In these two countries, local governments (state or provincial) tend to play a more active role relative to the national government in the support of plant breeding than elsewhere in the world.

Privately supported plant breeding is conducted by several different kinds of private industries, such as large general-seed companies, certain more specialized-seed firms (e.g., hybrid-seed-corn companies), large canning and packing concerns, and producers of sugar beet, sugar cane, pineapple, and other plant products. Frequently grower cooperatives support plant breeders, and of course many individuals engage in plant breeding as a hobby or part-time business.

The relative importance of publicly and privately supported plant breeding varies tremendously from crop to crop and from country to country. The breeding of self-pollinated field crops, such as wheat, oats, barley, rice, and flax, has been dominated by public agencies. This is natural because seed sales are small after the initial distribution of a new variety, and private industry can finance plant breeding only on the basis of large recurrent sales of seed. On the other hand, private industry has been the primary source of new varieties of crops for which farmers cannot produce the seed they need economically, such as many of the vegetable crops. Private industry has also been important in the development of new varieties in highly organized enterprises such as sugar-beet and pineapple production.

Technical advances have an influence also on the role of private industry in plant breeding. For example, the advent of hybrid varieties in corn made it uneconomical for the farmer to produce his own seed and paved the way for the development of a large seed industry. This in turn made it possible for seed-corn companies to support their own plant breeders, and changed corn breeding from a predominantly publicly supported undertaking to one in which private breeders participate to a much greater extent than they did previously. Similar developments now appear to be occurring in response to technical advances which have made hybrid sorghum varieties a possibility.

Seed-Certifying Agencies

In the United States, seed certification is a responsibility of each state; but in Canada and most other countries, seed is certified on a national basis. The organization of seed-certifying agencies differs in various countries and from state to state in the United States. In general, however, seed-certifying agencies have the following characteristics: (1) Seed growers, seed companies, and other parties interested in pure-seed production are eligible for membership. (2) The agency is governed by a board of directors elected by the members. (3) The agency has legal status under state or national legislation with authority to set standards for seed certification. (4) It has close working relations with seed growers and agricultural research, extension, and regulatory agencies.

The purpose of seed certification, according to the International Crop Improvement Association, is ". . . to maintain and make available to the public, through certification, high quality seeds and propagating materials of superior crop varieties so grown and distributed as to insure genetic identity and genetic purity. Only those varieties that are approved by a State or Governmental agricultural experiment station and accepted by the certifying agency shall be eligible for certification."

Varietal purity is the first consideration in seed certification, but factors such as weed seeds, seed-borne diseases, mechanical purity, viability, and other grading considerations are also important. Seed certification is therefore designed not only to maintain the genetic purity of superior varieties, but also to set reasonable standards of seed quality and condition.

The International Crop Improvement Association was organized in 1919 to coordinate the activities of certifying agencies in Canada and the United States. This organization has established minimum seed-certification standards for various crops for the guidance of its member agencies. It also played an important role in developing procedures whereby seed-certifying agencies in different areas can cooperate in the various field and cleaning-plant inspections necessary for certification. This made it possible for farmers to obtain planting seed produced in another area where seed production is more dependable and economical, and at the same time to obtain the benefits of seed certification.

Classes of Seed

Four classses of pure seed are recognized by the International Crop Improvement Association.

Breeder seed. Breeder seed is seed or vegetative propagating material produced by or under the direct control of the sponsoring plant breeder. It is the basis of the first and recurring increases of foundation seed.

Foundation seed. Foundation seed (including elite seed in Canada) is seed so designated by an agricultural experiment station. Production must be carefully supervised or approved by representatives of an agricultural experiment station. It is the source of all other certified-seed classes, either directly or through registered seed.

Registered seed. Registered seed is the progeny of foundation or registered seed. Its production and handling must be approved and certified by the certifying agency, and its quality must be such that it is suitable for the production of certified seed. Registered seed is used as the source of certified seed in some crops by some agencies. In other cases the registered-seed class is omitted.

Certified seed. Certified seed is the progeny of foundation, registered, or certified seed. It is the seed produced on a large scale by certified-seed growers for general farm sale. It must be produced and handled in such a way as to meet the standards set by the certifying agency.

Each seed-certifying agency establishes procedures for the production of each class of seed and the standards of purity and quality for each crop. The agency publishes these standards and enforces them through inspection at various stages in the growing and harvesting of the seed. It also sets up qualifications for seed producers, taking into account general adaptation of the farm and equipment, ability of the grower, his integrity, and other qualifications necessary for production of pure seed.

The Certification Process

Before a variety can be certified, it must be approved for certification by the appropriate review board of the certifying agency. Qualification for certification is ordinarily based on recommendation by the agricultural experiment station, or equivalent agency, that the variety is suitable for commercial production in the area. Areas that produce seed for export usually have two classes of certified varieties. One

FIGURE 36-1. Off-type plants in a foundation-seed field of grain sorghum. These plants could have arisen in any one of three ways. What are these possibilities?

group comprises those varieties recommended for commercial production in the area where the seed is produced. The other group is made up of varieties recommended only for seed production and not for general planting purposes. Seed of the second group is therefore shipped to its area of adaptation for sale.

The certification procedure follows this general pattern:

1. The seed grower must plant foundation, registered, or certified seed of an approved variety. Only one variety of the same crop can be grown for seed production on a farm, except on prior approval of the certifying agency.

2. The seed must be planted on fields which have not been planted for some specified period to other varieties of the same crop that might volunteer and affect the genetic purity of the certified seed. The

field should also not have been planted for some specified period, depending on the crop, to other similar crops that might affect purity. Freedom from noxious weeds is also required.

3. Appropriate isolation of the seed-producing field from other varieties of the same crop is required. Minimum distances recommended for different crops and different classes of seeds are given by the International Crop Improvement Association.

4. Field inspections are made by representatives of the certifying agency. Although the frequency and purpose of inspection vary from crop to crop, in general the intent is to check on the completeness of roguing of off-type plants, adequacy of isolation, thoroughness of detasseling or freedom from pollen shedding in male-sterile lines, and other factors that might influence genetic purity. Fields are also inspected for seed-borne diseases and weeds.

5. Seed inspections may be made at any time to observe harvesting, cleaning, storage, or labeling. A representative sample of each lot as it is offered for sale is obtained by the certifying agency and tested for impurities, germination, and other factors affecting quality.

6. All seed stocks sold as certified seed must be labeled with an official tag properly affixed to each container and appropriately sealed.

It will be noted that the responsibility for the growing, harvesting, processing, and marketing of certified seeds falls on the grower. The responsibility of the certifying agency is restricted to determining whether the grower has followed regulations and whether the seed meets the requirements established for crop and class of seed.

Release of New Varieties

The procedures by which new varieties are released to farmers vary widely. In general, the steps by which varieties developed by state agricultural experiment stations reach the farmer are orderly and efficient, and hence they will be described in some detail. The steps are: (1) decision to release; (2) naming the variety; (3) production of breeder seed; (4) production and distribution of foundation and registered seed; and (5) production and distribution of certified seed.

1. *Release of Improved Varieties.* Authority to release a new variety is ordinarily vested in a review board appointed by the agricultural experiment station to consider such proposals for release. The sponsoring breeder or agency must present evidence that the new type has a place in the agriculture of some specified area or areas. This evidence usually consists of observations of its response to vari-

ous environmental conditions, and precise data from comparative trials in which the performance of the variety has been compared with standard varieties over a period of years at a number of locations. Other important evidence concerning the new variety is provided by data on its response to races of diseases, its quality, and its suitability for mechanical handling or other factors of farm practice. Data from regional trials is helpful in determining the range of adaptation of the variety in areas outside the jurisdiction of the experiment station.

The grower usually has less information to guide him when he contemplates planting a new variety developed by private industry. Some companies breed varieties only for their own use or for the use of their contract growers, and such varieties are usually thoroughly tested before their production on a commercial scale. Many seed companies conduct their own evaluation trials or enter their varieties in tests conducted by public agencies. In the latter case a fee is paid for each variety, and the results are made available to the public. In general, the standards set by seed companies in North America are high because of the keen competition and increasing awareness on the part of farmers of the risks associated with inadequate testing. Nevertheless, the farmer must rely largely on the reputation and integrity of the company with which he deals. Some countries, particularly the smaller Northern European countries, have strict laws governing the release of all new varieties, whether developed by public or private breeders, and sale of seed is contingent on prior testing in official trials.

2. *Naming a Variety.* The sponsoring breeder suggests a name for the variety to facilitate its identification. One-word names are preferred or even required by most experiment stations.

3. *Production of Breeder Seed.* Once the decision to release has been made and the variety identified with a name, the plant breeder makes a limited increase of the new variety. This breeder seed is turned over to some agency, generally the foundation-seed department within the certifying agency, which uses it to produce foundation seed.

Either breeder or foundation seed is usually shared with other experiment stations which may also wish to release the new variety. Usually other stations have had a chance to test the variety in regional cooperative trials and are aware of the potential and limitations of the variety.

In the development of forage varieties, the increase of breeders seed may be accomplished in an area outside the region of adaptation. This is usually done under the auspices of the National Foundation

Seed Project, organized in 1948 to facilitate the rapid increase and maintenance of stocks of superior varieties of grasses and small-seeded legumes. Weather conditions in the midwestern and eastern states make seed production undependable, and there is consequently difficulty in increasing and maintaining seed stocks. However, seed of many crops can be produced dependably in the western states, where environmental conditions are favorable for high yields of superior-quality seed. Originally, forage varieties could be certified only in areas where they were adapted to avoid genetic changes from natural selection. But when it was found that production for one or a few generations outside the area of adaptation had little or no ill effect, the way was paved for rapid increase of seed stocks of new varieties and for economical large-scale production of certified seed. Details of the organization and operation of the National Foundation Seed Project are described by Garrison (1956).

4. *Distribution of Foundation Seed.* The next step is the distribution of foundation seed to seed growers. Since improved varieties developed by tax-supported agencies can be considered public property, it is desirable that no one be permitted to exploit a new variety by excessive charges for seed while the supply is limited. Consequently an ample supply of foundation seed is produced to assure generous supplies of registered (and subsequently certified) seed. The distribution of foundation seed is made to selected growers of proven ability and integrity. Sometimes control over distribution and price of the first increase of the foundation seed is retained by the certifying agency. Seed increased from foundation seed is classed as registered or certified seed, depending on the policy of the certifying agency and the circumstances of the particular case.

5. *Distribution of Registered Seed.* The final step in the release of a variety is the distribution of registered seed to certified-seed growers in each locality in the state and their use of this registered seed to produce certified seed, which is available without restriction to any grower.

Certification of Vegetatively Propagated Varieties

The general seed-certification standards of the International Crop Improvement Association also apply to vegetatively propagated materials. Certification procedures to meet the requirements of such diverse materials as vegetatively propagated grasses, bulb onions, potatoes, and sweet potatoes are given in publications of the International Crop Improvement Association.

Maintenance of Pure Seed Stocks

Despite the best attempts of certified seed growers and farmers, new varieties soon become objectionably contaminated with off-types, and periodic purifications are necessary to maintain varietal purity and identity. These periodic purifications are usually made cooperatively by the seed-certification agency and experiment-station plant breeders. Methods appropriate to various types of crops have already been discussed and need not be repeated here.

REFERENCES*

Beard, D. F., and E. A. Hollowell. 1952. The effect on performance when seed of forage crop species is grown under different environmental conditions. *Proc. Sixth Intern. Grasslands Congr.* pp. 860–866.

Beeson, K. E. 1952. The forage seed certification program in the United States. *Proc. Sixth Intern. Grasslands Congr.* pp. 1912–1917.

Garrison, C. S. 1956. *National Foundation Seed Project, Organization and Operational Policies.* Field Crops Research Branch, Agricultural Research Service, U. S. Dept. Agric., Beltsville, Maryland (Mimeograph).

Minimum Seed Certification Standards. Published by International Crop Improvement Association, Beltsville, Maryland.

Smith, Dale. 1955. Influence of area of seed production on the performance of Ranger alfalfa. *Agron. Jour.* 47: 201–205.

* The details of seed production in various crops are given in circulars published by Extension Services in various states, particularly the California, Oregon, Utah, and Washington Agricultural Extension Services, and are available upon request.

Glossary

ADAPTATION. The process by which individuals (or parts of individuals), populations, or species change in form or function in such a way to better survive under given environmental conditions. Also the result of this process.

ALLELE OR ALLELOMORPH. One of a pair or series of forms of a gene which are alternative in inheritance because they are situated at the same locus in homologous chromosomes.

ALLOPOLYPLOID. A polyploid containing genetically different sets of chromosomes, for example, sets from two or more different species.

AMPHIDIPLOID. A polyploid whose chromosome complement is made up of the entire somatic complements of two species.

ANEUPLOID. An organism whose somatic chromosome number is not an even multiple of the haploid number.

APOMIXIS. Reproduction in which sexual organs or related structures take part but fertilization does not occur, so that the resulting seed is vegetatively produced.

ASEXUAL REPRODUCTION. Reproduction which does not involve the union of gametes.

ASYNAPSIS. Failure of pairing of homologous chromosomes during meiosis.

AUTOGAMY. Self-fertilization.

AUTOPOLYPLOID. A polyploid arising through multiplication of the complete haploid set of a species.

BACKCROSS. A cross of a hybrid to either of its parents. In genetics, a cross of a heterozygote to a homozygous recessive. (*See* test cross.)

BACKCROSS BREEDING. A system of breeding whereby recurrent backcrosses are made to one of the parents of a hybrid, accompanied by selection for a specific character or characters.

BALANCE. The condition in which genetic components are adjusted in proportions that give satisfactory development. Balance applies to individuals and populations.

BASIC NUMBER. The number of chromosomes in ancestral diploid ancestors of polyploids, represented by x.

BIAS. A consistent and false departure of a statistic from its proper value.

BIOMETRY. The branch of science which deals with statistical procedures in biology.

BIOTYPE. A group of individuals with the same genotype. Biotypes may be homozygous or heterozygous.

BIVALENT. A pair of homologous chromosomes united in the first meiotic division.

465

BREEDER SEED. Seed produced by the agency sponsoring a variety and used to produce foundation seed.

BREEDING. The art and science of changing plants or animals genetically.

BULK BREEDING. The growing of genetically diverse populations of self-pollinated crops in a bulk plot with or without mass selection, followed by single-plant selection.

CERTIFIED SEED. Seed used for commercial crop production produced from foundation, registered, or certified seed under the regulation of a legally constituted agency.

CENTROMERE. (*See* kinetochore.)

CHARACTER. An attribute of an organism resulting from the interaction of a gene or genes with environment.

CHIASMA. An exchange of partners between paired chromatids in the first division of meiosis.

CHROMATID. One of two threadlike structures formed by the longitudinal division of a chromosome during meiotic prophase and known as a daughter chromosome during anaphase.

CHROMOSOMES. Structural units of the nucleus which carry the genes in linear order. Chromosomes undergo a typical cycle in which their morphology changes drastically in various phases of the life cycles of organisms.

CLONE. A group of organisms descended by mitosis from a common ancestor.

COMBINING ABILITY. *General,* average performance of a strain in a series of crosses. *Specific,* deviation from performance predicted on the basis of general combining ability.

COUPLING. Linked recessive alleles occur in one homologous chromosome and their dominant alternatives occur in the other chromosome. Opposed to repulsion in which one dominant and one recessive occur in each member of the pair of homologous chromosomes.

COVARIANCE. The mean of the product of the deviation of two variates from their individual means. A statistical measure of the interrelation between variables.

CROSSING OVER. The exchange of corresponding segments between chromatids of homologous chromosomes during meiotic prophase. Its genetic consequence is the recombination of linked genes.

CYTOPLASTIC INHERITANCE. Transmission of hereditary characters through the cytoplasm as distinct from transmission by genes carried by chromosomes. Detected by differing contribution of male and female parents in reciprocal crosses.

DEFICIENCY. The absence or deletion of a segment of chromosome.

DEGREES OF FREEDOM, NUMBER OF. The number of independent comparisons that can be made in a set of data.

DETASSEL. Remove the tassel (male inflorescence) as in maize.

DEVIATION. Departure of an observation from its expected value.

DIALLEL CROSS, COMPLETE. The crossing in all possible combinations of a series of genotypes.

DIHYBRID. Heterozygous with respect to two genes.

DIOECIOUS. Plants in which staminate and pistillate flowers occur on different individuals.

DIPLOID. An organism with two chromosomes of each kind.

DIPLOTENE. The stage of meiosis which follows pachytene and during which the four chromatids of each bivalent move apart in two pairs but remain attached in the region of chiasmata.

DISJUNCTION. The separation of chromosomes at anaphase.

DOMINANCE. Intra-allelic interaction such that one allele manifests itself more or less, when heterozygous, than its alternative allele.

DONOR PARENT. The parent from which one or a few genes are transferred to the recurrent parent in backcross breeding.

DOUBLE CROSS. A cross between two F_1 hybrids.

DRIFT. Changes in gene and genotypic frequencies in small populations due to random processes.

DUPLICATION. Occurrence of a segment of a chromosome twice in the haploid set.

DUPLEX. (*See* nulliplex.)

EMASCULATION. Removal of the anthers from a flower.

EPISTASIS. Dominance of one gene over a nonallelic gene. The gene suppressed is said to be hypostatic. More generally, the term epistasis is ued to describe all types of interallelic interaction whereby manifestation at any locus is affected by genetic phase at any or all other loci.

ENVIRONMENT. The sum total of the external conditions which affect growth and development of an organism.

ERROR VARIANCE. Variance arising from unrecognized or uncontrolled factors in an experiment with which the variance of recognized factors is compared in tests of significance.

EPIPHYTOTIC. An unarrested spread of a plant disease.

EXPRESSIVITY. The degree of manifestation of a genetic character.

F_1. The first generation of a cross.

F_2. The second filial generation obtained by self-fertilization or crossing *inter se* of F_1 individuals.

F_3. Progeny obtained by self-fertilizing F_2 individuals.

FACTOR. Same as gene.

FAMILY. A group of individuals directly related by descent from a common ancestor.

FERTILITY. Ability to produce viable offspring.

FERTILIZATION. Fusion of the nuclei of male and female gametes.

FOUNDATION SEED. Seed stock produced from breeder seed by or under the direct control of an agricultural experiment station. Foundation seed is the source of certified seed, either directly or through registered seed.

GAMETE. Cell of meiotic origin specialized for fertilization.

GENE. The unit of inheritance. Genes are located at fixed loci in chromosomes and can exist in a series of alternative forms called alleles.

GENE FREQUENCY. The proportion in which alternative alleles of a gene occur in a population.

GENE INTERACTION. Modification of gene action by a nonallelic gene or genes.

GERMPLASM. The sum total of the hereditary materials in a species.

GENOME. A set of chromosomes corresponding to the haploid set of a species.

GENOTYPE. The entire genetic constitution of an organism.

GENETIC EQUILIBRIUM. The condition in which successive generations of a population contain the same genotypes in the same proportions with respect to particular genes or combinations of genes.

HAPLOID. A cell or organism with the gametic chromosome number (n).

HERITABILITY. The proportion of observed variability which is due to heredity, the remainder being due to environmental causes. More strictly, the proportion of observed variability due to the additive effects of genes.

HERMAPHRODITISM. Reproductive organs of both sexes present in the same individual or in the same flower in higher plants.

HETEROSIS. Hybrid vigor such that an F_1 hybrid falls outside the range of the parents with respect to some character or characters. Usually applied to size, rate of growth, or general thriftiness.

HETEROCARYOSIS. The presence of two or more genetically different nuclei within single cells of a mycellium.

HETEROTHALLY. Haploid incompatibility in fungi (opposite of homothally).

HETEROZYGOUS. Having unlike alleles at one or more corresponding loci (opposite of homozygous).

HOMOLOGY OF CHROMOSOMES. Applied to whole chromosomes or parts of chromosome which synapse or pair in meiotic prophase.

HOMOZYGOUS. Having unlike alleles at corresponding loci on homologous chromosomes. An organism can be homozygous at one, several, or all loci.

HYBRID. The product of a cross between genetically unlike parents.

I_1, I_2, I_3, . . . Symbols used to designate first, second, etc. inbred generations. (See S_1, S_2, . . .)

INBRED LINE. A line produced by continued inbreeding. In plant breeding a nearly homozygous line usually originating by continued self-fertilization, accompanied by selection.

INBREEDING. The mating of individuals more closely related than individuals mating at random.

INBREEDING COEFFICIENT. A quantitative measure of the intensity of inbreeding.

INBRED-VARIETY CROSS. The F_1 cross of an inbred line with a variety.

INDEPENDENCE. The relationship between variables when the variation of each is uninfluenced by that of others, that is, correlation is zero.

INTERFERENCE. The effect of one crossover influencing the probability that another will occur in the immediate vicinity.

INVERSION. A rearrangement of a chromosome segment so that its genes are in reversed linear order.

IRRADIATION. Exposure of plants or plant parts to X-rays or other radiations to increase mutation rates.

ISOALLELES. Alleles indistinguishable except by special tests.

ISOGENIC LINES. Two or more lines differing from each other genetically at one locus only. Distinguished from clones, homozygous lines, identical twins, etc., which are identical at all loci.

ISOLATION. The separation of one group from another so that mating between or among groups is prevented.

KINETOCHORE. Spindle attachment. A localized region in each chromosome to which the "spindle fiber" appears to be attached and which seems to determine movement of the chromosomes during mitosis and meiosis.

LINE BREEDING. A system of breeding in which a number of genotypes which have been progeny tested in respect to some character or group of characters are composited to form a variety.

LINKAGE. Association of characters in inheritance due to location of genes in proximity on the same chromosome.

LINKAGE MAP. Map of position of genes in chromosomes determined by recombination relationships.

LINKAGE VALUE. Recombination fraction expressing the proportion of crossovers versus parental types in a progeny. The recombination fraction can vary from zero to one half.

LOCUS. The position occupied by a gene in a chromosome.

M_1, M_2, M_3, . . . Symbols used to designate first, second, third, . . . generations after treatment with a mutagenic agent.

MALE STERILITY. Absence or nonfunction of pollen in plants.

MASS-PEDIGREE METHOD. A system of breeding in which a population is propagated in mass until conditions favorable for selection occur, after which pedigree selection is practiced.

MASS SELECTION. A form of selection in which individual plants are selected and the next generation propagated from the aggregate of their seeds.

MATING SYSTEM. Any of a number of schemes by which individuals are assorted in pairs leading to sexual reproduction. *Random,* assortment of pairs is by chance. *Genetic assortative mating,* mating together of individuals more closely related than individuals mating at random. *Genetic disassortative mating,* mating together of individuals less closely related than individuals mating at random. *Phenotypic assortative mating,* mating individuals more alike in appearance than the average. *Phenotypic disassortative mating,* mating of individuals less alike in appearance than individuals mating at random.

MEAN. The arithmetic average of a series of observations.

MEDIAN. The value of the variate on each side of which there is an equal number of larger and smaller variates.

MEIOSIS. A double mitosis occurring in sexual reproduction which results in production of gametes with haploid (n) chromosome number.

METAPHASE. The stage of meiosis or mitosis at which the chromosomes lie on the spindle.

METAXENIA. Influence of pollen on maternal tissues of the fruit. (*See* xenia.)

MITOSIS. The process by which the nucleus is divided into two daughter nuclei with equivalent chromosome complements, usually accompanied by division of the cell containing the nucleus.

MODE. The value of variate in the class of greatest frequency in a frequency distribution.

MODIFYING GENES. Genes that affect the expression of a nonallelic gene or genes.

MONOECIOUS. Staminate and pistillate flowers borne separately on the same plant.

MONOHYBRID. Heterozygous with respect to one gene.

MONOPLOID. An organism with the basic (x) chromosome number . (*See* haploid.)

MONOSOME. An organism lacking one chromosome of the diploid complement, hence, having $2n - 1$ chromosomes.

MULTIPLE ALLELE. A member of a series of more than two alternative forms of a gene.

MULTIVALENTS. (*See* univalent.)

MUTATION. A sudden heritable variation in a gene or in chromosome structure.

NULL HYPOTHESIS. Hypothesis that there is no discrepancy between observation and expectation based on some set of postulates.

NULLIPLEX. The condition in which a polyploid is recessive in all chromosomes in respect to a particular gene. Simplex denotes recessiveness at all loci except one, duplex two, triplex three, quadriplex four, etc.

NULLISOME. An otherwise $2n$ plant that lacks both members of one specific pair of chromosomes, hence, with $2n - 2$ chromosomes.

OUTCROSS. A cross, usually natural, to a plant of different genotype.

P_1, P_2, P_3, . . . First, second, etc., generations from a parent. Also used to designate different parents used in making a hybrid or series of hybrids.

PACHYTENE. The double-thread stage of meiosis.

PANMIXIA. Random mating without restriction (usually extended to include random mating under the restrictions of sex or incompatibility).

PARTHENOGENESIS. Development of an organism from a sex cell but without fertilization.

PARAMETER. A numerical quantity which specifies a population in respect to some characteristic.

PEDIGREE. A record of the ancestry of an individual, family, or strain.

PEDIGREE BREEDING. A system of breeding in which individual plants are selected in the segregating generations from a cross on the basis of their desirability judged individually and on the basis of a pedigree record.

PENETRANCE. The frequency with which a gene produces a recognizable effect in individuals which carry it.

PHENOTYPE. Appearance of an individual as contrasted with its genetic make-up or genotype. Also used to designate a group of individuals with similar appearance but not necessarily identical genotypes.

PHYSIOLOGICAL RACES. Pathogens of the same species with similar or identical morphology but differing pathogenic capabilities.

POLYCROSS. Open pollination of a group of genotypes (generally selected) in isolation from other compatible genotypes in such a way as to promote random mating *inter se.*

POLYGENES. Genes whose effects are too slight to be identified individually but which, through similar and supplementary effects, can have important effects on total variability.

POLYMORPHISM. The occurrence together in the same population of two or more distinct forms at frequencies too great to be explained by recurrent mutation.

POLYPLOID. An organism with other than two basic sets of chromosomes, that is, monoploid, triploid, tetraploid, and various aneuploids.

POPULATION. In genetics, a community of individuals which share a common gene pool. In *statistics,* a hypothetical and infinitely large series of potential observations among which observations actually made constitute a sample.

PREPOTENCY. The capacity of a parent to impress characteristics on its offspring so they are more alike than usual.

PROBABILITY. The proportion of times in which an event occurs in an infinitely large and hypothetical series of cases, each capable of producing the event.

PROGENY TEST. A test of the value of a genotype based on the performance of its offspring produced in some definite system of mating.

PROTANDRY. Maturation of anthers before pistils.

PROTOGYNY. Reverse of protandry.

PURE LINE. A strain homozygous at all loci, ordinarily obtained by successive self-fertilizations in plant breeding.

QUADRIVALENT. (*See* univalent.)

QUADRIPLEX. (*See* nulliplex.)

QUALITATIVE CHARACTER. A character in which variation is discontinuous.

QUANTITATIVE CHARACTER. A character in which variation is continuous so that classification into discrete categories is not possible.

RANDOM. Arrived at by chance without discrimination.

RANDOMIZATION. Process of making assignments at random.

RECESSIVE. The member of an allelic pair which is not expressed when the other (dominant) member occupies the homologous chromosome.

RECIPROCAL CROSSES. Crosses in which the sources of male and female gametes are reversed.

RECOMBINATION. Formation of new combinations of genes as a result of segregation in crosses between genetically different parents. Also the rearrangement of linked genes due to crossing over.

RECURRENT PARENT. The parent to which successive backcrosses are made in backcross breeding.

RECURRENT SELECTION. A method of breeding designed to concentrate favorable genes scattered among a number of individuals by selecting in each generation among the progeny produced by matings *inter se* of the selected individuals (or their selfed progeny) of the previous generation.

REGISTERED SEED. The progeny of foundation seed normally grown to produce certified seed.

REGRESSION, COEFFICIENT OF. A numerical measure of the rate of change of the dependent on the independent variable.

ROGUE. A variation from the standard type of a variety or strain. *Roguing,* removal of undesirable individuals to purify the stock.

S_1, S_2, S_3, . . . Symbols for designating first, second, third, etc., selfed generations from an ancestral plant (S_0).

SAMPLE. A finite series of observations taken from a population.

SAMPLING ERROR. Deviation of a sample value from the true value owing to the limited size of sample.

SEGREGATION. Separation of paternal from maternal chromosomes at meiosis and consequent separation of genes leading to the possibility of recombination in the offspring.

SELECTION. In *genetics,* discrimination among individuals in the number of offspring contributed to the next generation. In *statistics,* discrimination in sampling leading to bias. Opposed to randomness.

SELF-FERTILIZATION. Fusion of male and female gametes from the same individual.

SELF-FERTILITY. Capability of producing seed upon self-fertilization.

SELF-INCOMPATIBILITY. Genetically controlled physiological hindrance to self-fruitfulness.

SIBS. Progeny of the same parents derived from different gametes. *Half sibs,* progeny with one parent in common.

SIB MATING. The mating of sibs.

SIGNIFICANCE, TEST OF. Statistical test designed to distinguish differences due to sampling error from differences due to discrepancy between observation and hypothesis.

SINGLE CROSS. A cross between two genotypes, usually two inbred lines, in plant breeding.

SOMATOPLASTIC STERILITY. Collapse of zygotes during embryonic stages due to disturbances in embryo-endosperm relationships.

SPECIES. The unit of taxonomic classification into which genera are subdivided. A group of similar individuals different from other similar arrays of individuals.

In sexually reproducing organisms, the maximum interbreeding group isolated from other species by barriers of sterility or reproductive incapacity.

STANDARD DEVIATION. A measure of variability. Mathematically, the distance along the abscissa from the mean to the point of inflection of a normal curve.

STANDARD ERROR. A statistic which is the estimated value of the standard deviation (a parameter).

STRAIN. A group of similar individuals within a variety.

STRAIN BUILDING. Improvement of cross-fertilizing plants by any one of a number of methods of selection.

SYNAPSIS. Conjugation at pachytene and zygotene of homologous chromosomes.

SYNTHETIC VARIETY. A variety produced by crossing *inter se* a number of geno-types selected for good combining ability in all possible hybrid combinations, with subsequent maintenance of the variety by open pollination.

STATISTIC. Estimate of a parameter made from a sample. Statistic is to sample what parameter is to population.

TELOPHASE. The last stage in cell division before the nucleus returns to a resting condition.

TEST CROSS. A cross of a double or multiple heterozygote to the corresponding multiple recessive to test for homozygosity or linkage.

TETRAPLOID. An organism with four basic (x) sets of chromosomes.

TOP CROSS. A cross between a selection, line, clone, etc., and a common pollen parent which may be a variety, inbred line, single cross, etc. The common pollen parent is called the top cross or tester parent. In corn, a top cross is commonly an inbred-variety cross.

TRANSGRESSIVE SEGREGATION. Appearance in segregating generations of individuals falling outside the parental range in respect to some character.

TRANSLOCATION. Change in position of a segment of a chromosome to another location in the same or a different chromosome.

TRIPLEX. (*See* nulliplex.)

TRIPLOID. An organism with three basic (x) sets of chromosomes.

TRISOMIC. An organism diploid except for one kind of chromosome which is present in triplicate, hence, having $2n + 1$ chromosomes.

UNIVALENT. An unpaired chromosome in meiosis. Bivalents, trivalents, quadri-valents, etc. are associations of 2, 3, 4 . . . homologous chromosomes held together by chiasmata.

VARIANCE. Mean squared deviation of a population of variates from their mean The square of the standard deviation. The corresponding statistic is the mean square.

VARIATE. A single observation or measurement.

VARIATION. The occurrence of differences among individuals due to differences in their genetic composition and/or the environment in which they were raised.

VARIETY. A subdivision of a species. A group of individuals within a species which are distinct in form or function from other similar arrays of individuals.

VIRULENCE. Capacity of a pathogen to incite a disease.

x. Basic number of chromosomes in a polyploid series.

X_1, X_2, X_3. Symbols denoting first, second, third . . . generations from an ir-radiated ancestral plant (X_0).

XENIA. Effect of pollen on the embryo and endosperm. (*See* metaxenia.)

ZYGOTE. Cell formed by the union of two gametes and the individual developing from this cell.

ZYGOTENE. A stage in meiotic prophase when the threadlike chromosomes pair.

Index